Fields Institute Communications

The Fields Institute for Research in Mathematical Sciences

Geometric Representation Theory and Extended Affine Lie Algebras

Erhard Neher
Alistair Savage
Weiqiang Wang
Editors

American Mathematical Society
Providence, Rhode Island

The Fields Institute
for Research in Mathematical Sciences
Toronto, Ontario

FIELDS

The Fields Institute
for Research in Mathematical Sciences

The Fields Institute is a center for mathematical research, located in Toronto, Canada. Our mission is to provide a supportive and stimulating environment for mathematics research, innovation and education. The Institute is supported by the Ontario Ministry of Training, Colleges and Universities, the Natural Sciences and Engineering Research Council of Canada, and seven Ontario universities (Carleton, McMaster, Ottawa, Toronto, Waterloo, Western Ontario, and York). In addition there are several affiliated universities in both Canada and the United States, and five Corporate Affiliate Members (Algorithmics, General Motors, QWeMA Group Inc., R2 Financial Technologies Inc., and Sigma Analysis and Management).

2000 *Mathematics Subject Classification.* Primary 16G20, 17B10, 17B35, 17B37, 17B50, 17B65, 17B67.

Library of Congress Cataloging-in-Publication Data

Geometric representation theory and extended affine Lie algebras / Erhard Neher, Alistair Savage, Weiqiang Wang.

p. cm. — (Fields Institute Communications, ISSN 1069-5265 ; 59)

Includes bibliographical references and index.

ISBN 978-0-8218-5237-8 (alk. paper)

1. Representations of rings (Algebra). 2. Nonassociative rings. 3. Representations of Lie algebras. I. Neher, Erhard, 1949– II. Savage, Alistair. III. Wang, Weiqiang, 1970–

QA252.3.G46 2011

512′.482—dc22

2010043432

∞ The paper used in this book is acid-free and falls within the guidelines
established to ensure permanence and durability.
This publication was prepared by the Fields Institute.
http://www.fields.utoronto.ca
Visit the AMS home page at http://www.ams.org/

10 9 8 7 6 5 4 3 2 1 16 15 14 13 12 11

Contents

Preface

Lie algebras and their representations form an extremely rich and important field of mathematics. This area can sometimes be technically abstract, but is also well known for its connections and applications to many other disciplines such as geometry, topology, mathematical physics, and algebraic combinatorics.

The interplay between algebra and geometry goes both ways. One uses geometric methods to construct or gain insight on representations of Lie groups and Lie algebras. The best known examples in this direction include the Borel-Weil-Bott construction, Springer theory of Weyl group representations, Kazhdan-Lusztig theory for category $\mathcal{O}$, and Nakajima and Lusztig quiver variety realizations of Kac-Moody algebras and their representations. On the other hand, one is often able to use the representation theory of semisimple or affine Lie algebras to organize and better understand the geometric invariants (such as homology) of various interesting varieties, such as Hilbert schemes, flag varieties, and affine Grassmannians. In addition, geometric representation theory has been found to be intimately related to the combinatorics of crystals and quivers.

Within the theory of Kac-Moody algebras, it is the finite-dimensional and affine Lie algebras that stand out, not only for their well-developed representation theory, but also for applications to number theory, combinatorics, and mathematical physics. Extended affine Lie algebras are a class of Lie algebras that encompasses these two important types of Lie algebras, as well as toroidal Lie algebras. The structure theory of extended affine Lie algebras (sometimes jokingly called Canadian algebras by friends) has just recently been completed, as the culmination of the last decade of work of a number of researchers, many of them Canadian.

It was before such a background that the University of Ottawa hosted a summer school in "Geometric Representation Theory and Extended Affine Lie Algebras", on June 15–27, 2009. The summer school was followed by a week-long conference on the same general area of representation theory. There were 3 lecture series in each of the two weeks of summer school, with one and half an hour lecture per weekday. These lecture series were as follows:

(i) *Introduction to geometric representation theory*, by Joel Kamnitzer.
(ii) *Introduction to quantum groups and crystals*, by Seok-Jin Kang.
(iii) *Geometric realizations of crystals*, by Alistair Savage.
(iv) *Nilpotent orbits and finite W-algebras*, by Weiqiang Wang.
(v) *Affine, toroidal and extended affine Lie algebras*, by Erhard Neher.
(vi) *Representation theory of affine and toroidal Lie algebras*, by Vyjayanthi Chari.

In his lectures, Kamnitzer explains three distinct geometric approaches of constructing irreducible rational representations of the Lie group GL_n, namely, the Borel-Weil construction, the Ginzburg construction, and the geometric Satake correspondence. While each of the approaches generalizes to more general semisimple or reductive Lie groups, the focus on type A helps to make things explicit and avoid introducing Langlands dual groups and further deeper geometric objects such as Nakajima quiver varieties.

Kang starts his lectures by reviewing the basics of crystal basis theory. He then continues to describe the theory of perfect crystals which has played an important role in providing a realization of crystals for irreducible integrable modules of quantum affine algebras. In the case of level one integrable modules, a combinatorial approach to crystals using Young walls is explained in the end.

Like the semisimple Lie algebras, extended affine Lie algebras admit root space decompositions. The associated root systems can be understood and studied in the framework of affine reflection systems, which, roughly speaking, consist of finite root systems together with additional extension data. The structure theory of extended affine Lie algebras serves as a good example of the unity of mathematics: One needs all the important classes of non-associative algebras to describe their structure (alternative algebras, Jordan algebras and structurable algebras). Neher's lectures provide a detailed exposition of the structure theory, and they complement Chari's lectures on representations.

Savage's lectures on realizations of crystals are geometric in nature, in contrast to the algebraic approach taken by Kang. The main geometric objects involved are partial flag manifolds, and Lusztig and Nakajima quiver varieties. The crystal graph corresponds to the set of irreducible components of these (or closely related) varieties. Along the way, the basics of quiver representations are also introduced. In the end, Savage describes the relationship between the geometric realization and combinatorial realization via tableaux.

The category of positive level integrable modules of affine Lie algebras with finite-dimensional weight spaces is semisimple and the irreducibles satisfy the Weyl-Kac character formula. Chari's lectures focus on the finite-dimensional level zero modules of the loop algebras. The category of such modules is not semisimple, and homological algebra and quivers have come to play an increasingly important role recently. Some important constructions such as Weyl modules are motivated through their connections to quantum affine algebras.

Finite W-algebras, which appeared as a counterpart to the affine W-algebras in the 1990's, are certain highly nonlinear algebras resulting from a (quantum) hamiltonian reduction. Alternatively, finite W-algebras over the complex numbers can be regarded as a quantization of Slodowy slices, i.e., transversal slices to nilpotent orbits for semisimiple Lie algebras. Finite W-algebras in prime characteristic also arise naturally in Premet's solution to the Kac-Weisfeiler conjecture for modular representations of Lie algebras. Wang's lectures provide an overview of some basic aspects of the structure and representation theory of finite W-algebras and their super generalizations.

More than 130 participants from 14 different countries took part in the summer school. In addition to many Canadian students, a particularly strong contingent of students came from Korea and, of course, the United States. We gratefully acknowledge the generous financial support of The Fields Institute for Research in the Mathematical Sciences (Canada) and the National Science Foundation (USA). We thank the University of Ottawa for providing excellent facilities for the summer school, and all the lecture series speakers for their fine contributions. We also thank all the participants for their active participation which made the summer school (and the workshop) such a success. Finally, we would like to thank the staff of the Fields Institute, particularly Debbie Iscoe, for their help in the preparation of this book.

Erhard Neher
Alistair Savage
Weiqiang Wang

Fields Institute Communications
Volume **59**, 2011

CHAPTER 1

Geometric Constructions of the Irreducible Representations of GL_n

JOEL KAMNITZER

Abstract. We give an exposition of three geometric constructions of the irreducible representations of GL_n. In particular, we discuss Borel-Weil theory, the Ginzburg construction, and the geometric Satake construction. We also explain how to deduce the Ginzburg construction from the geometric Satake equivalence.

Introduction

In these notes, we will give three different geometric constructions of the irreducible representations of GL_n.

First, we discuss the Borel-Weil theorem, which uses line bundles on flag varieties. The flag variety is a projective variety which parametrizes complete flags in a vector space. The flag variety has an action of GL_n and this group also acts on the space of algebraic sections of natural line bundles. These spaces of sections are the irreducible representations.

Second, we give Ginzburg's construction [77] of an action on the cohomology of n-step Springer fibres. A Springer fibre parametrizes flags in a vector space which are invariant under the action of a nilpotent linear operator. In this construction, we get an action of the generators of the Lie algebra of GL_n by convolution.

Finally, we describe the geometric Satake correspondence [163, 112, 169], which is a cornerstone of the geometric Langlands program. In this approach, GL_n acts on the intersection homology of subvarieties of the affine Grassmannian – which parametrizes certain infinite dimension subspaces of an infinite dimensional vector space.

All three of these constructions generalize to arbitrary reductive groups. The Borel-Weil construction works for the flag variety of any reductive group. Ginzburg's construction has been generalized by Nakajima [178] using quiver varieties (see Section 3.4). Finally, the geometric Satake construction generalizes using the affine Grassmannian of the Langlands dual group [112, 169].

In these lectures, we have chosen to focus on the case of GL_n for three reasons. First, we can discuss the results without using too much of the terminology of reductive groups and root systems. Second, flag varieties are much more concrete in this case. Third, we are able to avoid introducing quiver varieties and the Langlands dual group.

The outline of these notes is as follows. In the first section we give an overview of the basic facts concerning the representation theory of GL_n. In the second, third, and fourth sections we give the Borel-Weil, Ginzburg, and geometric Satake constructions. In the final section we explain how to deduce the Ginzburg construction from the geometric Satake construction. In general, we have endevoured to keep the exposition elementary. In particular, we have avoided using perverse sheaves until the final section.

Acknowledgments. I would like to thank all the participants of the summer school for their questions and comments. I especially thank Bruce Fontaine and Lucy Zhang for reading an early draft of these notes. I would also like to thank Erhard Neher and Alistair Savage for organizing this summer school.

1.1 Representation theory of GL_n

1.1.1 Preliminaries and examples. We will begin by reviewing the representation theory of the group $GL_n = GL_n(\mathbb{C})$. First, note that we will consider algebraic (=holomorphic) representations of GL_n. Namely, a representation of GL_n is a finite-dimensional complex vector space V along with a group homomorphism $GL_n \to GL(V)$ which is also required be a morphism of algebraic varieties.

This is equivalent to considering smooth representations of its maximal compact subgroup $U(n)$. This is also equivalent to studying finite-dimensional integral weight representations of its Lie algebra $\mathfrak{gl}_n$.

The material of this section is standard. Good references are [52], [125], and [104].

Example 1.1.1 The simplest representation of GL_n is the "standard representation" which is the action of GL_n on $\mathbb{C}^n$ (so $V = \mathbb{C}^n$ and the map $GL_n \to GL(V)$ is the identity).

Example 1.1.2 More generally, we can also consider actions of GL_n on vector spaces built out of $\mathbb{C}^n$. The first examples are the actions of GL_n on the kth symmetric powers, $\mathrm{Sym}^k\mathbb{C}^n$, and kth exterior powers, $\Lambda^k\mathbb{C}^n$. The action of GL_n on $\mathrm{Sym}^k\mathbb{C}^n$ is given by

$$g \cdot v_1 \cdots v_k = (g \cdot v_1) \cdots (g \cdot v_k)$$

and similarly for $\Lambda^k\mathbb{C}^n$.

Exercise 1.1.3 Identify $\mathrm{Sym}^k\mathbb{C}^n$ with the vector space of homogeneous polynomials of degree k in n variables. Consider the case $n = 2$. Describe the action of an invertible matrix $\left[\begin{smallmatrix} a & b \\ c & d \end{smallmatrix}\right]$ on a homogeneous polynomial $p(x, y)$.

Example 1.1.4 As a special case of the previous example, we may consider $V = \Lambda^n\mathbb{C}^n$. This is a 1-dimensional vector space. Let $e_1, \dots, e_n$ be the standard basis for $\mathbb{C}^n$. Then $e_1 \wedge \cdots \wedge e_n$ is a basis for $\Lambda^n\mathbb{C}^n$.

With respect to this basis, the action of GL_n becomes a group homomorphism $GL_n \to GL_1 = \mathbb{C}^\times$. This homomorphism is the map taking a matrix to its determinant. Hence we will refer to $\Lambda^n\mathbb{C}^n$ as the *determinant representation.*

1.1.2 Representations of tori. Our goal now is to classify all representations of GL_n.

Let us begin with the case of $GL_1 = \mathbb{C}^\times$. A one dimensional representation of $\mathbb{C}^\times$ is given by a (algebraic) group homomorphism $\mathbb{C}^\times \to \mathbb{C}^\times$. All of these are of the form $z \mapsto z^n$ for some integer n.

More generally, let V be a representation of $\mathbb{C}^\times$. We write

$$V_n = \{v \in V : z \cdot v = z^n v, \text{ for all } z \in \mathbb{C}^\times\}$$

for the subspace of vectors of weight n. We have the following result.

Proposition 1.1.5 *With the above notation, we have the direct sum decomposition $V = \oplus_{n\in\mathbb{Z}} V_n$.*

So we have written V as a direct sum of eigenspaces for the elements of $\mathbb{C}^\times$. This proposition captures two important facts about the representation theory of $\mathbb{C}^\times$. First, all representations are semisimple; meaning that they can be written as a direct sum of irreducible subrepresentations. Second, all irreducible representations of $\mathbb{C}^\times$ are 1-dimensional (this holds since $\mathbb{C}^\times$ is abelian). The terms V_n appearing the direct sum decomposition in Proposition 1.1.5 are the *isotypic* components of the representation V. By definition, V_n is the set of all vectors in V which lie in a subrepresentation isormorphic to the one-dimensional representation labelled by n.

More generally, let us consider a group $T = (\mathbb{C}^\times)^n$. Such a group is called a *torus*. As before, we begin by studying the one dimensional representations of this group. These are all of the form

$$z = (z_1, \dots, z_n) \mapsto z^\mu := z_1^{\mu_1} \cdots z_n^{\mu_n}$$

for some sequence of integers $\mu = (\mu_1, \dots, \mu_n) \in \mathbb{Z}^n$. We will call μ a *weight* of T and call the set $P := \mathbb{Z}^n$ the *weight lattice* of T.

Now let V be an arbitrary representation of T. For any weight μ, we can consider the subspace of vectors of weight μ

$$V_\mu := \{v \in V : z \cdot v = z^\mu v\}$$

Proposition 1.1.6 *We have the direct sum decomposition $V = \oplus_{\mu\in\mathbb{Z}^n} V_\mu$*

1.1.3 Weight spaces of representations of GL_n. Within GL_n, there is a large torus $T = (\mathbb{C}^\times)^n$ consisting of the invertible diagonal matrices. We will study representations of GL_n by restricting them to representations of this maximal torus.

Let V be a representation of GL_n. Then it is also a representation of the maximal torus T and so we get a weight decomposition $V = \oplus_\mu V_\mu$. We can record some numerical information about the representation by recording the dimensions of these weight spaces in the function

$$\chi_V : \mathbb{Z}^n \to \mathbb{N}, \ \mu \mapsto \dim(V_\mu).$$

This function is called the *character* of the representation.

Here is the first fundamental result about the representation theory of GL_n.

Theorem 1.1.7 *(i) Consider the action of the symmetric group S_n on $\mathbb{Z}^n$ given by*

$$w(\mu_1, \dots, \mu_n) = (\mu_{w(1)}, \dots, \mu_{w(n)}).$$

If χ_V is a character, then $\chi_V(\mu) = \chi_V(w\mu)$ for any $w \in S_n$.

(ii) If V, W are two representations and $\chi_V = \chi_W$, then $V \cong W$.

Thus the character completely determines the representation.

Exercise 1.1.8 Prove (i). Hint: consider the inclusion $S_n \hookrightarrow GL_n$ which takes permutations to permutation matrices.

Example 1.1.9 Consider the representation $\mathrm{Sym}^3\mathbb{C}^2$. It has a basis given by

$$e_1^3, e_1^2e_2, e_1e_2^2, e_2^3.$$

Note that if $z = \begin{bmatrix} z_1 & 0 \\ 0 & z_2 \end{bmatrix}$, then

$$z \cdot e_1^a e_2^b = (z_1e_1)^a(z_2e_2)^b = z_1^a z_2^b e_1^a e_2^b$$

and hence $e_1^a e_2^b$ has weight (a, b). Hence the non-zero weight spaces of $\mathrm{Sym}^3\mathbb{C}^2$ are $(3,0), (2,1), (1,2), (0,3)$ and each weight space has dimension 1.

Example 1.1.10 More generally consider the representation $\mathrm{Sym}^k\mathbb{C}^n$. It has a basis given by monomials $e_1^{a_1} \dots e_n^{a_n}$ of total degree k. Such a monomial has weight $(a_1, \dots, a_n)$. Hence the character is the function

$$\chi(\mu) = \begin{cases} 1 \text{ if } \mu_1 + \dots + \mu_n = k \\ 0 \text{ otherwise.} \end{cases}$$

In particular, the weights of this representation form a simplex.

Exercise 1.1.11 Find the character of the representation $\Lambda^k\mathbb{C}^n$.

1.1.4 Irreducible representations of GL_n.

A *subrepresentation* $W \subseteq V$ of a representation of GL_n is a subspace of $W \subseteq V$ which is invariant under the action of GL_n.

A representation V of GL_n is called *irreducible* if it has no subrepresentations (other than $0, V$). We have the following semisimplicity result.

Theorem 1.1.12 *Every representation V can be written as a direct sum of irreducible subresentations.*

There are a number of proofs of this theorem. One approach is to use the existence of a Haar measure on the maximal compact subgroup $U(n)$.

Example 1.1.13 Consider the representation of GL_n on $\mathbb{C}^n \otimes \mathbb{C}^n$. We have a direct sum decomposition into subrepresentations $\mathbb{C}^n \otimes \mathbb{C}^n = \Lambda^2\mathbb{C}^n \oplus \mathrm{Sym}^2\mathbb{C}^n$. Later we will see that $\Lambda^2\mathbb{C}^n$ and $\mathrm{Sym}^2\mathbb{C}^n$ are irreducible representations.

Now we would like to describe the irreducible representations of GL_n. For this it will be convenient to introduce a partial order on $\mathbb{Z}^n$. We say $\lambda \geq \mu$ iff we can write

$$\lambda - \mu = k_1\alpha_1 + \dots + k_{n-1}\alpha_{n-1}$$

for some non-negative integers $k_1, \dots, k_{n-1}$ where $\alpha_1 = (1, -1, 0, \dots, 0), \dots, \alpha_{n-1} = (0, \dots, 0, 1, -1)$. In this situation we write $\mathrm{ht}(\lambda - \mu) = k_1 + \dots + k_{n-1}$ and call this number the *height* of $\lambda - \mu$.

It will also be convenient to introduce the subset $P^+ = \mathbb{Z}_+^n$ of *dominant weights* which consists of weakly decreasing sequences, $\mu_1 \geq \dots \geq \mu_n$, with all $\mu_i \in \mathbb{Z}$.

A representation V is said to have *highest weight* λ if $V_\lambda \neq 0$ and whenever $V_\mu \neq 0$, then $\mu \leq \lambda$.

Example 1.1.14 Consider $\mathrm{Sym}^3\mathbb{C}^2$. As calculated in Example 1.1.9, the weights are $(3,0), (2,1), (1,2), (0,3)$. Hence this representation has highest weight $(3,0)$.

Let N denote the subgroup of GL_n consisting of those upper triangular matrices with ones on the diagonals. Let V be a representation of GL_n. A vector $v \in V$ is called a *highest weight vector* if it lies in V_λ for some λ and is N-invariant.

Here is the final main theorem concerning the representation theory of GL_n.

Theorem 1.1.15 *For each $\lambda \in \mathbb{Z}^n_+$, there exists a representation $V(\lambda)$ which can be characterized in one of the following two equivalent ways.*

- *It is irreducible and of highest weight λ.*
- *It has a 1-dimensional space of highest weight vectors and these vectors have weight λ.*

The set $\{V(\lambda)\}_{\lambda\in\mathbb{Z}^n_+}$ is the complete collection of the irreducible representations of GL_n.

These $V(\lambda)$ are the representations that we will describe geometrically.

An interesting problem is to determine the character χ_λ of V_λ. Using our geometric constructions, we will see that $\chi_\lambda(\mu)$ can be expressed as the number of components of a certain variety (see Sections 1.3.4 and 1.4.3).

Example 1.1.16 Take $\lambda = (k, 0, \dots, 0)$. Then $V(\lambda) = \mathrm{Sym}^k\mathbb{C}^n$. To see this note that $\mathrm{Sym}^k\mathbb{C}^n$ has a highest weight vector e_1^k. This vector is of weight $(k, 0, \dots, 0)$ and this is its only highest weight vector (up to scalar).

Exercise 1.1.17 Show that $\Lambda^k\mathbb{C}^n$ has a 1-dimensional space of highest weight vectors. Show that this vector has weight $(1, \dots, 1, 0 \dots, 0)$. We will write ω_k for this weight. Conclude that $\Lambda^k\mathbb{C}^n = V(\omega_k)$.

Example 1.1.18 Consider $\mathfrak{sl}_3$ the vector space of 3×3 trace 0 matrices. We have an action of GL_3 on this space by conjugation. This is the irreducible representation $V(1, 0, -1)$. In particular

$$\begin{bmatrix} 0 & 0 & 1 \\ 0 & 0 & 0 \\ 0 & 0 & 0 \end{bmatrix}$$

is a highest weight vector of weight $(1, 0, -1)$.

The weights of this representation are

$$(1,0,-1), (0,1,-1), (1,-1,0), (0,0,0), (0,-1,1), (-1,1,0), (-1,0,1).$$

All of these weight spaces are 1-dimensional with the exception of the $(0, 0, 0)$ weight space which consists of the diagonal matrices and hence is 2-dimensional.

1.1.5 Determinant representation and polynomial representations. The determinant representation described above is the irreducible representation of highest weight $(1, \dots, 1)$. Its dual representation has highest weight $(-1, \dots, -1)$. In this representation the action of g is given by $g \mapsto \det(g)^{-1}$. Tensoring with these representations moves us around the set of dominant weights in the sense that

$$V(1,\dots,1) \otimes V(\lambda) \cong V(\lambda_1 + 1, \dots, \lambda_n + 1)$$
$$V(-1,\dots,-1) \otimes V(\lambda) \cong V(\lambda_1 - 1, \dots, \lambda_n - 1).$$

Let $\mathbb{N}^n_+ = \{\lambda \in \mathbb{Z}^n_+ : \lambda_i \geq 0 \text{ for all } i\}$. We say that an irreducible representation $V(\lambda)$ is *polynomial*, if $\lambda \in \mathbb{N}^n_+$. More generally, a representation of GL_n is called polynomial if it is the direct sum of polynomial irreducible representations.

Starting with polynomial irreducible representations, we can get to all irreducible representations by repeatedly tensoring with $V(-1,\ldots,-1)$. For this reason it is enough to consider just these polynomial irreducible representations which we will do in later sections.

1.1.6 The Lie algebra action. For some of our geometric constructions, it will be useful to think about the Lie algebra $\mathfrak{gl}_n$ of GL_n. By definition $\mathfrak{gl}_n$ is the Lie algebra of $n \times n$ matrices.

Let $\mathfrak{h}$ denote the Lie subalgebra of diagonal matrices in $\mathfrak{gl}_n$. It is an abelian Lie algebra, usually called the Cartan subalgebra. Define the Chevalley generators E_i, F_i for $i = 1,\ldots,n-1$ by setting E_i to be the matrix with a 1 in the $(i, i+1)$ entry and 0 elsewhere and setting F_i to be the transpose of E_i. The Lie algebra $\mathfrak{gl}_n$ is generated by elements E_i, F_i and the subalgebra $\mathfrak{h}$.

Let V be a representation of GL_n. Then there exists an action of $\mathfrak{gl}_n$ on V by differentiation. This means that there is a Lie algebra morphism $\mathfrak{gl}_n \to \mathrm{End}(V)$.

Suppose that μ is a weight of V. Then μ gives us a Lie algebra morphism $\mathfrak{h} \to \mathbb{C}$ which takes $X = \begin{bmatrix} a_1 & & \\ & \ddots & \\ & & a_n \end{bmatrix}$ to $\mu(X) = \mu_1 a_1 + \cdots + \mu_n a_n$.

The relationship between the action of GL_n and its Lie algebra are summarized by the following result.

Proposition 1.1.19
- *If v is a vector in V of weight μ, then for $X \in \mathfrak{h}$ we have that $X \cdot v = \mu(X)v$.*
- *If $v \in V_\mu$ then $E_i \cdot v \in V_{\mu+\alpha_i}$ and $F_i \cdot v \in V_{\mu-\alpha_i}$.*
- *If $v \in V_\mu$ then $E_i F_i v - F_i E_i v = \langle \mu, \alpha_i \rangle v$ where $\langle , \rangle$ denotes the usual bilinear form (dot product) on $\mathbb{Z}^n$.*
- *A non-zero vector $v \in V_\mu$ is N-invariant (i.e. is a highest weight vector) iff $E_i \cdot v = 0$ for all i.*

Using this Proposition, we get a "picture" of the representation V_λ. We have a weight space decomposition, we move around the weight spaces using E_i, F_i and we can start with our highest weight vector and generate the whole representation by moving down with the F_i. This picture can be formalized through the notion of a crystal (see Section 2.4).

1.2 Borel-Weil theory

The Borel-Weil construction is the most "classical" of the geometric constructions we will consider and hence it can be found in many sources, for example, [77, Section 6.1.13], or [132, Section II.5].

The geometry that will be used in this section concerns line bundles on projective varieties. We begin by recalling some general facts.

1.2.1 Line bundles on projective varieties. Let $\mathbb{P}^n$, *projective space*, denote the set of one dimensional subspaces of $\mathbb{C}^{n+1}$. It is naturally an algebraic variety.

A *projective variety* X is a subset of projective space defined by the vanishing of some homogeneous polynomial equations. A *line bundle* L on a projective variety X is an algebraic variety $\pi : L \to X$ over X such that all fibres carry the structure of a one dimensional complex vector space. Moreover we require that π be locally trivial in the Zariski topology, compatible with the vector space structure on each fibre. Usually we will construct a line bundle by describing the fibre of π over each point in X. If this is done in a natural way, then these fibres will glue together uniquely to form a variety L.

Given the line bundle L, we may consider the set $\Gamma(X, L)$ of all sections, ie $s : X \to L$ such that $\pi \circ s = \text{id}$. The set $\Gamma(X, L)$ forms a vector space using the vector space structure on each fibre.

Example 1.2.1 A simple example of a line bundle is the trivial line bundle $L = X \times \mathbb{C}$. In this case, $\Gamma(X, L)$ will be the vector space of regular functions on X (usually denoted $\mathcal{O}(X)$). A well-known result in algebraic geometry is that the only regular functions on a connected projective variety are constant.

Example 1.2.2 Take $Y = \mathbb{P}^1$ and consider the line bundle L_2 whose fibre at a point $W \in \mathbb{P}^1$ is $\mathbb{C}^2/W$ (one should keep in mind that a point $W \in \mathbb{P}^1$ is a 1-dimensional subspace of $\mathbb{C}^2$).

There is an isomorphism of vector spaces (and later GL_2 representations)

$$\mathbb{C}^2 \to \Gamma(\mathbb{P}^1, L_2)$$
$$v \mapsto [v]_W$$

where for $v \in \mathbb{C}^2$, we let $[v]_W$ denote its image in $\mathbb{C}^2/W$.

We have the following fundamental result.

Theorem 1.2.3 *Let X be a projective variety and L be a line bundle. Then $\Gamma(X, L)$ is a finite dimensional vector space.*

Now suppose that X is a projective variety with an action of an algebraic group G. An *equivariant line bundle* L is a line bundle together an action of G on L such that π is equivariant for these actions.

In this case $\Gamma(X, L)$ will be a representation of G. This action is given as follows. Let $s \in \Gamma(X, L)$ and $g \in G$. Then $(g \cdot s)(x) = g \cdot s(g^{-1} \cdot x)$.

We will construct the irreducible representations of GL_n in this manner for certain line bundles on a projective variety called the flag variety.

1.2.2 The case of $\mathbb{P}^1$. We will begin with $X = \mathbb{P}^1$. To understand the structure of $\mathbb{P}^1$, note that almost every line in $\mathbb{C}^2$ is the span of $(x, 1)$ for some $x \in \mathbb{C}$. The only line not of this form is the span of $(1, 0)$. This allows us to think of $\mathbb{P}^1$ as the union of $\mathbb{C}$ with a "point at infinity".

Note there is an action of GL_2 on $\mathbb{P}^1$ coming from the action of GL_2 on $\mathbb{C}^2$.

On $\mathbb{P}^1$ we have two natural line bundles, L_1 and L_2. Recall that a point in $\mathbb{P}^1$ is a one dimensional subspace W of $\mathbb{C}^2$. Then the fibre of L_1 at W is W and the fibre of L_2 at W is $\mathbb{C}^2/W$ (as described above).

In other words, the total space of the line bundle L_1 is the variety of pairs

$$\{(W, w) : w \in \mathbb{C}^2, W \subseteq \mathbb{C}^2, \dim W = 1, \text{ and } w \in W\}.$$

From this description it is clear that L_1 carries an action of GL_2 compatible with its action on $\mathbb{P}^1$. A similar analysis holds for L_2.

From these line bundles L_1, L_2 we can construct more by tensor product. In particular, for $(a, b) \in \mathbb{N}^2$ we define the line bundle $\mathcal{L}(a, b) = L_1^{\otimes b} \otimes L_2^{\otimes a}$.

The following theorem is the first case of our construction.

Theorem 1.2.4 *Assume $a \geq b$. We have an isomorphism of GL_2 representations $\Gamma(\mathbb{P}^1, \mathcal{L}(a, b)) \cong Sym^{a-b}\mathbb{C}^2 \otimes V(1, 1)^b$.*

From our description of $V(a,0)$ (Example 1.1.16) and our observation about tensoring with the determinant representation (Section 1.1.5), we see that Theorem 1.2.4 implies that $\Gamma(\mathbb{P}^1, \mathcal{L}(a,b)) \cong V(a,b)$.

Proof Let us assume that $b = 0$. There is a map $\mathrm{Sym}^a\mathbb{C}^2 \to \mathrm{Sym}^a(\mathbb{C}^2/W) = (\mathbb{C}^2/W)^{\otimes a}$ for any W. This gives us a map $\mathrm{Sym}^a\mathbb{C}^2 \to \Gamma(\mathbb{P}^1, \mathcal{L}(a,0))$. This map is clearly injective. Showing that it is surjective requires a bit more machinery.

The case of $b \neq 0$ follows immediately since $\mathcal{L}(1,1)$ is isomorphic to the trivial line bundle except with non-trivial GL_2 action (GL_2 acts by the determinant representation). □

1.2.3 The flag variety. Now, we will generalize this construction for $n \geq 2$. For this we will need to define a variety called the flag variety.

A *flag* in $\mathbb{C}^n$ is a sequence of subspaces $0 = V_0 \subseteq V_1 \subseteq \cdots \subseteq V_n = \mathbb{C}^n$ where $\dim V_i = i$.

Example 1.2.5 Let $e_1, \ldots, e_n$ be the standard basis for $\mathbb{C}^n$.

The standard flag is

$$0 \subseteq \langle e_1 \rangle \subseteq \langle e_1, e_2 \rangle \subseteq \cdots \subseteq \mathbb{C}^n$$

and the opposite flag is

$$0 \subseteq \langle e_n \rangle \subseteq \langle e_n, e_{n-1} \rangle \subseteq \cdots \subseteq \mathbb{C}^n.$$

The set of all flags in $\mathbb{C}^n$ forms a projective variety denoted $\mathrm{Fl}(\mathbb{C}^n)$. Note that $\mathrm{Fl}(\mathbb{C}^2)$ is the same thing as $\mathbb{P}^1$.

Exercise 1.2.6 Find an embedding $\mathrm{Fl}(\mathbb{C}^n) \hookrightarrow \mathbb{P}^N$ for some N. Hint: use the Plücker embedding.

The group GL_n acts on the flag variety $\mathrm{Fl}(\mathbb{C}^n)$ by acting on each piece in the flag. In fact, this action is transitive. To see this just note that GL_n acts transitively (in fact simply transitively) on the set of all ordered bases of $\mathbb{C}^n$ and there is a surjective map from ordered bases to flags.

Consider now the set of flags $\mathrm{Fl}(\mathbb{C}^n)^T$ which are invariant under the action of the maximal torus T of GL_n.

Proposition 1.2.7 *There is a bijection between* $\mathrm{Fl}(\mathbb{C}^n)^T$ *and the symmetric group* S_n.

Proof If $V_\bullet \in \mathrm{Fl}(\mathbb{C}^n)^T$, then each V_i is invariant under the action of T. This means that each V_i is a coordinate subspace of $\mathbb{C}^n$. From this it is immediate that $V_\bullet$ can be constructed by taking the standard basis $e_1, \ldots, e_n$ for $\mathbb{C}^n$, putting it in some order and then constructing a flag by adding one basis vector at a time. This ordering gives us an element of S_n.

Conversely if $\sigma \in S_n$, then the corresponding T-fixed flag is

$$0 \subseteq \langle e_{\sigma(1)} \rangle \subseteq \langle e_{\sigma(1)}, e_{\sigma(2)} \rangle \subseteq \cdots \subseteq \mathbb{C}^n.$$

□

The standard and opposite flags are both T-fixed and correspond to the identity and longest permutations, respectively.

Now we will consider the action of the group N of upper triangular matrices with 1s on the diagonal.

Proposition 1.2.8 *The N orbit through the opposite flag is dense in* $\mathrm{Fl}(\mathbb{C}^n)$

Exercise 1.2.9 Show that in the case of $\mathbb{P}^1$, the N-orbit through the opposite flag is precisely the open subset $\mathbb{C}$ in $\mathbb{P}^1$.

Exercise 1.2.10 Prove this Proposition.

1.2.4 Line bundles on the flag variety. For each $\lambda \in \mathbb{N}^n_+$, we now define a line bundle $\mathcal{L}(\lambda)$ on the flag variety. The fibre of $\mathcal{L}(\lambda)$ at a point $V_\bullet$ is

$$(V_1/V_0)^{\otimes \lambda_n} \otimes \cdots \otimes (V_n/V_{n-1})^{\otimes \lambda_1}.$$

Notice that this is a GL_n equivariant line bundle, so we have an action of GL_n on $\Gamma(\mathrm{Fl}(\mathbb{C}^n), \mathcal{L}(\lambda))$. In particular, if p is a T-fixed point, then T acts on the fibre $\mathcal{L}(\lambda)_p$ over p. The reason for the "backwards" order in the definition of $\mathcal{L}(\lambda)$ is that under this definition, the action of T on the fibre over the opposite flag is by weight λ.

Theorem 1.2.11 $\Gamma(\mathrm{Fl}(\mathbb{C}^n), \mathcal{L}(\lambda))$ *is the irreducible representation of* GL_n *of highest weight* λ.

Proof It suffices to show that $\mathcal{L}(\lambda)$ has exactly one N-invariant section (up to scalar) and that this section has weight λ.

Let s be an N-invariant section. Let $p \in \mathrm{Fl}(\mathbb{C}^n)$ denote the opposite flag. Since s is N-invariant, its values on the N orbit through p are determined by its value at p. Since this N orbit is dense, this means that its values on the whole flag variety is determined by its value at p. Put another way, the map $\Gamma(\mathrm{Fl}(\mathbb{C}^n), \mathcal{L}(\lambda))^N \to \mathcal{L}(\lambda)_p$ taking a section to its value at p, is a surjective map. Hence the space of N-invariant sections is at most one dimensional.

To see that there actually is a non-zero N-invariant section requires more detailed understanding of the geometry of the flag variety and this is where we use that $\lambda \in \mathbb{N}^n_+$. (For a good exposition, see [162, Proof of Theorem 2].)

Finally note that if s is an N-invariant section, then $(t \cdot s)(p) = t^\lambda s(p)$ since p is a T-fixed point and the action of T on the fibre over p is by weight λ. Hence if $s(p) \neq 0$, then s has weight λ as desired. □

Example 1.2.12 Let us consider the case $\lambda = (1, 0, \cdots, 0) = \omega_1$. There is a map $\mathrm{Fl}(\mathbb{C}^n) \to \mathbb{P}^{n-1}$ which takes $(V_0, \ldots, V_n)$ to V_{n-1} (here we regard $\mathbb{P}^{n-1}$ as the variety of hyperplanes in $\mathbb{C}^n$). The line bundle $\mathcal{L}(\omega_1)$ is the pullback of the quotient line bundle $\mathcal{O}(1)$ on $\mathbb{P}^{n-1}$. Hence we get a map

$$\Gamma(\mathbb{P}^{n-1}, \mathcal{O}(1)) \to \Gamma(\mathrm{Fl}(\mathbb{C}^n), \mathcal{L}(\omega_1)).$$

This map is an isomorphism and as in the proof of Theorem 1.2.4, $\Gamma(\mathbb{P}^{n-1}, \mathcal{O}(1)) \cong \mathbb{C}^n$. Hence we see that

$$\mathbb{C}^n \cong \Gamma(\mathbb{P}^{n-1}, \mathcal{O}(1)) \cong \Gamma(\mathrm{Fl}(\mathbb{C}^n), \mathcal{L}(\omega_1)) \cong V(\omega_1).$$

Exercise 1.2.13 Carry out a similar analysis when $\lambda = \omega_k$ and when $\lambda = (k, 0, \ldots, 0)$.

1.3 Ginzburg construction

In the Borel-Weil construction, we started with a variety with an action of GL_n and obtained a representation of GL_n. In the Ginzburg (and geometric Satake) constructions we will start with a variety that does not carry an action of GL_n and so the construction of a representation of GL_n is more indirect. Actually instead of getting an action of GL_n, we will get a vector space with a weight decomposition and an action of the generators E_i, F_i of the Lie algebra $\mathfrak{gl}_n$.

Ginzburg's construction was developed by Ginzburg in [77, Chapter 4], building on earlier work of Beilinson-Lusztig-MacPherson [23].

1.3.1 Partial flag varieties, nilpotent operators and Springer fibres. Let us fix two positive integers n, N. We will now consider the variety $\mathrm{Fl}_n(\mathbb{C}^N)$ of all n-step partial flags in $\mathbb{C}^N$

$$\mathrm{Fl}_n(\mathbb{C}^N) = \{0 = V_0 \subseteq V_1 \subseteq \cdots \subseteq V_n = \mathbb{C}^N\}.$$

Note that we do not put any restrictions on the dimensions of V_i, however it is clear that $\mathrm{Fl}_n(\mathbb{C}^N)$ breaks into different connected pieces according to the dimensions of the V_i.

Let $\mu \in \mathbb{N}^n$ with $|\mu| = \mu_1 + \cdots + \mu_n = N$. Then we can consider the connected component $\mathrm{Fl}_\mu(\mathbb{C}^N)$ of those partial flags in $\mathbb{C}^N$ with $\dim(V_i/V_{i-1}) = \mu_i$. In particular when $n = N$, then $\mathrm{Fl}_{(1,\ldots,1)}(\mathbb{C}^n)$ is our flag variety $\mathrm{Fl}(\mathbb{C}^n)$ from the previous section.

We now consider the cotangent bundle $T^*\mathrm{Fl}_n(\mathbb{C}^N)$ of this n-step flag variety. We will think of a point of $T^*\mathrm{Fl}_n(\mathbb{C}^N)$ as a pair $(X, V_\bullet)$ such that $X \in \mathrm{End}(\mathbb{C}^N)$, $V_\bullet \in \mathrm{Fl}_n(\mathbb{C}^N)$ and $XV_i \subseteq V_{i-1}$.

Notice that those X which occur in $T^*\mathrm{Fl}_n(\mathbb{C}^N)$ will necessarily have $X^n = 0$. Hence there is a projection $\pi : T^*\mathrm{Fl}_n(\mathbb{C}^N) \to \mathcal{N}_n(\mathbb{C}^N)$ where $\mathcal{N}_n(\mathbb{C}^N)$ is the variety of $N \times N$ matrices X with $X^n = 0$. The fibres of this map are called *n-step Springer fibres*. We will write $\mathrm{Fl}_n(\mathbb{C}^N)^X$ for the fibre over the nilpotent X.

Example 1.3.1 Let $N = 3, n = 3$. Let $X = \left[\begin{smallmatrix} 0&1&0\\0&0&0\\0&0&0 \end{smallmatrix}\right]$ be an operator on $\mathbb{C}^3$. Let $\mu = (1,1,1)$. Then if $V_\bullet \in \mathrm{Fl}_\mu(\mathbb{C}^3)^X$, we see that either

$$0 \subseteq V_1 = \langle e_1 \rangle \subseteq V_2 \subseteq \mathbb{C}^3 \text{ or } 0 \subseteq V_1 \subseteq V_2 = \langle e_1, e_3 \rangle \subseteq \mathbb{C}^3$$

This shows that $\mathrm{Fl}_\mu(\mathbb{C}^3)^X$ has two irreducible components each of which is isomorphic to $\mathbb{P}^1$. These $\mathbb{P}^1$s are glued together at the point $0 \subseteq \langle e_1 \rangle \subseteq \langle e_1, e_3 \rangle \subseteq \mathbb{C}^3$.

Recall from linear algebra, that every $N \times N$ matrix is similar to a Jordan form matrix. If we just consider nilpotent linear operators (i.e. all eigenvalues are 0) then there is a bijection between Jordan form matrices and decreasing sequences $\nu_1 \geq \cdots \geq \nu_m$ of positive integers with $|\nu| = N$ — here the ν_i are the sizes of the blocks. If X is a nilpotent linear operator on $\mathbb{C}^m$, then we will speak of its *Jordan type*, which will be such a sequence.

Let $X \in \mathcal{N}_n(\mathbb{C}^N)$ be a nilpotent operator with Jordan type ν. Let λ be the conjugate of ν — this means that if we write ν as a Young diagram where the ν_i are the lengths of the rows, then λ_i will be the lengths of the columns. In other words λ_i is the number of j such that $\nu_j \geq i$. Since $X^n = 0$, $\nu_i \leq n$ for all i and this means that $\lambda_k = 0$ for $k > n$. In particular, λ can be regarded as a dominant weight for GL_n. Also note that $|\lambda| = N$.

Exercise 1.3.2 Let X, λ be as above. Suppose that the Springer fibre $\mathrm{Fl}_\mu(\mathbb{C}^N)^X$ is non-empty. Show that for each k,

$$\mu_1 + \cdots + \mu_k \leq \lambda_1 + \cdots + \lambda_k.$$

(Hint: first show that $\lambda_1 + \cdots + \lambda_k = \dim(\ker X^k)$.)

Conclude that as weights of GL_n, we have $\mu \leq \lambda$.

Show that $\mathrm{Fl}_\lambda(\mathbb{C}^N)^X$ is a single point.

The intuition behind the Ginzburg construction is that we will construct a representation of $\mathfrak{gl}_n$ on $\mathrm{Fl}_n(\mathbb{C}^N)^X$ where the weight spaces are the $\mathrm{Fl}_\mu(\mathbb{C}^N)^X$ and where E_i acts on a flag $V_\bullet$ to give the sum of all the flags $V'_\bullet$ with $V_i \subseteq V'_i$ and $V_j = V'_j$ for $j \neq i$. Of course, this description does not really make sense since $\mathrm{Fl}_n(\mathbb{C}^N)^X$ is a variety, not a vector space. So we will need a way to get a vector space out of a variety and for this we will use homology.

1.3.2 Borel-Moore homology. For the purpose of this construction, we will need to consider *Borel-Moore homology* of certain varieties.

For a well-behaved topological space Y (for example a complex algebraic variety), its Borel-Moore homology is the homology of the complex of locally finite, but possibly unbounded, i-chains in Y. This means that $H_i(Y)$ is the group of (formal $\mathbb{C}$-linear combinations of) (possibly singular and unbounded) closed i-dimensional subspaces of Y modulo those which are boundaries of $i+1$-dimensional subspaces.

The important properties that we will need for Borel-Moore homology are

(i) Pullback: if $f : X \to Y$ is a locally trivial fibre bundle with smooth fibre of dimension d, then there is a pull-back $f^* : H_i(Y) \to H_{i+d}(X)$.

(ii) Pushforward: if $f : X \to Y$ is a proper map (ie the pull back of a compact set is compact), then there is a push forward $f_* : H_i(X) \to H_i(Y)$.

(iii) Intersection pairing: if M is a smooth manifold of real dimension m and Z, Z', Z'' are three closed (well-behaved) subsets such that Z'' contains $Z \cap Z'$, then there is an intersection product $\cap_M : H_i(Z) \times H_j(Z') \to H_{i+j-m}(Z'')$.

(iv) Fundamental class: if Y is a algebraic variety of complex dimension n, then we have a fundamental class $[Y] \in H_{2n}(Y)$. If Y has multiple irreducible components $Y_1, \dots, Y_k$ then the fundamental classes of these components $[Y_1], \dots, [Y_k]$ form a basis for $H_{2n}(Y)$. In this circumstance we will write $H_{\mathrm{top}}(Y) = H_{2n}(Y)$.

1.3.3 The main construction. Fix n, N as above.

For each $i = 1, \dots, n-1$ consider the following subvariety $Z_i \subseteq T^*\mathrm{Fl}_n(\mathbb{C}^N) \times T^*\mathrm{Fl}_n(\mathbb{C}^N)$ defined by

$$\begin{aligned} Z_i = \{(X, V_\bullet, V'_\bullet) &: (X, V_\bullet), (X, V'_\bullet) \in T^*\mathrm{Fl}_n(\mathbb{C}^N), \\ &V_j = V'_j \text{ for } j \neq i, \text{ and } V_i \subseteq V'_i \text{ with } \dim(V'_i) = \dim(V_i) + 1\} \end{aligned}$$

Let π_1, π_2 denote the two projections $T^*\mathrm{Fl}_n(\mathbb{C}^N) \times T^*\mathrm{Fl}_n(\mathbb{C}^N) \to T^*\mathrm{Fl}_n(\mathbb{C}^N)$. For any $X \in \mathcal{N}_n(\mathbb{C}^N)$ we define $E_i : H_*(\mathrm{Fl}_n(\mathbb{C}^N)^X) \to H_*(\mathrm{Fl}_n(\mathbb{C}^N)^X)$ by

$$\begin{aligned} H_*(\mathrm{Fl}_n(\mathbb{C}^N)^X) &\xrightarrow{\pi_1^*} H_*(\mathrm{Fl}_n(\mathbb{C}^N)^X \times T^*\mathrm{Fl}_n(\mathbb{C}^N)) \\ &\xrightarrow{\cap_M [Z_i]} H_*(\mathrm{Fl}_n(\mathbb{C}^N)^X \times \mathrm{Fl}_n(\mathbb{C}^N)^X) \xrightarrow{\pi_{2*}} H_*(\mathrm{Fl}_n(\mathbb{C}^N)^X) \end{aligned} \tag{1.1}$$

where $M = T^*\mathrm{Fl}_n(\mathbb{C}^N) \times T^*\mathrm{Fl}_n(\mathbb{C}^N)$ is the ambient variety where the intersection takes place.

We also define $F_i : H_*(\mathrm{Fl}_n(\mathbb{C}^N)^X) \to H_*(\mathrm{Fl}_n(\mathbb{C}^N)^X)$ in the same manner, by interchanging the roles of π_1 and π_2 in (1.1).

Note that by the very definition, we see that if $(X, V, V') \in Z_i$ and $V \in \mathrm{Fl}_\mu(\mathbb{C}^N)$ then $V' \in \mathrm{Fl}_{\mu+\alpha_i}(\mathbb{C}^N)$, thus we see that

$$Z_i = \cup_\mu Z_{i\mu}, \quad \text{with } Z_{i\mu} \subseteq T^*\mathrm{Fl}_\mu(\mathbb{C}^N) \times T^*\mathrm{Fl}_{\mu+\alpha_i}(\mathbb{C}^N)$$

This shows us that $E_i : H_*(\mathrm{Fl}_\mu(\mathbb{C}^N)^X) \to H_*(\mathrm{Fl}_{\mu+\alpha_i}(\mathbb{C}^N)^X)$ for each μ, and vice versa for F_i.

Theorem 1.3.3 *This defines a representation of GL_n with weight decomposition $H_*(\mathrm{Fl}_n(\mathbb{C}^N)^X) = \oplus_\mu H_*(\mathrm{Fl}_\mu(\mathbb{C}^N)^X)$ and action of Chevalley generators given by E_i, F_i.*

1.3.4 The irreducible representation. In general this representation will be reducible. Now let us describe an irreducible subrepresentation.

Theorem 1.3.4 *Consider $H_{\mathrm{top}}(\mathrm{Fl}_n(\mathbb{C}^N)^X) = \oplus_\mu H_{\mathrm{top}}(\mathrm{Fl}_\mu(\mathbb{C}^N)^X)$. This subspace is invariant under the action of $\mathfrak{gl}_n$. Moreover it is an irreducible representation of highest weight λ (where λ is the conjugate of the Jordan type of X). In particular there is a basis for $V(\lambda)_\mu$ labelled by irreducible components of* $\mathrm{Fl}_\mu(\mathbb{C}^N)^X$.

Carefully examining the homological degrees in (1.1) shows that $H_{\mathrm{top}}(\mathrm{Fl}_n(\mathbb{C}^N)^X)$ is invariant under the action of $\mathfrak{gl}_n$ (and hence GL_n). Also by Exercise 1.3.2, we see that λ is the highest weight of the representation $H_{\mathrm{top}}(\mathrm{Fl}_n(\mathbb{C}^N)^X)$. So to complete the proof of this theorem, we would need to know that $H_{\mathrm{top}}(\mathrm{Fl}_n(\mathbb{C}^N))^X$ is irreducible, but this is not as easy.

One nice feature of this Ginzburg construction is that it provides us with a basis of $V(\lambda)_\mu$. This allows us to answer combinatorial questions — for example we can determine $\chi_\lambda(\mu)$ by counting the irreducible components of $\mathrm{Fl}_\mu(\mathbb{C}^N)^X$.

Example 1.3.5 Consider the case $n = 2$ and $\lambda = (k, 0)$. So $N = k$. Then we see that $X = 0$ and for $\mu = (a, b)$ we see that $\mathrm{Fl}_\mu(\mathbb{C}^k)^X = \mathrm{Gr}(a, \mathbb{C}^k)$, the Grassmannian of a dimensional subspaces of $\mathbb{C}^n$. In particular it is an irreducible variety and so we see that $\chi_\lambda(\mu) = 1$ (which we already knew from our description $V_\lambda = \mathrm{Sym}^k\mathbb{C}^2$).

Exercise 1.3.6 We continue from Example 1.3.1 (so $\lambda = (2, 1, 0)$ and $\mu = (1, 1, 1)$). We saw that $\mathrm{Fl}_\mu(\mathbb{C}^3)^X$ has two irreducible components. On the another hand we already knew that $V(\lambda)_\mu$ was 2-dimensional from Example 1.1.18.

Exercise 1.3.7 Continue with $\lambda = (2, 1, 0)$. We have defined the operator

$$E_1 : H_{\mathrm{top}}(\mathrm{Fl}_{(1,1,1)}(\mathbb{C}^3)^X) \to H_{\mathrm{top}}(\mathrm{Fl}_{(2,0,1)}(\mathbb{C}^3)^X).$$

From the previous exercise, $H_{\mathrm{top}}(\mathrm{Fl}_{(1,1,1)}(\mathbb{C}^3)^X)$ is 2-dimensional with a basis given by the two irreducible components. On the other hand $\mathrm{Fl}_{(2,0,1)}(\mathbb{C}^3)^X)$ is a point (as in Section 1.3.2). So with respect to this basis E_1 can be written as 1×2 matrix. Show that it is [2 1].

1.4 Geometric Satake correspondence

The following construction is part of the geometric Satake correspondence which is due to Lusztig [163], Ginzburg [112], and Mirkovic-Vilonen [169].

1.4.1 The varieties Gr^λ. For each polynomial dominant weight $\lambda \in \mathbb{N}^n_+$, we will construct a variety Gr^λ.

Consider the (infinite-dimensional) $\mathbb{C}$-vector space $\mathbb{C}[z]\otimes\mathbb{C}^n$. So if $\mathbb{C}^n$ has basis $e_1, \dots, e_n$, then $\mathbb{C}[z] \otimes \mathbb{C}^n$ has a basis $\{z^k e_i\}$ where $k = 0, 1, \dots$ and $i = 1, \dots, n$. Define a linear operator

$$X : \mathbb{C}[z] \otimes \mathbb{C}^n \to \mathbb{C}[z] \otimes \mathbb{C}^n$$

$$z^k e_i \mapsto \begin{cases} z^{k-1} e_i \text{ if } k \geq 1 \\ 0 \text{ if } k = 0 \end{cases}$$

Let $\mathrm{Gr}^{\geq} = \{L \subseteq \mathbb{C}[z]\otimes\mathbb{C}^n : XL \subseteq L\}$ be the ind-variety of all finite-dimensional subspaces of this vector space which are X invariant. This is called the positive part of the *affine Grassmannian.*

Let $L \in \mathrm{Gr}^{\geq}$. We can consider the restriction of X to L. This will be a nilpotent operator and hence it will have a Jordan type. Note that since $\dim\ker(X) = n$, it follows that $\dim\ker(X|_L) \leq n$ which implies that $X|_L$ has at most n Jordan blocks. Hence the Jordan type of $X|_L$ can be regarded as an element of $\mathbb{N}^n_+$.

We define

$$\mathrm{Gr}^\lambda := \{L \subseteq \mathbb{C}[z] \otimes \mathbb{C}^n : XL \subseteq L, \text{ and } X|_L \text{ has Jordan type } \lambda\}.$$

It is a locally closed subset in $\mathrm{Gr}^{\geq}$ and we may take its closure to obtain $\overline{\mathrm{Gr}^\lambda}$.

It is easy to see exactly which subspaces we pick up in the closure — in the closure the Jordan type of $X|_L$ becomes more "special", i.e. more like the zero matrix. This can be expressed as follow.

Lemma 1.4.1

$$\overline{\mathrm{Gr}^\lambda} = \bigcup_{\mu\in\mathbb{N}^n_+ : \mu\leq\lambda} \mathrm{Gr}^\mu$$

Example 1.4.2 Suppose that $\lambda = \omega_k = (1, \dots, 1, 0, \dots, 0)$. Having Jordan type ω_k means that $X|_L$ is 0 and that $\dim L = k$. Hence we see that Gr^λ is isomorphic to the Grassmannian of k dimensional subspaces of $\ker(X) = \mathbb{C}^n$. Note that in this case, $\mathrm{Gr}^\lambda = \overline{\mathrm{Gr}^\lambda}$.

Exercise 1.4.3 Describe $\overline{\mathrm{Gr}^{(2,0)}}$.

It is not immediately obvious that these varieties Gr^λ are non-empty. Let us demonstrate this now. For each $\mu \in \mathbb{N}^n$, we define a subspace $L_\mu \subseteq \mathbb{C}[z] \otimes \mathbb{C}^n$ as the subspace generated by the action of X on $\{z^{\mu_1-1}e_1, \dots, z^{\mu_n-1}e_n\}$.

Exercise 1.4.4 $L_\lambda \in \mathrm{Gr}^\lambda$. More generally $L_{w\lambda} \in \mathrm{Gr}^\lambda$ for all $w \in S_n$.

We will construct the representation $V(\lambda)$ of GL_n using these varieties $\overline{\mathrm{Gr}^\lambda}$.

1.4.2 Intersection homology and MV cycles. Recall that the homology of smooth manifolds obeys Poincaré duality. However, the variety $\overline{\mathrm{Gr}^\lambda}$ is usually singular. For well-behaved compact topological spaces (for example complex projective varieties), there is an invariant, *intersection homology*, which does obey Poincaré duality. It is defined by restricting the possible intersections between the chains and the singular strata.

More precisely, suppose that Y admits a stratification $Y = Y_n \supseteq Y_{n-1} \supseteq \cdots Y_0$, which means that the Y_k are closed subsets such that $Y_k \smallsetminus Y_{k-1}$ is a manifold of dimension $2k$ (and some other conditions). Then $IH_*(Y)$ is the homology of the complex of i-chains $Z \subseteq Y$ such that $\dim(Z \cap Y_{n-k}) \leq i - k$.

We will now describe a basis for the intersection homology of $\overline{\mathrm{Gr}^\lambda}$ using a type of Bialynicki-Birula decomposition.

Consider the action of the torus $T = (\mathbb{C}^\times)^n$ on $\mathbb{C}[z] \otimes \mathbb{C}^n$ (acting on the second tensor factor). This action commutes with the operator X and hence gives us an action on the variety of X-invariant subspaces. In particular it acts on $\overline{\mathrm{Gr}^\lambda}$.

Exercise 1.4.5 The fixed points of this T-action are precisely the subspaces L_μ described above.

Now, define a map $\mathbb{C}^\times \hookrightarrow (\mathbb{C}^\times)^n$ by $s \mapsto \begin{bmatrix} s^{n-1} & & \\ & \ddots & \\ & & 1 \end{bmatrix}$. Then given a point $L \in \overline{\mathrm{Gr}^\lambda}$, we can consider $\lim_{s \to 0} s \cdot L$. This limit will exist since $\overline{\mathrm{Gr}^\lambda}$ is proper. For each μ such that $L_\mu \in \overline{\mathrm{Gr}^\lambda}$, we consider the subvariety

$$\{L \in \overline{\mathrm{Gr}^\lambda} : \lim_{s \to 0} s \cdot L = L_\mu\}$$

The components of this subvariety are called *Mirkovic-Vilonen cycles* for $\overline{\mathrm{Gr}^\lambda}$ of weight μ.

Theorem 1.4.6 *The set of all MV cycles for* $\overline{\mathrm{Gr}^\lambda}$ *forms a basis for* $IH_*(\overline{\mathrm{Gr}^\lambda})$. *In particular the MV cycles of weight* μ *contribute to* $IH_{2k}(\overline{\mathrm{Gr}^\lambda})$ *where* $k = \mathrm{ht}(\lambda - \mu)$.

This theorem allows us to formally define a grading on $IH_*(\overline{\mathrm{Gr}^\lambda})$ by the weight lattice of GL_n. Namely, we set

$$IH_\mu(\overline{\mathrm{Gr}^\lambda}) = \text{ span of MV cycles of weight } \mu.$$

Example 1.4.7 Let us examine this theorem in the example discussed in Section 1.4.2. So $\lambda = \omega_k$ and $\mathrm{Gr}^\lambda = \mathrm{Gr}(k, \mathbb{C}^n)$. The T-fixed points in $\mathrm{Gr}(k, \mathbb{C}^n)$ are naturally labelled by subsets S of $\{1, \dots, n\}$ of size k. These subsets give us coordinate subspaces $W_S \subseteq \mathbb{C}^n$ (namely W_S is the span of $\{e_i : i \in S\}$). For such a subset S, the attracting set

$$\{W \in \mathrm{Gr}(k, \mathbb{C}^n) : \lim_{s \to 0} s \cdot W = W_S\}$$

is called a *Schubert cell*.

These Schubert cells (which are irreducible and isomorphic to an affine space) give a cell decomposition of $\mathrm{Gr}(k, \mathbb{C}^n)$. Thus there is only one MV cycle for each μ and it is the closure of this Schubert cell (usually called a Schubert variety). Thus in this case $IH_*(\overline{\mathrm{Gr}^\lambda}) = H_*(\mathrm{Gr}^\lambda)$ has a basis given by MV cycles which are labelled by k-element subsets of $\{1, \dots, n\}$.

1.4.3 The representation. To define the action of $\mathfrak{gl}_n$ on $IH_*(\overline{\mathrm{Gr}^\lambda})$ we need one more geometric ingredient. Let Y be a projective variety with a very ample line bundle $\mathcal{L}$. The first Chern class of $\mathcal{L}$ gives a map

$$c(\mathcal{L}) : IH_k(Y) \to IH_{k-2}(Y)$$

One may think of this map as taking a k-chain and intersecting with the zero set of a generic section of $\mathcal{L}$, or equivalently, intersecting with a hyperplane inside the projective embedding coming from $\mathcal{L}$.

For our variety $\overline{\mathrm{Gr}^\lambda}$, we define a line bundle $\mathcal{L}$ whose fibre at the point L is given by $\mathcal{L}_L := \det(L)^*$, where det of a vector space denotes the top exterior power.

This $\mathcal{L}$ gives us a map $c(\mathcal{L}) : IH_k(\overline{\mathrm{Gr}^\lambda}) \to IH_{k-2}(\overline{\mathrm{Gr}^\lambda})$. Using our grading by the weight lattice, we may split this $c(\mathcal{L})$ into matrix coefficients $c(\mathcal{L})_{\mu\nu} : IH_\mu(\overline{\mathrm{Gr}^\lambda}) \to IH_\nu(\overline{\mathrm{Gr}^\lambda})$, where $ht(\lambda - \nu) = ht(\lambda - \mu) - 1$. A refined analysis shows that this matrix coefficient vanishes unless $\nu = \mu + \alpha_i$ for some $i = 1, \dots, n-1$. Thus, we can split $c(\mathcal{L})$ as a direct sum $n-1$ of operators E_i, with

$$E_i : IH_\mu(\overline{\mathrm{Gr}^\lambda}) \to IH_{\mu+\alpha_i}(\overline{\mathrm{Gr}^\lambda}).$$

Theorem 1.4.8 *The vector space $IH_*(\overline{\mathrm{Gr}^\lambda})$ with its weight space decomposition and E_i defined as above is a representation of $\mathfrak{gl}_n$. In particular it is the irreducible representation of highest weight λ.*

One might wonder where the F_i come from. A deep result in the topology of algebraic varieties, called the Hard Lefschetz theorem, states that for any projective variety Y, there exists an operator $c(\mathcal{L})^*$ such that the pair $c(\mathcal{L}), c(\mathcal{L})^*$ generate an action of $\mathfrak{sl}_2$ on $IH_*(Y)$ compatible with the grading. To get our F_i, we then split $c(\mathcal{L})^*$ into F_i, just as we split our $c(\mathcal{L})$ into E_i (except we actually have to divide by certain coeffiecients when defining the F_i).

As a corollary of Theorems 1.4.6 and 1.4.8, we find that the number of MV cycles for $\overline{\mathrm{Gr}^\lambda}$ of weight μ equals the dimension of the μ weight space in $V(\lambda)$.

Example 1.4.9 Continuing from Example 1.4.7, where we saw that $IH_*(Gr^{\omega_k})$ has a basis labelled by k element subset of $\{1, \dots, n\}$. Also $V(\omega_k) = \Lambda^k \mathbb{C}^n$ has a basis labelled by k elements subsets of $\{1, \dots, n\}$, so this isomorphism is clear in that case.

Exercise 1.4.10 Consider $\overline{\mathrm{Gr}^{(2,1,0)}}$. Show that there are two MV cycles of weight $(1,1,1)$.

1.5 Geometric skew Howe duality

We will discuss a "duality" relation between the Ginzburg and geometric Satake constructions. In particular, this will allow us to deduce the Ginzburg construction from the geometric Satake correspondence, following ideas of Braverman, Gaitsgory, Mirkovic, and Vybornov (see [43, 42, 170]).

We will begin by giving a more refined statement of the geometric Satake correspondence.

1.5.1 Geometric Satake. Let $Y = \cup Y_\alpha$ be a stratified space. Here each Y_α is a locally closed submanifold of Y. Also we require that the closure of any Y_α is the union of various Y_β. To simplify the following discussion, let us assume that all of the Y_α are simply-connected.

In this case, there is a category $\mathcal{P}(Y)$ of perverse sheaves on Y which are constructible with respect to this stratification. Perverse sheaves are a subset of the set of all complexes of constructible sheaves on Y which obey certain vanishing conditions with respect to their stalks and costalks. For more information on the definition and basic properties of perverse sheaves, see [77, Section 8.4].

We will need the following properties.

Proposition 1.5.1
- *The category $\mathcal{P}(Y)$ is an abelian category.*
- *The irreducible objects of $\mathcal{P}(Y)$ are the IC sheaves $IC(\overline{Y_\alpha})$ of the strata. These can be characterized as the "simplest" perverse sheaves which are supported on $\overline{Y_\alpha}$ and whose restriction to Y_α is the (shifted) constant sheaf.*
- *The push-forward of a perverse sheaf under a semi-small map is perverse (see* [77, Section 8.9].

Given a perverse sheaf $\mathcal{F}$ on Y, we may consider the cohomology of Y with coefficients in $\mathcal{F}$. In particular we have $H^*(Y, IC(\overline{Y_\alpha})) = IH(\overline{Y_\alpha})$.

Recall that $\mathrm{Gr}^{\geq} = \cup_{\lambda \in \mathbb{N}^n_+} \mathrm{Gr}^\lambda$ is the space of all subspaces $L \subseteq \mathbb{C}[z] \otimes \mathbb{C}^m$ such that $XL \subseteq L$. We will consider it as a stratified space with strata Gr^λ.

Let $\mathrm{Rep}^{\geq}(GL_m)$ denote the category of polynomial representations of GL_m. Now we can state the geometric Satake correspondence.

Theorem 1.5.2 *There is an equivalence of categories*

$$\mathcal{P}(\mathrm{Gr}^{\geq}) \cong \mathrm{Rep}^{\geq}(GL_m).$$

This equivalence is compatible with the functors to the category of vector spaces (denoted $Vect$) on both sides. One functor is $\mathcal{F} \mapsto H^*(\mathrm{Gr}^{\geq}, \mathcal{F})$ and the other functor is the forgetful functor.

Under this equivalence $IC(\overline{\mathrm{Gr}^\lambda})$ is taken to $V(\lambda)$.

Using the compatibility with the functors to $Vect$, we get our previous statement that $IH(\overline{\mathrm{Gr}^\lambda}) \cong V(\lambda)$.

The above equivalence is actually an equivalence of tensor categories with respect to a tensor structure on $\mathcal{P}(\mathrm{Gr}^{\geq})$. Instead of giving a full definition of this tensor structure, we will explain how it works for the IC sheaves.

Let $\lambda^1, \dots, \lambda^n$ be a sequence of dominant weights of GL_m. Then we define the variety

$$\begin{aligned}\overline{\mathrm{Gr}^{\lambda^1,\dots,\lambda^n}} := \{0 = &L_0 \subseteq L_1 \subseteq \cdots \subseteq L_n \subseteq \mathbb{C}[z] \otimes \mathbb{C}^m : \\ &XL_i \subseteq L_i \text{ and } X|_{L_i/L_{i-1}} \text{ has Jordan type } \leq \lambda^i\}\end{aligned}$$

There is an obvious map $m_{\lambda^1,\dots,\lambda^n} : \overline{\mathrm{Gr}^{\lambda^1,\dots,\lambda^n}} \to \mathrm{Gr}$ taking $(L_0, \dots, L_n)$ to L_n. Actually its image is in $\overline{\mathrm{Gr}^{\lambda^1+\cdots+\lambda^n}}$.

Theorem 1.5.3 *Under the geometric Satake equivalence, $m_*(IC(\overline{\mathrm{Gr}^{\lambda_1,\dots,\lambda_n}}))$ corresponds to $V(\lambda^1) \otimes \cdots \otimes V(\lambda^n)$.*

Corollary 1.5.4 *Let $\mu, \lambda^1, \dots, \lambda^n \in \mathbb{N}^m_+$. Then*

$$\begin{aligned}\mathrm{Hom}_{GL_m}\left(V(\mu), V(\lambda^1) \otimes \cdots \otimes V(\lambda^n)\right) &\cong \mathrm{Hom}_{\mathcal{P}(\mathrm{Gr})}(IC(\overline{\mathrm{Gr}^\mu}), m_*(IC(\overline{\mathrm{Gr}^{\lambda_1,\dots,\lambda_n}})) \\ &\cong H_{\mathrm{top}}(m^{-1}_{\lambda^1,\dots,\lambda^n}(L_\mu)).\end{aligned}$$

For the second isomorphism, we take the stalk of $m_*(IC(\overline{\mathrm{Gr}^{\lambda_1,\dots,\lambda_n}}))$ at the point L_μ and use the fact that $m_{\lambda^1,\dots,\lambda^n}$ is strictly semismall (for a general statement of this type, see [113, Theorem 5.4]).

1.5.2 Skew-Howe duality. Fix n, m. Consider the vector space $\mathbb{C}^m \otimes \mathbb{C}^n$. This vector space has commuting actions of GL_n, GL_m.

Let V be a vector space with actions of two reductive groups G, H. These actions are called *Howe dual* if the actions commute and the image of G generates the $\mathrm{Hom}_H(V, V)$ (i.e., $U(\mathfrak{g}) \to \mathrm{Hom}_H(V, V)$ is onto) and vice versa.

Proposition 1.5.5 *Under this situation, $V \cong \oplus_{i=1}^k U_i \otimes W_i$ where U_i is an irreducible representation of G, W_i is an irreducible representation of H. In particular $\mathrm{Hom}_G(W_i, V) \cong U_i$ as H-representations.*

Exercise 1.5.6 Prove this proposition.

Now we fix N and consider $\Lambda^N(\mathbb{C}^n \otimes \mathbb{C}^m)$. This vector space has actions of GL_n and GL_m coming from their actions on $\mathbb{C}^n \otimes \mathbb{C}^m$. In fact these actions of GL_n and GL_m are Howe dual. We have the following result.

Theorem 1.5.7

$$\Lambda^N(\mathbb{C}^n \otimes \mathbb{C}^m) \cong \bigoplus_{\lambda \in \mathbb{N}_m^+, |\lambda| = N, \lambda_i \leq n} V(\lambda^\vee) \otimes V(\lambda)$$

Here $\lambda^\vee$ is the conjugate of λ as defined in Section 1.3.1. The Young diagrams of these λ fit inside a $n \times m$ box.

In particular this theorem shows that

$$\mathrm{Hom}_{GL_m}(V(\lambda), \Lambda^N(\mathbb{C}^n \otimes \mathbb{C}^m)) \cong V(\lambda^\vee) \tag{1.2}$$

as representations of GL_n.

Example 1.5.8 Consider $\Lambda^2(\mathbb{C}^2 \otimes \mathbb{C}^3)$. By the theorem we have

$$\Lambda^2(\mathbb{C}^2 \otimes \mathbb{C}^3) \cong V(2,1) \otimes V(2,1,0) \oplus V(3,0) \otimes V(1,1,1)$$

Alternatively, we can see that as GL_3 representations

$$\begin{aligned}\Lambda^2(\mathbb{C}^2 \otimes \mathbb{C}^3) &\cong \Lambda^3(\mathbb{C}^3 \oplus \mathbb{C}^3) \\ &\cong \Lambda^3\mathbb{C}^3 \oplus (\Lambda^2\mathbb{C}^3 \otimes \mathbb{C}^3) \oplus (\mathbb{C}^3 \otimes \Lambda^2\mathbb{C}^3) \oplus \Lambda^3\mathbb{C}^3 \\ &\cong V(1,1,1) \oplus V(2,1,0) \oplus V(1,1,1) \oplus V(2,1,0) \oplus V(1,1,1) \oplus V(1,1,1).\end{aligned}$$

So we see $4 = \dim V_{(3,0)}$ copies of $V_{(1,1,1)}$ and $2 = \dim V_{(2,1)}$ copies of $V_{(2,1,0)}$ as expected.

1.5.3 Deriving the Ginzburg construction. We begin by noting that as GL_m representations, we have

$$\Lambda^N(\mathbb{C}^n \otimes \mathbb{C}^m) \cong \Lambda^N(\mathbb{C}^m \oplus \cdots \oplus \mathbb{C}^m)$$

and now we can expand out the right hand side using a "binomial formula" to deduce that

$$\Lambda^N(\mathbb{C}^n \otimes \mathbb{C}^m) \cong \bigoplus_{\mu \in \mathbb{N}^n, |\mu| = N} \Lambda^{\mu_1}\mathbb{C}^m \otimes \cdots \otimes \Lambda^{\mu_n}\mathbb{C}^m.$$

Combining this with (1.2), we learn that there is an isomorphism of GL_n representations

$$\bigoplus_{\mu \in \mathbb{N}^n, |\mu| = N} \mathrm{Hom}_{GL_m}(V_\lambda, \Lambda^{\mu_1}\mathbb{C}^m \otimes \cdots \otimes \Lambda^{\mu_n}\mathbb{C}^m) \cong V(\lambda^\vee).$$

Moreover, we note that the direct sum decomposition on the left hand side matches the weight decomposition of the right hand side.

Now, we apply Corollary 1.5.4 to deduce that

$$V(\lambda^\vee) \cong \bigoplus_\mu H_{\text{top}}(m^{-1}_{(\omega_{\mu_1},\dots,\omega_{\mu_n})}(L_\lambda))$$

(here we use that $\Lambda^k\mathbb{C}^m = V(\omega_k)$ and we also use the convention that $\omega_0 = 0$).

Let us carefully examine the last variety

$$m^{-1}_{(\omega_{\mu_1},\dots,\omega_{\mu_n})}(L_\lambda) = \{0 = L_0 \subseteq L_1 \subseteq \cdots \subseteq L_n = L_\lambda : \\ XL_i \subseteq L_i, X|_{L_i/L_{i-1}} \text{ has Jordan type } \omega_{\mu_i}\}$$

However, since $\omega_k = (1,\dots,1,0,\dots,0)$, having Jordan type ω_k is the same thing as being the zero matrix on a space of dimension k. So we see that we can rewrite the above condition as $XL_i \subseteq L_{i-1}$ and $\dim L_i/L_{i-1} = \mu_i$. Hence we conclude that

$$m^{-1}_{(\omega_{\mu_1},\dots,\omega_{\mu_n})}(L_\lambda) = \mathrm{Fl}_\mu(L_\lambda)^X$$

Thus using the geometric Satake correspondence and the skew Howe duality we have proven the following.

Theorem 1.5.9 *Identifying L_λ with $\mathbb{C}^N$, we find that*

$$\bigoplus_{\mu\in\mathbb{N}^n,|\mu|=N} H_{\text{top}}(\mathrm{Fl}_\mu(\mathbb{C}^N)^X)$$

has an action of GL_n and is isomorphic to $V(\lambda^\vee)$.

Since the Jordan type of X on L_λ is λ, this reproves the main theorem of the Ginzburg construction.

One might ask if these two constructions are the same. We have constructed two (possibly different) actions of GL_n on $\bigoplus_{\mu\in\mathbb{N}^n,|\mu|=N} H_{\text{top}}(\mathrm{Fl}_\mu(\mathbb{C}^N)^X)$. In the first, we defined the action of the Chevalley generators using the varieties Z_i. In the second we worked more abstractly using the geometric Satake correspondence and skew Howe duality to define an action of the entire GL_n at once.

Exercise 1.5.10 Prove that these two action coincide. Hint: first, we can see easily that the weight decompositions coincide. So it remains to see that in our second version of the story, the Chevalley generators act using the varieties Z_i. This can be done by expressing the action of E_i on

$$\mathrm{Hom}_{GL_m}(V(\lambda), \Lambda^{\mu_1}\mathbb{C}^m \otimes \cdots \otimes \Lambda^{\mu_n}\mathbb{C}^m) \cong V_{\lambda^\vee}(\mu)$$

as the composition with basic GL_m-equivariant maps

$$\Lambda^k\mathbb{C}^m \otimes \Lambda^l\mathbb{C}^m \to \Lambda^{k-1}\mathbb{C}^m \otimes \mathbb{C}^m \otimes \Lambda^l\mathbb{C}^m \to \Lambda^{k-1}\mathbb{C}^m \otimes \Lambda^{l+1}\mathbb{C}^m,$$

and then expressing these basic maps using the geometric Satake correspondence.

To conclude, in this section, we have seen a geometric incarnation of skew-Howe duality, where one group acts by the Ginzburg construction and the other group acts by the geometric Satake construction[1].

Joel Kamnitzer
Department of Mathematics
University of Toronto
Toronto, ON, Canada M5S 2E4
jkamnitz@math.toronto.edu

---

[1] A geometric manifestation of symmetric-Howe duality was achieved earlier by Wang [215], using the Ginzburg construction on both sides.

Fields Institute Communications
Volume **59**, 2011

CHAPTER 2

Introduction to Crystal Bases

SEOK-JIN KANG

Abstract. We give an introduction to the theory of quantum groups and crystal bases. We first review the definition of crystal bases, the tensor product rule, and the theory of abstract crystals. We then explain the theory of perfect crystals for quantum affine algebras including the fundamental crystal isomorphism theorem. By applying the fundamental isomorphism theorem repeatedly, we obtain a path realization of irreducible highest weight crystals for quantum affine algebras. Finally, we describe combinatorics of Young walls and give an explicit construction of the level 1 irreducible highest weight crystals for all classical quantum affine algebras in terms of reduced proper Young walls.

Introduction

The purpose of this article is to give a brief introduction to the theory of quantum groups and crystal bases. The *quantum groups*, introduced by Drinfel'd and Jimbo in [88, 135], are certain families of Hopf algebras that are deformations of universal enveloping algebras of Kac-Moody algebras. In [164], it was shown by Lusztig that the representation theory of quantum groups can also be deformed in the sense that the integrable highest weight modules over Kac-Moody algebras can be deformed to those over quantum groups in such a way that the characters are invariant under deformation. In other words, the *classical limit* of quantum groups and their integrable representations are universal enveloping algebras of Kac-Moody algebras and their integrable representations. In Section 2.3, we will discuss the process of taking classical limit.

In [141], Kashiwara introduced the notion of *crystal bases* for quantum groups, which can be viewed as bases at $q = 0$, and proved several fundamental theorems including the tensor product rule, the existence and uniqueness of crystal/global bases, and the characterization of global bases. Since its discovery, the crystal basis theory has played a central role in combinatorial representation theory. On the other hand, in [165], Lusztig took a geometric approach to yield the *canonical basis theory*, which gives rise to a lot of exciting developments as well.

In Section 2.4 and Section 2.5, we will review the definition of crystal bases, the tensor product rule, and the theory of *abstract crystals*. We will also discuss the *recognition theorems* that identify certain abstract crystals with the highest weight crystals $B(\infty)$ and $B(\lambda)$ $(\lambda \in P^+)$.

Our main objective is to explain the theory of *perfect crystals* for quantum affine algebras which was developed in [140]. After giving some examples of perfect crystals and non-perfect crystals, we will deal with the *fundamental crystal isomorphism theorem* for perfect crystals. By applying the fundamental isomorphism theorem repeatedly, we obtain a *path realization* of irreducible highest weight crystals for quantum affine algebras.

Finally, we will describe combinatorics of *Young walls* introduced in [139]. The Young walls consist of colored blocks with various shapes and can be viewed as generalizations of Young diagrams and Young tableaux. Using combinatorics of Young walls, we are able to construct an explicit combinatorial realization of the level 1 highest weight crystals $B(\Lambda)$ for *all* classical quantum affine algebras in terms of *reduced proper Young walls* (see also [123]).

Acknowledgments. This work was supported by KRF Grant #2007-341-C00001 and NRF Grant #2010-0010753. This article is based on the lectures given at University of Ottawa in June, 2009. The author should confess that he enjoyed his own lectures a lot more than the audience did. The audience created a lively, exciting, inspiring and stimulating atmosphere throughout the workshop. The author would like to express his sincere gratitude to the organizers and the audience for giving him a beautiful and unforgettable memory.

2.1 Lie algebras

Let F be a field of characteristic 0 and let G be an algebraic group over F. Thus G is an algebraic variety over F with a group structure where the group multiplication and the inverse map are morphisms of algebraic varieties. We denote by $F[G]$ the algebra of regular functions on G.

For each $x \in G$, let $T_x(G)$ be the tangent space to G at x and let $D_x(G)$ be the algebra of linear functionals $d : F[G] \to F$ such that

$$d(fg) = d(f)g(x) + f(x)d(g) \quad \text{for all } f, g \in F[G].$$

For a tangent vector X to G at x, define a linear functional $d_X : F[G] \to F$ by

$$d_X(f) = \nabla f(x) \cdot X \quad (f \in F[G]),$$

where $\nabla f(x)$ is the gradient of f at x. Then the map $X \mapsto d_X$ yields a natural identification $T_x(G) \xrightarrow{\sim} D_x(G)$.

Consider the algebra $\mathfrak{L}(G)$ consisting of linear maps $D : F[G] \to F[G]$ such that

$$D(fg) = D(f)g + fD(g), \quad D\lambda(x) = \lambda(x)\mathcal{D} \quad \text{for all } x \in G,$$

where $(\lambda(x)f)(y) = f(xy)$ for all $x, y \in G$. The elements of $\mathfrak{L}(G)$ are called the *left invariant derivations* of $F[G]$. For $X \in T_1(G)$, define a left invariant derivation $D_X : F[G] \to F[G]$ by

$$D_X(f)(x) = d_X(\lambda(x)f) \quad \text{for } f \in F[G], \ x \in G.$$

It is easy to show that every left invariant derivation D has the form $D = D_X$ for a unique $X \in T_1(G)$. Thus we get a natural identification $T_1(G) \xrightarrow{\sim} \mathfrak{L}(G)$ given by $X \mapsto D_X$.

Moreover, one can show that the *bracket* $[D_X, D_Y] = D_X D_Y - D_Y D_X$ of left invariant derivations is again a left invariant derivation. That is, $[D_X, D_Y] = D_Z$

for a uniquely determined $Z \in T_1(G)$. Hence we may define the bracket $[\ ,\ ]$ on $T_1(G)$ via

$$[D_X, D_Y] = D_{[X,Y]} \quad \text{for } X, Y \in T_1(G).$$

The tangent space $T_1(G)$ with the bracket defined above is called the *Lie algebra* of G and is denoted by $Lie(G)$. It is straightforward to verify that the bracket operation satisfies:

$$[X, X] = 0 \ \text{ for all } X \in T_1(G),$$
$$[X, [Y, Z]] + [Y, [Z, X]] + [Z, [X, Y]] = 0 \ \text{ for all } X, Y, Z \in T_1(G).$$

The point of the above discussion is that, to study the structure and properties of an algebraic group G, a *geometric* object, we would rather study its Lie algebra $Lie(G) = T_1(G)$, an *algebraic* object, because it is easier to deal with $Lie(G)$ and the structure and properties of G are reflected on those of $Lie(G)$. We begin by giving an algebraic definition of Lie algebras.

Definition 2.1.1 A *Lie algebra* over F is a vector space L over F with a bilinear product $[\ ,\ ] : L \times L \to L$, called the *bracket*, satisfying

$$\begin{aligned} &[x, x] = 0 \quad \text{for all } x \in L, \\ &[x, [y, z]] + [y, [z, x]] + [z, [x, y]] = 0 \quad \text{for all } x, y, z \in L. \end{aligned} \tag{2.1}$$

Example 2.1.2
(a) Let $L = \mathbb{R}^3$ and define $[x, y] = x \times y$, the cross product of x and y, for $x, y \in \mathbb{R}^3$. Then L is a Lie algebra.

(b) Let A be an associative algebra over F and define

$$[x, y] = xy - yx \quad \text{for all } x, y \in A.$$

Then A becomes a Lie algebra.

(c) For a vector space V over F, consider the algebra $\mathrm{End}_F(V)$ of all linear endomorphisms of V. The induced Lie algebra is called the *general linear Lie algebra* of V and is denoted by $\mathfrak{gl}(V)$. When $V = F^n$, we use the notation $\mathfrak{gl}_n(F)$.

(d) The subspace $\mathfrak{sl}_n(F) = \{x \in \mathfrak{gl}_n(F) \mid \mathrm{tr}(x) = 0\}$ of $\mathfrak{gl}_n(F)$ is a Lie algebra, called the *special linear Lie algebra.*

Definition 2.1.3 Let L and L' be Lie algebras.
(a) A subspace L_1 of L is called a *Lie subalgebra* of L if L_1 is a Lie algebra with the bracket induced from L.
(b) An *ideal* I of L is a subalgebra such that $[x, y] \in I$ for all $x \in L$, $y \in I$.
(c) A nontrivial Lie algebra L is *simple* if the only ideals of L are 0 and L itself.
(d) A linear map $\phi : L \to L'$ is a *Lie algebra homomorphism* if

$$[\phi(x), \phi(y)] = [x, y] \quad \text{for all } x, y \in L.$$

If $\phi : L \to L'$ is a Lie algebra homomorphism, $\mathrm{Im}\phi$ is a Lie subalgebra of L' and $\ker \phi$ is an ideal of L. We define the notion of *monomorphism*, *epimorphism* and *isomorphism* in the usual way.

Definition 2.1.4 Let L be a Lie algebra and V be a vector space over F.

(a) A *representation* of L is a Lie algebra homomorphism $\rho : L \to \mathfrak{gl}(V)$.

(b) V is called an *L-module* if there exists a bilinear map $L \times V \to V$, denoted by $(x, v) \mapsto x \cdot v$, such that

$$[x, y] \cdot v = x \cdot (y \cdot v) - y \cdot (x \cdot v) \quad \text{for all } x, y \in L,\ v \in V.$$

It is an easy exercise to see that (a) is equivalent to (b) via

$$x \cdot v = \rho(x)(v) \quad \text{for } x \in L,\ v \in V.$$

Thus the representation theory of L is the study of structure and properties of L that are reflected on various L-modules.

Definition 2.1.5 Let L be a Lie algebra and let V, W be L-modules.

(a) A subspace U of V is called a *L-submodule* of V if $x \cdot u \in U$ for all $x \in L$, $u \in U$.

(b) A nonzero L-module V is *irreducible* if the only L-submodules of V are 0 and V itself.

(c) A linear map $\phi : V \to W$ is an *L-module homomorphism* if

$$\phi(x \cdot v) = x \cdot \phi(v) \quad \text{for all } x \in L,\ v \in V.$$

If $\phi : V \to W$ is an L-module homomorphism, $\ker \phi$ is a submodule of V and $\mathrm{Im}\phi$ is a submodule of W. We define the notion of *monomorphism*, *epimorphism* and *isomorphism* of L-modules in the usual way.

Definition 2.1.6 Let L be a Lie algebra and let U be an associative algebra. Suppose there exists a linear map $\iota : L \to U$ satisfying

$$\iota([x, y]) = \iota(x)\iota(y) - \iota(y)\iota(x) \quad \text{for all } x, y \in L.$$

The pair (U, ι) is called a *universal enveloping algebra of L* if for every associative algebra A and a linear map $j : L \to A$ satisfying

$$j([x, y]) = j(x)j(y) - j(y)j(x) \quad \text{for all } x, y \in L,$$

there exists a unique algebra homomorphism $\phi : U \to A$ such that $\phi \circ \iota = j$.

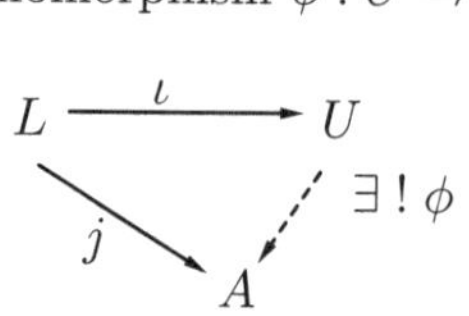

The uniqueness of a universal enveloping algebra can be proved by a standard argument. For the existence, let $T(L) = \bigoplus_{k=0}^{\infty} L^{\otimes k}$ be the tensor algebra of L and set $U(L) = T(L)/I$, where I is the two-sided ideal of $T(L)$ generated by $x \otimes y - y \otimes x - [x, y]$ $(x, y \in L)$. Consider the composite of linear maps $\iota : L \hookrightarrow T(L) \to U(L)$. Then it is straightforward to show that $(U(L), \iota)$ is a universal enveloping algebra of L. Moreover, we have the following important theorem (see, for example, [126, Theorem 17.3]).

Theorem 2.1.7 (Poincaré-Birkhoff-Witt)

(a) The linear map $\iota : L \to U(L)$ is injective.

(b) Let $B = \{x_\alpha \mid \alpha \in \Omega\}$ be an ordered basis of L. Then the set $\mathbb{B} = \{x_{\alpha_1} x_{\alpha_2} \cdots x_{\alpha_r} \mid r \geq 0,\ \alpha_1 \leq \alpha_2 \leq \cdots \leq \alpha_r,\ \alpha_k \in \Omega\}$ forms a basis of $U(L)$.

As an immediate corollary, one can see that V is an L-module if and only if V is a $U(L)$-module. Moreover, there exists an algebra homomorphism $\Delta : U(L) \to U(L) \otimes U(L)$ and an anti-involution $S : U(L) \to U(L)$ defined by

$$\Delta(x) = x \otimes 1 + 1 \otimes x, \quad S(x) = -x \quad \text{for all } x \in L.$$

Thus if V and W are L-modules, then $V \otimes W$ and V^* have the structure of L-modules given by

$$\begin{aligned} x \cdot (v \otimes w) &= x \cdot v \otimes w + v \otimes x \cdot w \quad \text{for all } x \in L, \ v \in V, \ w \in W, \\ (x \cdot f)(v) &= -f(x \cdot v) \quad \text{for all } x \in L, \ v \in V, \ f \in V^*. \end{aligned}$$

The homomorphism Δ and the anti-involution S are called the *comultiplication* and the *antipode* of $U(L)$, respectively.

2.2 Kac-Moody algebras

Let $L = \mathfrak{sl}_{n+1}(F)$ be the special linear Lie algebra. Then it is spanned by the matrices E_{ij} $(i \neq j)$, $E_{ii} - E_{i+1,i+1}$ $(i = 1, \cdots n)$, where E_{ij} denotes the matrix whose (i,j)-entry is 1 and the other entries are 0. Set $e_i = E_{i,i+1}$, $f_i = E_{i+1,i}$, $h_i = E_{ii} - E_{i+1,i+1}$ $(i = 1, \cdots, n)$. Then, as a Lie algebra, L is generated by e_i, f_i, h_i $(i = 1, \cdots, n)$. Moreover, they satisfy the following relations:

$$\begin{aligned} & [h_i, h_j] = 0, \quad [e_i, f_j] = \delta_{ij} h_i, \\ & [h_i, e_j] = \begin{cases} 2e_i & \text{if } j = i, \\ -e_j & \text{if } j = i \pm 1, \\ 0 & \text{otherwise}, \end{cases} \\ & [h_i, f_j] = \begin{cases} -2f_i & \text{if } j = i, \\ f_j & \text{if } j = i \pm 1, \\ 0 & \text{otherwise}, \end{cases} \\ & [e_i[e_i, e_j]] = [f_i, [f_i, f_j]] = 0 \ \text{ if } j = i \pm 1, \\ & [e_i, e_j] = 0 \ \text{ otherwise}. \end{aligned} \tag{2.2}$$

Let

$$A = (a_{ij})_{1 \le i,j \le n} = \begin{pmatrix} 2 & -1 & 0 & \cdots & 0 & 0 \\ -1 & 2 & -1 & \cdots & 0 & 0 \\ \vdots & \vdots & \ddots & \vdots & \vdots & \\ 0 & 0 & 0 & \cdots & 2 & -1 \\ 0 & 0 & 0 & \cdots & -1 & 2 \end{pmatrix}.$$

Then the relations (2.2) can be written as

$$\begin{aligned} & [h_i, h_j] = 0, \quad [e_i, f_j] = \delta_{ij} h_i, \\ & [h_i, e_j] = a_{ij} e_j, \quad [h_i, f_j] = -a_{ij} f_j, \\ & (\operatorname{ad} e_i)^{1-a_{ij}}(e_j) = (\operatorname{ad} f_i)^{1-a_{ij}} = 0 \quad (i \neq j), \end{aligned}$$

where $\operatorname{ad} x(y) = [x, y]$ for $x, y \in L$. The matrix A is called the *Cartan matrix* of L.

We will consider a class of (usually infinite dimensional) Lie algebras defined by generators and relations associated with certain square matrices. Let I be a finite index set. A square matrix $A = (a_{ij})_{i,j \in I}$ is called a *generalized Cartan matrix* if

(i) $a_{ii} = 2$ for all $i \in I$,
(ii) $a_{ij} \in \mathbb{Z}_{\le 0}$ for $i \neq j$,
(iii) $a_{ij} = 0$ if and only if $a_{ji} = 0$.

We assume that A is symmetrizable; i.e., there is a diagonal matrix $D = \mathrm{diag}(s_i \in \mathbb{Z}_{>0} \mid i \in I)$ such that DA is symmetric. Set

$$P^\vee = \left(\bigoplus_{i\in I} \mathbb{Z}h_i \right) \oplus \left(\bigoplus_{j=1}^{\operatorname{corank} A} \mathbb{Z}d_j \right),$$

$\mathfrak{h} = \mathbb{C} \otimes_{\mathbb{Z}} P^\vee$, $P = \{\lambda \in \mathfrak{h}^* \mid \lambda(P^\vee) \subseteq \mathbb{Z}\}$, and let $\Pi = \{\alpha_i \mid i \in I\}$ be a linearly independent subset of P such that $\langle h_i, \alpha_j \rangle = a_{ij}$ for all $i, j \in I$. The quadruple $(A, P^\vee, P, \Pi)$ is called a *Cartan datum*. We denote by $Q = \bigoplus_{i\in I} \mathbb{Z}\alpha_i$, $Q^+ = \sum_{i\in I} \mathbb{N}\alpha_i$, $Q^- = -Q^+$ and $P^+ = \{\lambda \in P \mid \lambda(h_i) \ge 0 \text{ for all } i \in I\}$. The elements of Π and P^+ are called the *simple roots* and *dominant integral weights*, respectively. There is a partial ordering on $\mathfrak{h}^*$ defined by $\lambda \ge \mu$ if and only if $\lambda - \mu \in Q^+$ for $\lambda, \mu \in \mathfrak{h}^*$.

Definition 2.2.1 The *Kac-Moody algebra* $\mathfrak{g}$ associated with $(A, P^\vee, P, \Pi)$ is the Lie algebra generated by e_i, f_i $(i \in I)$ and $h \in \mathfrak{h}$ with defining relations

$$\begin{aligned}
&[h, h'] = 0 \quad \text{for all } h, h' \in \mathfrak{h},\\
&[e_i, f_j] = \delta_{ij} h_i \quad \text{for all } i, j \in I,\\
&[h, e_j] = \alpha_j(h) e_j, \quad [h, f_j] = -\alpha_j(h) f_j \quad \text{for } h \in \mathfrak{h},\ j \in I,\\
&(\operatorname{ad} e_i)^{1-a_{ij}}(e_j) = (\operatorname{ad} f_i)^{1-a_{ij}}(f_j) = 0 \quad \text{for } i \neq j.
\end{aligned} \tag{2.3}$$

We denote by U^+ (resp. U^- and U^0) the subalgebra of $U(\mathfrak{g})$ generated by the e_i's (resp. f_i's and $h \in \mathfrak{h}$). By the Poincaré-Birkhoff-Witt Theorem, we have the *triangular decomposition*:

$$U(\mathfrak{g}) \cong U^- \otimes U^0 \otimes U^+.$$

Example 2.2.2

(a) It is known that finite dimensional complex simple Lie algebras are Kac-Moody algebras of *finite type* (see, for example, [126, 137]).

(b) Let $\mathfrak{g}$ be a finite dimensional simple Lie algebra and let

$$\widehat{\mathfrak{g}} = \mathfrak{g} \otimes \mathbb{C}[t, t^{-1}] \oplus \mathbb{C}c \oplus \mathbb{C}d,$$

where the Lie bracket is defined by

$$\begin{aligned}
&[x^m \otimes, y \otimes t^n] = [x, y] \otimes t^{m+n} + m\delta_{m+n,0}\, c,\\
&[c, x \otimes t^m] = [c, d] = 0, \quad [d, x \otimes t^m] = mx \otimes t^m \quad (m, n \in \mathbb{Z}).
\end{aligned}$$

Then $\widehat{\mathfrak{g}}$ is a Kac-Moody algebra of *affine type*.

Definition 2.2.3 We define the *category* $\mathcal{O}_{int}$ to be the category of $\mathfrak{g}$-modules M satisfying the following properties:

(i) $M = \bigoplus_{\lambda\in P} M_\lambda$, where $M_\lambda = \{m \in M \mid h \cdot m = \lambda(h)m \text{ for all } h \in \mathfrak{h}\}$ and $\dim M_\lambda < \infty$ for all $\lambda \in P$;

(ii) there are finitely many $\lambda_1, \cdots, \lambda_s \in P$ such that $\mathrm{wt}(M) := \{\lambda \in P \mid M_\lambda \neq 0\} \subseteq \bigcup_{j=1}^{s} (\lambda_j - Q^+)$;

(iii) e_i, f_i $(i \in I)$ are locally nilpotent on M.

For a $\mathfrak{g}$-module $M = \bigoplus_{\lambda \in P} M_\lambda$ in $\mathcal{O}_{int}$, we define the *character* of M to be

$$\operatorname{ch}(M) = \sum_{\lambda \in P} (\dim M_\lambda) e^\lambda,$$

where e^λ are (multiplicative) basis elements of the group algebra $\mathbb{C}[P]$. One of the most important problems in combinatorial representation theory is to understand the characters in various ways, because the characters are the first clue to the understanding of $\mathfrak{g}$-modules, they are algebraic invariants, and they often stand for important and interesting mathematical quantities such as symmetric functions, modular forms, 1 point functions, etc.

A $\mathfrak{g}$-module V is called a *highest weight module* with highest weight $\lambda \in P$ if there exists a nonzero vector v_λ in V such that

(i) $hv_\lambda = \lambda(h)v_\lambda$ for all $h \in \mathfrak{h}$,
(ii) $e_i v_\lambda = 0$ for all $i \in I$,
(iii) $V = U(\mathfrak{g})v_\lambda$.

The vector v_λ is called a *highest weight vector* of V. By the triangular decomposition, every highest weight module V can be written as $V = U(\mathfrak{g})v_\lambda = U^- v_\lambda = \bigoplus_{\mu \leq \lambda} V_\mu$, which justifies the term *highest weight module.*

For $\lambda \in P$, let $J(\lambda)$ be the left ideal of $U(\mathfrak{g})$ generated by e_i $(i \in I)$, $h - \lambda(h)1$ $(h \in \mathfrak{h})$ and set $M(\lambda) = U(\mathfrak{g})/J(\lambda)$. Then via left multiplication, $M(\lambda)$ becomes a $U(\mathfrak{g})$-module called a *Verma module.*

Proposition 2.2.4 ([137, Chapter 9])
(a) $M(\lambda)$ is a highest weight module with highest weight λ.
(b) $M(\lambda)$ is a free U^--module of rank 1.
(c) Every highest $\mathfrak{g}$-module with highest weight λ is a homomorphic image of $M(\lambda)$.
(d) $M(\lambda)$ has a unique maximal submodule $R(\lambda)$.

The irreducible quotient $V(\lambda) := M(\lambda)/R(\lambda)$ is called the *irreducible highest weight module* with highest weight λ. The basic properties of $\mathfrak{g}$-modules in the category $\mathcal{O}_{int}$ are summarized in the following proposition.

Proposition 2.2.5 ([137, Chapters 9 and 10])
(a) Every irreducible $\mathfrak{g}$-module in the category $\mathcal{O}_{int}$ is isomorphic to $V(\lambda)$ for some $\lambda \in P^+$.
(b) Every $\mathfrak{g}$-module in the category $\mathcal{O}_{int}$ is completely reducible.
(c) For each $i \in I$, every $\mathfrak{g}$-module in the category $\mathcal{O}_{int}$ is a direct sum of finite dimensional irreducible U_i submodules, where U_i is the subalgebra of $U(\mathfrak{g})$ generated by e_i, f_i, h_i.

Remark. The main ingredient of the proof is the *Weyl-Kac character formula:*

$$\operatorname{ch} V(\lambda) = \frac{\sum_{w \in W} \det(w) e^{w(\lambda+\rho)-\rho}}{\prod_{\alpha \in \Phi^+} (1 - e^{-\alpha})^{\dim \mathfrak{g}_\alpha}},$$

where W is the Weyl group, Φ^+ is the set of positive roots of $\mathfrak{g}$, and $\mathfrak{g}_\alpha = \{x \in \mathfrak{g} \mid [h, x] = \alpha(h)x \text{ for all } h \in \mathfrak{h}\}$.

2.3 Quantum groups

In this section, we construct the quantum deformation $U_q(\mathfrak{g})$ of $U(\mathfrak{g})$, the *quantum group*. Let q be an indeterminate. For $i \in I$ and $m, n \in \mathbb{Z}$, set

$$q_i = q^{s_i}, \quad [n]_i = \frac{q_i^n - q_i^{-n}}{q_i - q_i^{-1}}, \qquad [n]_i! = \prod_{k=1}^{n} [k]_i, \quad \begin{bmatrix} m \\ n \end{bmatrix}_i = \frac{[m]_i!}{[n]_i![m-n]_i!}.$$

Definition 2.3.1 The *quantum group* or a *quantized universal enveloping algebra* associated with a Cartan datum $(A, P^\vee, P, \Pi)$ is the associative algebra over $\mathbb{C}(q)$ with 1 generated by the elements e_i, f_i $(i \in I)$ and q^h $(h \in P^\vee)$ with defining relations

$$\begin{aligned}
&q^0 = 1, \quad q^h q^{h'} = q^{h+h'} \ \text{ for } h, h' \in P^\vee, \\
&q^h e_j q^{-h} = q^{\alpha_j(h)} e_j, \quad q^h f_j q^{-h} = q^{-\alpha_j(h)} f_j \ \text{ for } h \in P^\vee,\, j \in I, \\
&e_i f_j - f_j e_i = \delta_{ij} \frac{K_i - K_i^{-1}}{q_i - q_i^{-1}}, \quad \text{where } K_i = q^{s_i h_i}, i, j \in I, \\
&\sum_{k=0}^{1-a_{ij}} (-1)^k \begin{bmatrix} 1-a_{ij} \\ k \end{bmatrix}_i e_i^{1-a_{ij}-k} e_j e_i^k = 0 \quad (i \neq j), \\
&\sum_{k=0}^{1-a_{ij}} (-1)^k \begin{bmatrix} 1-a_{ij} \\ k \end{bmatrix}_i f_i^{1-a_{ij}-k} f_j f_i^k = 0 \quad (i \neq j).
\end{aligned} \tag{2.4}$$

The *comultiplication* $\Delta : U_q(\mathfrak{g}) \to U_q(\mathfrak{g}) \otimes U_q(\mathfrak{g})$ and the *antipode* $S : U_q(\mathfrak{g}) \to U_q(\mathfrak{g})$ are defined by

$$\begin{aligned}
&\Delta(q^h) = q^h \otimes q^h \quad \text{for } h \in P^\vee, \\
&\Delta(e_i) = e_i \otimes K_i^{-1} + 1 \otimes e_i \quad \Delta(f_i) = f_i \otimes 1 + K_i \otimes f_i \quad (i \in I), \\
&S(q^h) = q^{-h}, \quad S(e_i) = -e_i K_i, \quad S(f_i) = -K_i^{-1} f_i \quad (h \in P^\vee,\, i \in I).
\end{aligned}$$

We denote by U_q^+ (resp. U_q^- and U_q^0) the subalgebra of $U_q(\mathfrak{g})$ generated by e_i's (resp. f_i's and q^h's). One cannot apply the Poincaré-Birkhoff-Witt Theorem to $U_q(\mathfrak{g})$ in a straightforward manner. Still, one can prove the quantum version of *triangular decomposition*:

$$U_q(\mathfrak{g}) \cong U_q^- \otimes U_q^0 \otimes U_q^+$$

(see, for example, [122, Thorem 3.1.5]).

We define the quantum version of the category $\mathcal{O}_{int}$, characters, and highest weight modules as follows.

Definition 2.3.2 The *category* $\mathcal{O}_{int}^q$ consists of $U_q(\mathfrak{g})$-modules M^q and $U_q(\mathfrak{g})$-module homomorphisms satisfying the following properties:

(i) $M^q = \bigoplus_{\lambda \in P} M_\lambda^q$, where $M_\lambda^q = \{m \in M^q \mid q^h \cdot m = q^{\lambda(h)} m \ \text{ for all } h \in P^\vee\}$ and $\dim M_\lambda^q < \infty$ for all $\lambda \in P$;

(ii) there are finitely many $\lambda_1, \cdots, \lambda_s \in P$ such that $\mathrm{wt}(M^q) := \{\lambda \in P \mid M_\lambda^q \neq 0\} \subseteq \bigcup_{j=1}^{s} (\lambda_j - Q^+)$;

(iii) e_i, f_i $(i \in I)$ are locally nilpotent on M^q.

For a $U_q(\mathfrak{g})$-module $M^q = \bigoplus_{\lambda \in P} M^q_\lambda$ in $\mathcal{O}^q_{int}$, we define the *character* of M to be

$$\mathrm{ch}(M^q) = \sum_{\lambda \in P} (\dim_{\mathbb{C}(q)} M^q_\lambda) e^\lambda,$$

where e^λ are (multiplicative) basis elements of the group algebra $\mathbb{C}[P]$.

A $U_q(\mathfrak{g})$-module V^q is called a *highest weight module* with highest weight $\lambda \in P$ if there exists a nonzero vector v_λ in V^q such that

(i) $q^h v_\lambda = q^{\lambda(h)} v_\lambda$ for all $h \in P^\vee$,
(ii) $e_i v_\lambda = 0$ for all $i \in I$,
(iii) $V^q = U_q(\mathfrak{g}) v_\lambda$.

The vector v_λ is called a *highest weight vector* of V^q. By the quantum version of triangular decomposition, we have $V^q = U_q(\mathfrak{g}) v_\lambda = U_q^- v_\lambda = \bigoplus_{\mu \le \lambda} V^q_\mu$.

For $\lambda \in P$, let $J^q(\lambda)$ be the left ideal of $U_q(\mathfrak{g})$ generated by e_i $(i \in I)$, $q^h - q^{\lambda(h)} 1$ $(h \in P^\vee)$ and set $M^q(\lambda) = U_q(\mathfrak{g})/J^q(\lambda)$. Then via left multiplication, $M^q(\lambda)$ becomes a $U_q(\mathfrak{g})$-module called a *Verma module*.

Proposition 2.3.3 ([122, Chapter 3])
(a) $M^q(\lambda)$ is a highest weight module with highest weight λ.
(b) $M^q(\lambda)$ is a free U_q^--module of rank 1.
(c) Every highest $U_q(\mathfrak{g})$-module with highest weight λ is a homomorphic image of $M^q(\lambda)$.
(d) $M^q(\lambda)$ has a unique maximal submodule $R^q(\lambda)$.

The irreducible quotient $V^q(\lambda) := M^q(\lambda)/R^q(\lambda)$ is called the *irreducible highest weight module* with highest weight λ.

Let $\mathbb{A}_1 = \{f/g \in \mathbb{C}(q) \mid f, g \in \mathbb{C}[q],\ g(1) \ne 0\}$. We define $U^{\mathbb{A}_1}(\mathfrak{g})$ to be the $\mathbb{A}_1$-subalgebra generated by e_i, f_i $(i \in I)$, q^h $(h \in P^\vee)$, $\dfrac{q^h - 1}{q - 1}$ $(h \in P^\vee)$, and set $V^{\mathbb{A}_1}(\lambda) = U^{\mathbb{A}_1}(\mathfrak{g}) v_\lambda$. Let $\mathbb{J}_1$ be the maximal ideal of $\mathbb{A}_1$ generated by $q - 1$. Then there is a canonical isomorphism $\mathbb{A}_1/\mathbb{J}_1 \xrightarrow{\sim} \mathbb{C}$ given by $f + \mathbb{J}_1 \mapsto f(1)$. Set

$$U^1(\mathfrak{g}) := \mathbb{A}_1/\mathbb{J}_1 \otimes_{\mathbb{A}_1} U^{\mathbb{A}_1}(\mathfrak{g}) \cong U^{\mathbb{A}_1}(\mathfrak{g})/\mathbb{J}_1 U^{\mathbb{A}_1}(\mathfrak{g}),$$
$$V^1(\lambda) := \mathbb{A}_1/\mathbb{J}_1 \otimes_{\mathbb{A}_1} V^{\mathbb{A}_1}(\lambda) \cong V^{\mathbb{A}_1}(\lambda)/\mathbb{J}_1 V^{\mathbb{A}_1}(\lambda).$$

We call $U^1(\mathfrak{g})$ and $V^1(\lambda)$ the *classical limit* of $U_q(\mathfrak{g})$ and $V^q(\lambda)$, respectively. The following following shows that the quantum group is a deformation of $U(\mathfrak{g})$ and that the $\mathfrak{g}$-modules in $\mathcal{O}_{int}$ can be deformed to those over $U_q(\mathfrak{g})$ in such a way that the characters are invariant under deformation.

Theorem 2.3.4 ([122, Theorem 3.4.4], [164, Theorem 4.12]) (a) The algebra $U^1(\mathfrak{g})$ is isomorphic to $U(\mathfrak{g})$.
(b) As a $U(\mathfrak{g})$-module, $V^1(\lambda)$ is isomorphic to $V(\lambda)$.
(c) $\dim_{\mathbb{C}(q)} V^q(\lambda)_\mu = \mathrm{rank}_{\mathbb{A}_1} V^{\mathbb{A}_1}(\lambda)_\mu = \dim_{\mathbb{C}} V(\lambda)_\mu$ for all $\mu \in P$.

Using the classical limit theorem, one can prove the following fundamental theorem for $U_q(\mathfrak{g})$-modules in the category $\mathcal{O}^q_{int}$.

Theorem 2.3.5 ([122, Corollary 3.4.8])
(a) Every irreducible $U_q(\mathfrak{g})$-module in the category $\mathcal{O}^q_{int}$ is isomorphic to $V^q(\lambda)$ for some $\lambda \in P^+$.
(b) Every $U_q(\mathfrak{g})$-module in the category $\mathcal{O}^q_{int}$ is completely reducible.

(c) For each $i \in I$, every $U_q(\mathfrak{g})$-module in the category $\mathcal{O}^q_{int}$ is a direct sum of finite dimensional irreducible U_i^q-submodules, where U_i^q is the subalgebra of $U_q(\mathfrak{g})$ generated by e_i, f_i, $K_i^{\pm 1}$.

2.4 Crystal bases

Let $M = \bigoplus_{\lambda \in P} M_\lambda$ be a $U_q(\mathfrak{g})$-module in the category $\mathcal{O}^q_{int}$. For each $i \in I$, every $u \in M_\lambda$ can be written uniquely as

$$u = \sum_{k \geq 0} f_i^{(k)} u_k, \quad \text{where } e_i u_k = 0, \quad f_i^{(k)} = f_i^k/[k]_i!. \tag{2.5}$$

The decomposition (2.5) is called the *i-string decomposition* of u. We define the *Kashiwara operators* on M by

$$\tilde{e}_i u = \sum_{k \geq 1} f_i^{(k-1)} u_k, \quad \tilde{f}_i u = \sum_{k \geq 0} f_i^{(k+1)} u_k.$$

Let $\mathbb{A}_0 = \{f/g \in \mathbb{C}(q) \mid f, g \in \mathbb{C}[q],\ g(0) \neq 0\}$.

Definition 2.4.1 A *crystal basis* of M is a pair (L, B), where
(i) L is a free $\mathbb{A}_0$-submodule of M such that $M = \mathbb{C}(q) \otimes_{\mathbb{A}_0} L$;
(ii) B is a $\mathbb{C}$-basis of $L/qL \cong \mathbb{C} \otimes_{\mathbb{A}_0} L$;
(iii) $L = \bigoplus_{\lambda \in P} L_\lambda$, where $L_\lambda = L \cap M_\lambda$;
(iv) $B = \bigsqcup_{\lambda \in P} B_\lambda$, where $B_\lambda = B \cap (L_\lambda / qL_\lambda)$;
(v) $\tilde{e}_i L \subseteq L$, $\tilde{f}_i L \subseteq L$ for all $i \in I$;
(vi) $\tilde{e}_i B \subseteq B \cup \{0\}$, $\tilde{f}_i B \subseteq B \cup \{0\}$ for all $i \in I$;
(vii) for all $b, b' \in B$ and $i \in I$, $\tilde{f}_i b = b'$ if and only if $b = \tilde{e}_i b'$.

The set B is given a colored oriented graph structure by defining $b \xrightarrow{i} b'$ if and only if $\tilde{f}_i b = b'$. The graph B is called the *crystal graph* or simply the *crystal* of M.

Example 2.4.2 Let $U_q(\mathfrak{sl}_2(F)) = \langle e_i, f_i, K_i^{\pm 1} \rangle$ and let $V = U_q(\mathfrak{sl}_2(F)) v_0$ be the $(m+1)$-dimensional irreducible $U_q(\mathfrak{sl}_2(F))$-module. The crystal graph B of V is given as follows.

$$v_0 \xrightarrow{i} f_i v_0 \xrightarrow{i} \cdots \xrightarrow{i} f_i^{(k)} v_0 \xrightarrow{i} \cdots \xrightarrow{i} f_i^{(m)} v_0$$

The point is that almost all combinatorial features of M are reflected on the crystal B. For instance, we have

$$\dim_{\mathbb{C}(q)} M_\lambda = \operatorname{rank}_{\mathbb{A}_0} L_\lambda = \dim_{\mathbb{C}} L_\lambda / qL_\lambda = \# B_\lambda.$$

Moreover, the crystals behave very nicely with respect to taking the tensor product as we can see in the following discussion. We first introduce some notations. For an element $b \in B$, define

$$\varepsilon_i(b) = \max\{k \geq 0 \mid \tilde{e}_i^k b \neq 0\}, \quad \varphi_i(b) = \max\{k \geq 0 \mid \tilde{f}_i^k b \neq 0\}.$$

An element $b \in B$ is called a *maximal vector* if $\tilde{e}_i b = 0$ for all $i \in I$.

Let B_j be the crystal of a $U_q(\mathfrak{g})$-module M_j $(j = 1, 2)$ in the category $\mathcal{O}^q_{int}$. Set $L = L_1 \otimes_{\mathbb{A}_0} L_2$, $B = B_1 \otimes B_2 = B_1 \times B_2$. Then we have the *tensor product rule*:

Theorem 2.4.3 ([141, Theorem 1]) The pair (L, B) is a crystal basis of $M_1 \otimes_{\mathbb{C}(q)} M_2$, where the Kashiwara operators on B is given by

$$\begin{aligned} \tilde{e}_i(b_1 \otimes b_2) &= \begin{cases} \tilde{e}_i b_1 \otimes b_2 & \text{if } \varphi_i(b_1) \geq \varepsilon_i(b_2), \\ b_1 \otimes \tilde{e}_i b_2 & \text{if } \varphi_i(b_1) < \varepsilon_i(b_2), \end{cases} \\ \tilde{f}_i(b_1 \otimes b_2) &= \begin{cases} \tilde{f}_i b_1 \otimes b_2 & \text{if } \varphi_i(b_1) > \varepsilon_i(b_2), \\ b_1 \otimes \tilde{f}_i b_2 & \text{if } \varphi_i(b_1) \leq \varepsilon_i(b_2). \end{cases} \end{aligned} \tag{2.6}$$

Example 2.4.4 Let B be the crystal of the 3-dimensional irreducible $U_q(\mathfrak{sl}_2(F))$-module $V = \mathbb{C}(q)u \oplus \mathbb{C}(q)fu \oplus \mathbb{C}(q)f^{(2)}u$. Then the crystal $B \otimes B$ is given as follows.

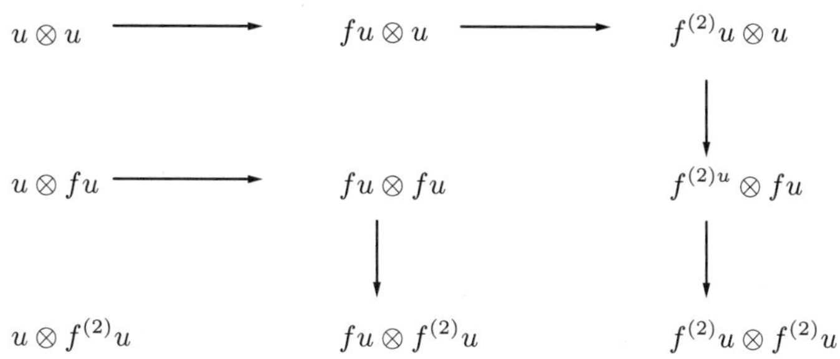

As an immediate corollary, we may describe the maximal vectors in the tensor product of crystals.

Corollary 2.4.5
(a) Let B_j be the crystal of M_j $(j = 1, 2)$. Then $b_1 \otimes b_2$ is a maximal vector in $B_1 \otimes B_2$ if and only if b_1 is a maximal vector and $\varepsilon_i(b_2) \leq \langle h_i \operatorname{wt}(b_1) \rangle$ for all $i \in I$.
(b) Let B_j be the crystal of M_j $(j = 1, \cdots, r)$. Then $b_1 \otimes \cdots \otimes b_r$ is a maximal vector in $B_1 \otimes \cdots \otimes B_r$ if and only if $b_1 \otimes \cdots \otimes b_k$ is a maximal vector for all $k = 1, \cdots, r$.

The existence and uniqueness of crystal basis is given in the following theorem.

Theorem 2.4.6 ([141, Theorem 2]) Let $V(\lambda) = U_q(\mathfrak{g})v_\lambda$ be the irreducible highest weight module with highest weight $\lambda \in P^+$. Let $L(\lambda)$ be the free $\mathbb{A}_0$-submodule of $V(\lambda)$ spanned by the vectors $\tilde{f}_{i_1} \cdots \tilde{f}_{i_r} v_\lambda$ $(r \geq 0, i_k \in I)$ and set

$$B(\lambda) = \{\tilde{f}_{i_1} \cdots \tilde{f}_{i_r} v_\lambda + qL(\lambda) \mid r \geq 0, i_k \in I\} \setminus \{0\}.$$

Then $(L(\lambda), B(\lambda))$ is a unique crystal basis of $V(\lambda)$.

One of the most interesting problems in crystal basis theory is to find an explicit combinatorial realization of $B(\lambda)$. For instance, let $U_q(\mathfrak{g})$ be the quantum group corresponding to the general linear Lie algebra $\mathfrak{gl}_n(\mathbb{C})$ and let λ be a dominant integral weight which can be identified with a partition with at most n rows. Then the crystal graph $B(\lambda)$ can be realized as the set of semistandard Young tableaux of shape λ with entries from $\{1, \cdots, n\}$ See, for example, [122, Chapter 7], [143] or Section 3.5.

Example 2.4.7 For $\lambda = (2, 1, 0)$, the $U_q(\mathfrak{gl}_3(F))$-crystal $B(\lambda)$ has the following realization.

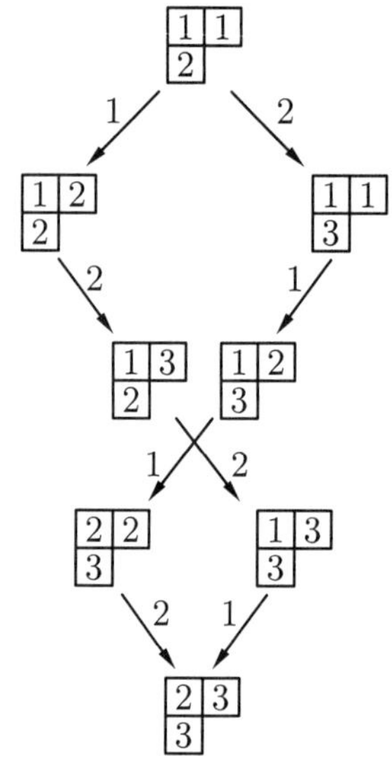

We now consider the crystal basis of $U_q^-(\mathfrak{g})$. For each $i \in I$ and $u \in U_q^-(\mathfrak{g})$, by induction on the degree of u, we can show that there exist unique u' and u'' in $U_q^-(\mathfrak{g})$ such that

$$e_i u - u e_i = \frac{K_i u'' - K_i^{-1} u'}{q_i - q_i^{-1}}.$$

We define a linear map $e_i' : U_q^-(\mathfrak{g}) \to U_q^-(\mathfrak{g})$ by $e_i'(u) = u'$. Then every u has a unique *i-string decomposition*:

$$u = \sum_{k \geq 0} f_i^{(k)} u_k, \quad \text{where } e_i' u_k = 0.$$

The *Kashiwara operators* on $U_q^-(\mathfrak{g})$ are defined by

$$\tilde{e}_i u = \sum_{k \geq 1} f_i^{(k-1)} u_k, \quad \tilde{f}_i u = \sum_{k \geq 0} f_i^{(k+1)} u_k.$$

Definition 2.4.8 A *crystal basis* of $U_q^-(\mathfrak{g})$ is a pair (L, B), where
(i) L is a free $\mathbb{A}_0$-submodule of $U_q^-(\mathfrak{g})$ such that $U_q^-(\mathfrak{g}) = \mathbb{C}(q)_{\mathbb{A}_0} L$;
(ii) B is a $\mathbb{C}$-basis of $L/qL \cong \mathbb{C} \otimes_{\mathbb{A}_0} L$;
(iii) $L = \bigoplus_{\alpha \in Q^+} L_{-\alpha}$, where $L_{-\alpha} = L \cap U_q^-(\mathfrak{g})_{-\alpha}$;
(iv) $B = \bigsqcup_{\alpha \in Q^+} B_{-\alpha}$, where $B_{-\alpha} = B \cap (L_{-\alpha}/qL_{-\alpha})$;
(v) $\tilde{e}_i L \subseteq L$, $\tilde{f}_i L \subseteq L$ for all $i \in I$;
(vi) $\tilde{e}_i B \subseteq B \cup \{0\}$, $\tilde{f}_i B \subseteq B$ for all $i \in I$;
(vii) for all $b, b' \in B$ and $i \in I$, $\tilde{f}_i b = b'$ if and only if $b = \tilde{e}_i b'$.

Theorem 2.4.9 ([141, Theorem 4]) Let $L(\infty)$ be the free $\mathbb{A}_0$-submodule of $U_q^-(\mathfrak{g})$ spanned by the vectors $\tilde{f}_{i_1} \cdots \tilde{f}_{i_r} 1$ ($r \geq 0, i_k \in I$) and set

$$B(\infty) = \{\tilde{f}_{i_1} \cdots \tilde{f}_{i_r} 1 + qL(\infty) \mid r \geq 0, i_k \in I\} \setminus \{0\}.$$

Then $(L(\infty), B(\infty))$ is a unique crystal basis of $U_q^-(\mathfrak{g})$.

We would also like to find an explicit combinatorial realization of the crystal graph $B(\infty)$. We will look into the basic properties of $B(\lambda)$ and $B(\infty)$ in the next section.

2.5 Abstract crystals

By extracting the properties of crystal graphs of $U_q(\mathfrak{g})$-modules in the category $\mathcal{O}^q_{int}$, we introduce the notion of *abstract crystals.*

Definition 2.5.1 An *abstract* $U_q(\mathfrak{g})$*-crystal* or simply a *crystal* is a set B together with the maps $wt : B \to P$, $\varphi_i, \varepsilon_i : B \to \mathbb{Z} \sqcup \{-\infty\}$, $\tilde{e}_i, \tilde{f}_i : B \to B \sqcup \{0\}$ $(i \in I)$ satisfying the following conditions: for all $b, b' \in B$ and $i \in I$,
(i) $\mathrm{wt}(\tilde{e}_i b) = \mathrm{wt}(b) + \alpha_i$, $\mathrm{wt}(\tilde{f}_i b) = \mathrm{wt}(b) - \alpha_i$,
(ii) if $\tilde{e}_i b \in B$, then $\varepsilon_i(\tilde{e}_i b) = \varepsilon_i(b) - 1$, $\varphi_i(\tilde{e}_i b) = \varphi_i(b) + 1$,
(iii) if $\tilde{f}_i b \in B$, then $\varepsilon_i(\tilde{f}_i b) = \varepsilon_i(b) + 1$, $\varphi_i(\tilde{f}_i b) = \varphi_i(b) - 1$,
(iv) $\varphi_i(b) = \varepsilon_i(b) + \langle h_i, wt(b)\rangle$,
(v) $\tilde{f}_i b = b'$ if and only if $b = \tilde{e}_i b'$,
(vi) if $\varphi_i(b) = -\infty$, then $\tilde{e}_i b = \tilde{f}_i b = 0$.

Example 2.5.2

(a) Let $B(\lambda)$ be the crystal graph of $V(\lambda)$ with $\lambda \in P^+$. For $b = \tilde{f}_{i_1} \cdots \tilde{f}_{i_r} v_\lambda$, define

$$\begin{aligned} \mathrm{wt}(b) &= \lambda - (\alpha_{i_1} + \cdots + \alpha_{i_r}), \\ \varepsilon_i(b) &= \max\{k \geq 0 \mid \tilde{e}_i^k b \neq 0\}, \\ \varphi_i(b) &= \max\{k \geq 0 \mid \tilde{f}_i^k b \neq 0\}. \end{aligned}$$

Then $B(\lambda)$ is an abstract crystal.

(b) Let $B(\infty)$ be the crystal graph of $U_q^-(\mathfrak{g})$. For $b = \tilde{f}_{i_1} \cdots \tilde{f}_{i_r} 1$, define

$$\begin{aligned} \mathrm{wt}(b) &= -(\alpha_{i_1} + \cdots + \alpha_{i_r}), \\ \varepsilon_i(b) &= \max\{k \geq 0 \mid \tilde{e}_i^k b \neq 0\}, \\ \varphi_i(b) &= \varepsilon_i(b) + \langle h_i, \mathrm{wt}(b)\rangle. \end{aligned}$$

Then $B(\infty)$ is an abstract crystal.

(c) For $\lambda \in P$, let $T_\lambda = \{t_\lambda\}$ and define

$$\mathrm{wt}(t_\lambda) = \lambda, \quad \varepsilon_i(t_\lambda) = \varphi_i(t_\lambda) = -\infty, \quad \tilde{e}_i t_\lambda = \tilde{f}_i t_\lambda = 0.$$

Then T_λ is an abstract crystal.

(d) Let $C = \{c\}$ and define

$$\mathrm{wt}(c) = 0, \quad \varepsilon_i(c) = \varphi_i(c) = 0, \quad \tilde{e}_i c = \tilde{f}_i c = 0.$$

Then C is an abstract crystal.

(e) Let $B_i = \{b_i(-n) \mid n \geq 0\}$ and define

$$\begin{aligned}
&\mathrm{wt}(b_i(-n)) = -n\alpha_i, \quad \varepsilon_i(b_i(-n)) = n, \quad \varphi_i(b_i(-n)) = -n,\\
&\varepsilon_j(b_i(-n)) = \varphi_j(b_i(-n)) = -\infty \quad (j \neq i),\\
&\tilde{e}_i(b_i(-n)) = b_i(-n+1), \quad \tilde{f}_i(b_i(-n)) = b_i(-n-1),\\
&\tilde{e}_j(b_i(-n)) = \tilde{f}_j(b_i(-n)) = 0 \quad (j \neq i).
\end{aligned}$$

Then B_i becomes an abstract crystal which is called the *i-th elementary crystal.*

Definition 2.5.3 Let B_1, B_2 be abstract crystals.

(a) A map $\psi : B_1 \to B_2$ is called a *crystal morphism* if it satisfies the following conditions: for all $b \in B_1$ and $i \in I$,

(i)$\mathrm{wt}(\psi(b)) = \mathrm{wt}(b)$, $\varepsilon_i(\psi(b)) = \varepsilon_i(b)$, $\varphi_i(\psi(b)) = \varphi_i(b)$,

(ii) if $\tilde{f}_i b \in B_1$, then $\psi(\tilde{f}_i b) = \tilde{f}_i \psi(b)$.

(b) A crystal morphism $\psi : B_1 \to B_2$ is *strict* if

$$\psi \circ \tilde{e}_i = \tilde{e}_i \circ \psi, \quad \psi \circ \tilde{f}_i = \tilde{f}_i \circ \psi \quad \text{for all } i \in I.$$

Here, we understand $\psi(0) = 0$.

(c) A crystal morphism $\psi : B_1 \to B_2$ is called an *embedding* if ψ is injective. In this case, we say that B_1 is a *subcrystal* of B_2. If ψ is a strict embedding, we say that B_1 is a *full subcrystal* of B_2.

We now define the tensor product of abstract crystals. Let B_1, B_2 be abstract crystals. Set $B_1 \otimes B_2 = B_1 \times B_2$ and define

$$\begin{aligned}
&\mathrm{wt}(b_1 \otimes b_2) = \mathrm{wt}(b_1) + \mathrm{wt}(b_2),\\
&\varepsilon_i(b_1 \otimes b_2) = \max(\varepsilon_i(b_1), \varepsilon_i(b_2) - \langle h_i, \mathrm{wt}(b_1)\rangle),\\
&\varphi_i(b_1 \otimes b_2) = \max(\varphi_i(b_2), \varphi_i(b_1) + \langle h_i, \mathrm{wt}(b_2)\rangle),\\
&\tilde{e}_i(b_1 \otimes b_2) = \begin{cases} \tilde{e}_i b_1 \otimes b_2 & \text{if } \varphi_i(b_1) \geq \varepsilon_i(b_2),\\ b_1 \otimes \tilde{e}_i b_2 & \text{if } \varphi_i(b_1) < \varepsilon_i(b_2),\end{cases}\\
&\tilde{f}_i(b_1 \otimes b_2) = \begin{cases} \tilde{f}_i b_1 \otimes b_2 & \text{if } \varphi_i(b_1) > \varepsilon_i(b_2),\\ b_1 \otimes \tilde{f}_i b_2 & \text{if } \varphi_i(b_1) \leq \varepsilon_i(b_2).\end{cases}
\end{aligned} \tag{2.7}$$

Theorem 2.5.4 ([142, Section 1]) $B_1 \otimes B_2$ is an abstract crystal.

Hence the category of abstract crystals becomes a tensor category. The crystals $B(\lambda)$ and $B(\infty)$ are closely related as can be seen in the following proposition.

Proposition 2.5.5 ([142]) There exists an injective map $\psi_\lambda : B(\lambda) \to B(\infty)$ such that

(i) $\psi_\lambda(v_\lambda) = 1$,

(ii) $\psi_\lambda(\tilde{f}_i b) = \tilde{f}_i \psi_\lambda(b)$ whenever $\tilde{f}_i b \neq 0$,

(iii) $\psi_\lambda(\tilde{e}_i b) = \tilde{e}_i \psi_\lambda(b)$ for all $b \in B(\lambda)$,

(iv) $\mathrm{wt}(\psi_\lambda(b)) = \mathrm{wt}(b) - \lambda$, $\varepsilon_i(\psi_\lambda(b)) = \varepsilon_i(b)$ for all $b \in B(\lambda)$.

The following crystal embedding theorem plays a crucial role in proving the recognition theorem for $B(\infty)$.

Theorem 2.5.6 ([142, Theorem 2.2.1]) For each $i \in I$, there exists a strict crystal embedding

$$\Psi_i : B(\infty) \to B(\infty) \otimes B_i \quad \text{given by } 1 \longmapsto 1 \otimes b_i(0).$$

Now we are ready to state the *recognition theorems* for $B(\infty)$ and $B(\lambda)$ ($\lambda \in P^+$).

Theorem 2.5.7 ([144, Proposition 3.2.3]) Suppose B is an abstract crystal satisfying the following conditions:
(i) $\mathrm{wt}(B) \subseteq -Q^+$,
(ii) there exists an element $b_0 \in B$ with $\mathrm{wt}(b_0) = 0$,
(iii) for any $b \neq b_0$, there exists an $i \in I$ such that $\tilde{e}_i b \neq 0$,
(iv) for each $i \in I$, there exists a strict crystal embedding $\Psi_i : B \to B \otimes B_i$.

Then there exists a crystal isomorphism $B \xrightarrow{\sim} B(\infty)$ sending b_0 onto 1.

Theorem 2.5.8 ([202, Proposition 2.3.1]) The crystal $B(\lambda)$ is isomorphic to the connected component of $1 \otimes t_\lambda \otimes c$ in $B(\infty) \otimes T_\lambda \otimes C$. That is, there exists a strict crystal embedding

$$B(\lambda) \hookrightarrow B(\infty) \otimes T_\lambda \otimes C \quad \text{such that } v_\lambda \longmapsto 1 \otimes t_\lambda \otimes c.$$

Remark. These recognition theorems can be applied in various ways. For example, in [144] and [202], Kashiwara and Saito constructed abstract crystals consisting of irreducible components of certain Lagrangian subvarieties. They identified these crystals with $B(\infty)$ and $B(\lambda)$, respectively, using these recognition theorems. (See Sections 3.3 and 3.4 for more details.) On the other hand, Kleshchev constructed abstract crystals on the Grothendieck groups of affine Hecke algebras and their cyclotomic quotients, and identified them with $B(\infty)$ and $B(\lambda)$, respectively, for the quantum affine algebra of type $U_q(A_n^{(1)})$ using these recognition theorems [147]. The same philosophy works well for the construction of abstract crystals on the Grothendieck groups of Khovanov-Lauda-Rouquier algebras [151].

2.6 Perfect crystals

Let $I = \{0, 1, \cdots, n\}$ be the index set and let $A = (a_{ij})_{i,j \in I}$ be a generalized Cartan matrix of affine type. That is, corank$A = 1$ and there exists $u = (d_0, d_1, \cdots, d_n)^t$ with $d_i \in \mathbb{Z}_{>0}$ such that $Au = 0$ and $\gcd(d_0, d_1, \cdots, d_n) = 1$. The affine Cartan matrices are classified in [137]. The *affine Cartan datum* $(A, P^\vee, P, \Pi)$ associated with A is given as follows:

$$\begin{aligned} P^\vee &= \mathbb{Z}h_0 \oplus \mathbb{Z}h_1 \oplus \cdots \oplus \mathbb{Z}h_n \oplus \mathbb{Z}d, \\ \Pi &= \{\alpha_0, \alpha_1, \cdots \alpha_n\}, \\ P &= \mathbb{Z}\Lambda_0 \oplus \mathbb{Z}\Lambda_1 \oplus \cdots \oplus \mathbb{Z}\Lambda_n \oplus \mathbb{Z}\frac{1}{d_0}\delta, \end{aligned} \tag{2.8}$$

where

$$\begin{aligned} &\alpha_j(h_i) = a_{ij}, \quad \alpha_j(d) = \begin{cases} 1 & \text{for } j = 0, \\ 0 & \text{for } j \neq 0, \end{cases} \\ &\Lambda_i(h_j) = \delta_{ij}, \quad \Lambda_i(d) = 0, \\ &\delta = d_0\alpha_0 + d_1\alpha_1 + \cdots + d_n\alpha_n. \end{aligned}$$

The quantum group $U_q(\mathfrak{g})$ associated with an affine Cartan datum is called a *quantum affine algebra*. The elements of P^+ are called the *affine dominant integral weights*. If A is an affine Cartan matrix, then so is A^t and hence there is

$v = (c_0, c_1, \cdots, c_n)$ with $c_i \in \mathbb{Z}_{>0}$ such that $vA = 0$ and $\gcd(c_0, c_1, \cdots, c_n) = 1$. The *canonical central element* is defined to be

$$c = c_0 h_0 + c_1 h_1 + \cdots + c_n h_n.$$

We say that $\lambda \in P$ has level l if $\lambda(c) = l$.

We denote by $U_q'(\mathfrak{g})$ the subalgebra of $U_q(\mathfrak{g})$ generated by e_i, f_i, $K_i^{\pm 1}$ $(i \in I)$. Let

$$\overline{P}^{\vee} = \mathbb{Z}h_0 \oplus \mathbb{Z}h_1 \oplus \cdots \oplus \mathbb{Z}h_n \quad \text{and} \quad \overline{\mathfrak{h}} = \mathbb{C} \otimes_{\mathbb{Z}} \overline{P}^{\vee}.$$

Consider α_i and Λ_i as linear functionals on $\overline{\mathfrak{h}}$ and set

$$\overline{P} = \mathbb{Z}\Lambda_0 \oplus \mathbb{Z}\Lambda_1 \oplus \cdots \oplus \mathbb{Z}\Lambda_n.$$

Thus $\delta = 0$ in $\overline{P}$ and Π is not linearly independent. We call the quadruple $(A, \overline{P}^{\vee}, \overline{P}, \Pi)$ a *classical Cartan datum.* The subalgebra $U_q'(\mathfrak{g})$ can be viewed as the quantum group associated with $(A, \overline{P}^{\vee}, \overline{P}, \Pi)$.

Let $\mathrm{cl} : P \to \overline{P}$ denote the projection map and fix an embedding $\mathrm{aff} : \overline{P} \to P$ such that $\mathrm{cl} \circ \mathrm{aff} = \mathrm{id}$, $\mathrm{aff} \circ \mathrm{cl}(\alpha_i) = \alpha_i$ for $i \neq 0$. Set $\overline{P}^{+} = \mathrm{cl}(P^+)$. The elements of $\overline{P}^{+}$ are called the *classical dominant integral weights.* The *level* of an element $\lambda \in \overline{P}$ is defined to be $\lambda(c)$.

We say that B is an *affine crystal* or a $U_q(\mathfrak{g})$*-crystal* if it is an abstract crystal associated with an affine Cartan datum $(A, P^{\vee}, P, \Pi)$. Similarly, B is called a *classical crystal* or a $U_q'(\mathfrak{g})$*-crystal* if it is an abstract crystal associated with a classical Cartan datum $(A, \overline{P}^{\vee}, \overline{P}, \Pi)$. For an element b in a finite classical crystal B, we define

$$\varepsilon(b) = \sum_{i \in I} \varepsilon_i(b)\Lambda_i \quad \varphi(b) = \sum_{i \in I} \varphi_i(b)\Lambda_i.$$

Definition 2.6.1 A finite classical crystal B is called a *perfect crystal of level* l if

(i) there exists a finite dimensional $U_q'(\mathfrak{g})$-module V whose crystal graph is isomorphic to B,

(ii) $B \otimes B$ is connected,

(iii) there exists a classical weight $\lambda_0 \in \overline{P}$ such that $\#(B_{\lambda_0}) = 1$ and

$$\mathrm{wt}(B) \subseteq \lambda_0 + \frac{1}{d_0} \sum_{i \neq 0} \mathbb{N}\alpha_i,$$

(iv) $\langle c, \varepsilon(b) \rangle \geq l$ for all $b \in B$,

(v) for each $\lambda \in \overline{P}^{+}$ with $\lambda(c) = l$, there exist uniquely determined b^{λ} and b_{λ} in B such that $\varepsilon(b^{\lambda}) = \lambda$, $\varphi(b_{\lambda}) = \lambda$.

Example 2.6.2 (see [122, Example 10.5.2] for details)

(a) Let $\mathfrak{g} = A_2^{(1)}$ and let B be the $U_q'(\mathfrak{g})$-crystal given below. Then B is a perfect crystal of level 1.

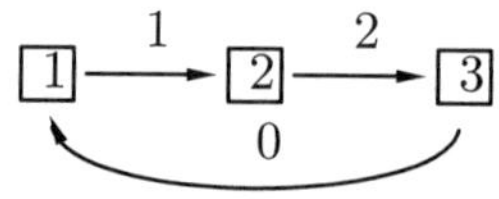

(b) Let $\mathfrak{g} = C_2^{(1)}$ and let B be the $U_q'(\mathfrak{g})$-crystal given below.

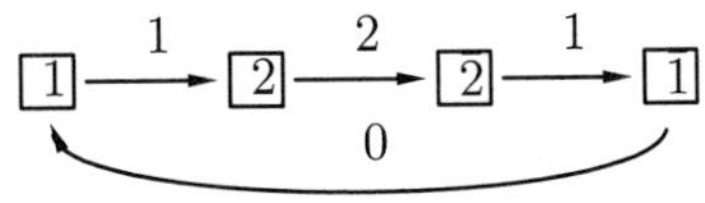

Then B is *not* a perfect crystal of level 1, since we have

$$\varepsilon(\boxed{2}) = \varepsilon(\boxed{\bar{1}}) = \Lambda_1, \qquad \varphi(\boxed{1}) = \varphi(\boxed{\bar{2}}) = \Lambda_1,$$

(c) Let $\mathfrak{g} = C_2^{(1)}$ and let B be the $U_q'(\mathfrak{g})$-crystal given below. Then B is a perfect crystal of level 1.

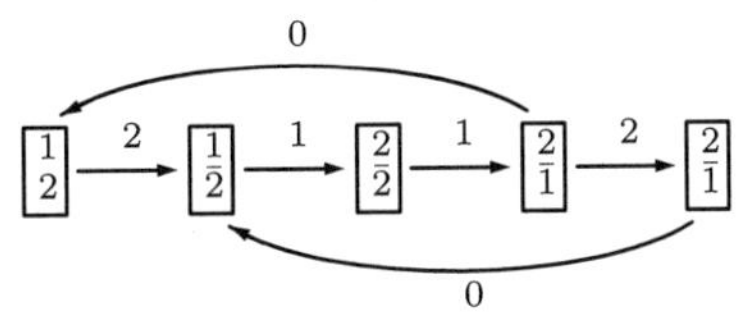

(d) Let $\mathfrak{g} = A_2^{(1)}$ and let B be the $U_q'(\mathfrak{g})$-crystal given below. Then B is a perfect crystal of level 2.

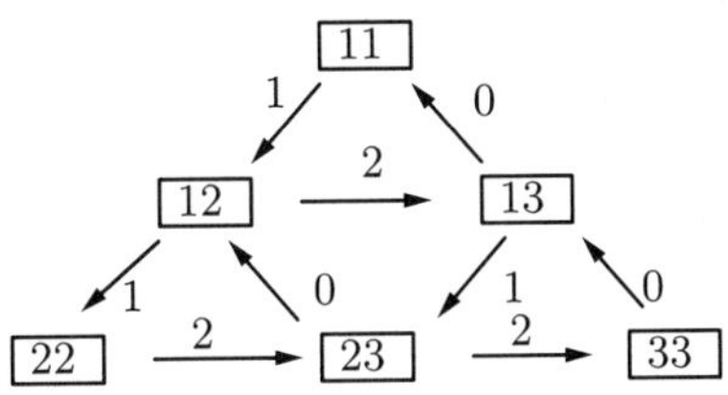

Here, we use the semistandard tableaux of shape $\boxed{}\boxed{}$ with entries in $\{1, 2, 3\}$ to describe the crystal $B(2\varpi_1)$ for $U_q(A_2)$.

The following *fundamental isomorphism theorem* plays a central role in the theory of perfect crystals.

Theorem 2.6.3 ([140, Theorem 4.4.1]) Let B be a perfect crystal of level l. Then for any classical dominant integral weight λ of level l, there exists a unique $U_q'(\mathfrak{g})$-crystal isomorphism

$$B(\lambda) \xrightarrow{\sim} B(\varepsilon(b_\lambda)) \otimes B \quad \text{given by } u_\lambda \mapsto u_{\varepsilon(b_\lambda)} \otimes b_\lambda,$$

where u_λ denotes the highest weight vector in $B(\lambda)$ and b_λ is the unique element in B such that $\varphi(b_\lambda) = \lambda$.

Idea of Proof. The main idea of the proof is that $u_{\varepsilon(b_\lambda)} \otimes b_\lambda$ is the only maximal vector in $B(\varepsilon(b_\lambda)) \otimes B$ and its weight is λ. □

Set

$$\lambda_0 = \lambda, \quad b_0 = b_\lambda; \quad \lambda_{k+1} = \varepsilon(b_k), \quad b_{k+1} = b_{\lambda_{k+1}} \ (k \geq 0).$$

By applying the fundamental isomorphism theorem repeatedly, we get a series of crystal isomorphisms

$$B(\lambda) \xrightarrow{\sim} B(\lambda_1) \otimes B \xrightarrow{\sim} \cdots \xrightarrow{\sim} B(\lambda_{k+1}) \otimes B \otimes \cdots \otimes B$$

given by

$$u_\lambda \longmapsto u_{\lambda_1} \otimes b_0 \longmapsto \cdots \longmapsto u_{\lambda_{k+1}} \otimes b_k \otimes \cdots \otimes b_0.$$

Definition 2.6.4

(a) The sequence $\mathbf{b} = (b_k)_{k\geq 0} = \cdots b_{k+1} \otimes b_k \otimes \cdots \otimes b_0$ is called the *ground-state path of weight λ in B.*

(b) The sequence $\mathbf{p} = (p_k)_{k\geq 0} = \cdots \otimes p_{k+1} \otimes p_k \cdots \otimes p_0$ is called a *λ-path in B* if $p_k \in B$ for all $k \geq 0$ and $p_k = b_k$ for all $k \gg 0$.

Let $\mathcal{P}(\lambda)$ be the set of all λ-paths in B. For $\mathbf{p} = (p_k)_{k\geq 0} \in \mathcal{P}(\lambda)$, let $N \geq 0$ be the smallest nonnegative integer such that $p_k = b_k$ for all $k \geq N$, and define

$$\begin{aligned}
\mathrm{wt}(\mathbf{p}) &= \lambda_N + \sum_{k=0}^{N-1} \mathrm{wt}(p_k), \\
\tilde{e}_i\mathbf{p} &= \cdots \otimes p_{N+1} \otimes \tilde{e}_i(p_N \otimes \cdots \otimes p_0), \\
\tilde{f}_i\mathbf{p} &= \cdots \otimes p_{N+1} \otimes \tilde{f}_i(p_N \otimes \cdots \otimes p_0), \\
\varepsilon_i(\mathbf{p}) &= \max(\varepsilon_i(\mathbf{p}') - \varphi_i(b_N), 0), \\
\varphi_i(\mathbf{p}) &= \varphi_i(\mathbf{p}') + \max(\varphi_i(b_N) - \varepsilon_i(\mathbf{p}'), 0),
\end{aligned} \tag{2.9}$$

where $\mathbf{p}' = p_{N-1} \otimes \cdots \otimes p_0$. Then we obtain the *path realization* of $B(\lambda)$ in terms of λ-paths.

Theorem 2.6.5 ([140, Proposition 4.6.4])
(a) The maps wt, $\tilde{e}_i$, $\tilde{f}_i$, ε_i, φ_i $(i \in I)$ define a $U_q'(\mathfrak{g})$-crystal structure on $\mathcal{P}(\lambda)$.
(b) There exists a unique $U_q'(\mathfrak{g})$-crystal isomorphism

$$\Psi : B(\lambda) \xrightarrow{\sim} \mathcal{P}(\lambda) \quad \text{given by} \quad u_\lambda \mapsto \mathbf{b}.$$

Example 2.6.6
(a) Let $\mathfrak{g} = A_2^{(1)}$ and consider the perfect crystal B of level 1 given below.

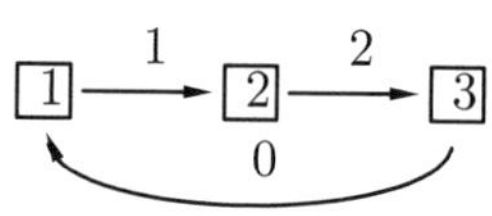

The ground-state of weight Λ_0 in B is $\mathbf{b} = (\cdots\cdots, 1, 2, 3, 1, 2, 3)$ and the top part of the crystal $\mathcal{P}(\Lambda_0)$ is given as follows.

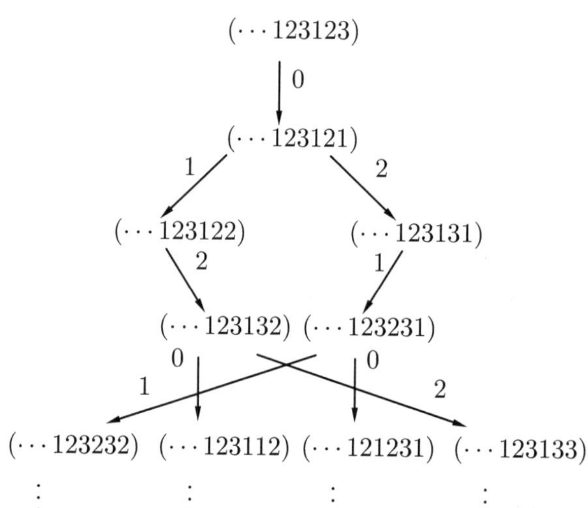

(b) Let $\mathfrak{g} = A_4^{(2)}$ and consider the perfect crystal B of level 1 given below.

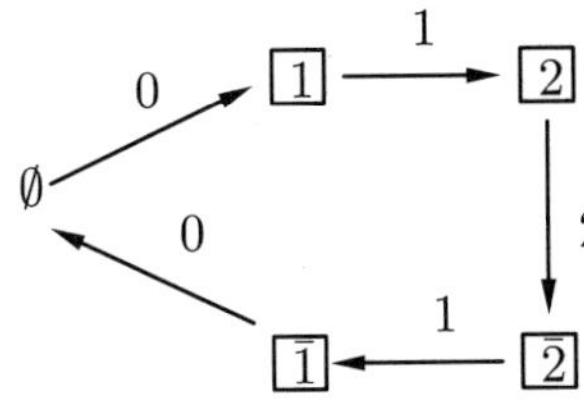

The ground-state of weight Λ_0 in B is $\mathbf{b} = (\cdots\cdots, \phi, \phi, \phi, \phi)$ and the top part of the crystal $\mathcal{P}(\Lambda_0)$ is given as follows.

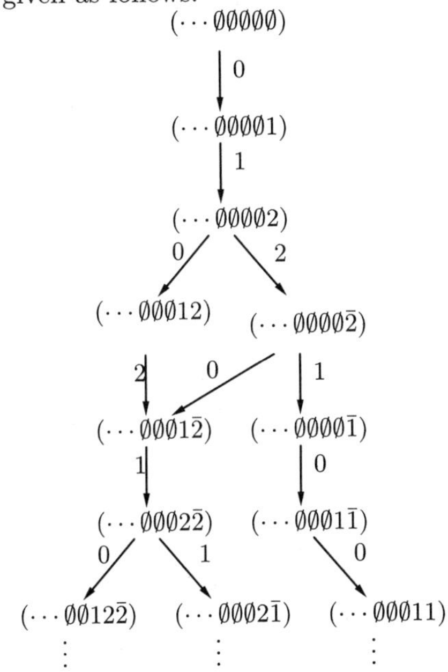

(c) Let $\mathfrak{g} = A_5^{(2)}$ and consider the perfect crystal B of level 1 given below.

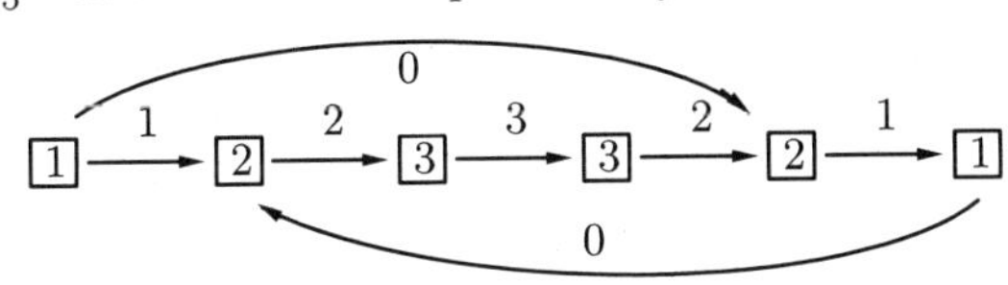

The ground-state of weight Λ_0 in B is $\mathbf{b} = (\cdots\cdots, 1, \bar{1}, 1, \bar{1}, 1, \bar{1})$ and the top part of the crystal $\mathcal{P}(\Lambda_0)$ is given as follows.

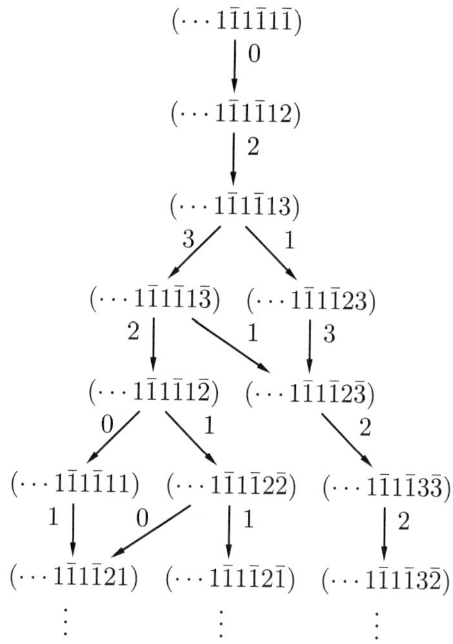

2.7 Combinatorics of Young walls

In [171], Misra and Miwa gave a realization of the level 1 highest weight crystal $B(\Lambda_0)$ over $U_q(A_{n-1}^{(1)})$ in terms of *colored Young diagrams*. More precisely, we build a Young diagram $Y = (y_k)_{k\geq 0}$ with each box colored by $0, 1, \cdots, n-1$ following the pattern given below.

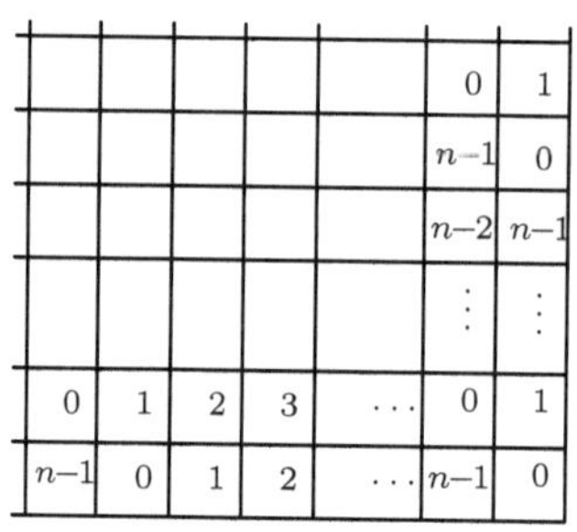

Note that each column of the pattern is in fact the *affinization* of the perfect crystal given in Example 2.6.6(a) (see [122, Chapter 10] for details). Let y_k $(k \geq 0)$ denote the k-th column of Y reading from right to left. Since Y is a Young diagram, we have $y_{k+1} \leq y_k$ for all $k \geq 0$ and $y_k = 0$ for all $k \gg 0$. A colored Young diagram is called *n-reduced* if $y_k - y_{k+1} < n$ for all $k \geq 0$. We denote by $\mathcal{Z}(\Lambda_0)$ (resp. $\mathcal{Y}(\Lambda_0)$) the set of all colored Young diagrams (resp. n-reduced Young diagrams). We define the Kashiwara operators as follows.

Fix an index $i \in I$. To each column y_k of Y, we assign its *i-signature* by

$$i\text{-signature of } y_k = \begin{cases} + & \text{if the top of } y_k \text{ is } i-1, \\ - & \text{if the top of } y_k \text{ is } i, \\ \cdot & \text{otherwise.} \end{cases}$$

Here, "·" means we don't put anything. Then we get an infinite sequence of +'s and −'s. From this infinite sequence, we cancel all the $(+,-)$-pairs to obatin a finite sequence of −'s followed by +'s:

$$i\text{-signature of } Y = (-,\cdots,-,+,\cdots,+),$$

which is called the *i-signature* of Y. Now we define $\tilde{e}_i Y$ (resp. $\tilde{f}_i Y$) to be the colored Young diagram obtained from Y by removing the i-box from the column (resp. by adding an i-box to the column) corresponding to the rightmost − (resp. leftmost +) in the i-signature of Y.

Theorem 2.7.1 ([171, Theorem 4.7])
(a) The Kashiwara operators define a $U_q(A_{n-1}^{(1)})$-crystal structure on $\mathcal{Z}(\Lambda_0)$.
(b) The set $\mathcal{Y}(\Lambda_0)$ is the connected component of $\mathcal{Z}(\Lambda_0)$ containing the empty diagram ϕ. Moreover, there exists a crystal isomorphism

$$B(\Lambda_0) \xrightarrow{\sim} \mathcal{Y}(\Lambda_0) \quad \text{given by} \quad u_{\Lambda_0} \longmapsto \phi.$$

Example 2.7.2 Let $\mathfrak{g} = A_2^{(1)}$. The top part of the crystal $\mathcal{Y}(\Lambda_0)$ consisting of 3-reduced Young diagrams is given as follows.

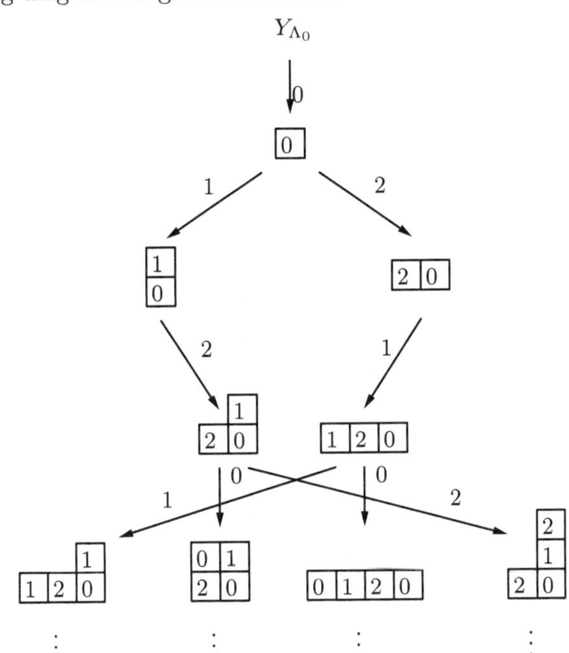

In [139] (see also [123]), we introduced a new family of combinatorial objects, called the *Young walls*, and gave a realization of the level 1 highest weight crystals $B(\Lambda)$ for all classical quantum affine algebras in terms of *reduced proper Young walls*. The Young walls consists of colored blocks of three different shapes shown below.

Type	Shape	Width	Thickness	Height	Volume
I	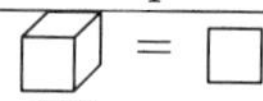	1	1	1	1
II	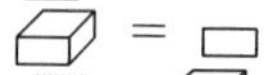	1	1	$\frac{1}{2}$	$\frac{1}{2}$
III	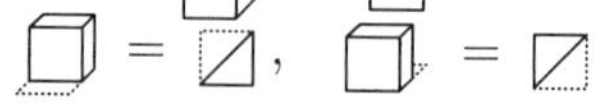	1	$\frac{1}{2}$	1	$\frac{1}{2}$

For each dominant integral weight Λ of level 1, we fix a frame Y_Λ called the *ground-state wall of weight* Λ, and on this frame, we build a wall of thickness less than or equal to one unit which extends infinitely to the left. The rules for building the walls are as follows:

(1) The colored blocks should be stacked in columns. No block can be placed on top of a column of half-unit thickness.

(2) Except for the right-most column, there should be no free space to the right of any block.

(3) The colored blocks should be stacked in a specified pattern which is determined by the type of the quantum affine algebra $U_q(\mathfrak{g})$ and the level 1 dominant integral weight Λ.

The coloring of blocks, the description of ground-state walls and the patterns for building the walls are given in [139] (see also [123]).

Example 2.7.3

(a) If $\mathfrak{g} = A_4^{(2)}$, we use the colored blocks

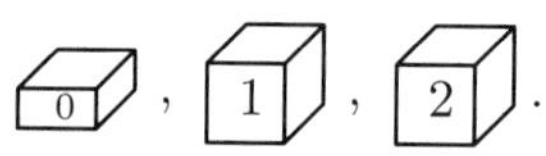

The walls are built on the ground-state wall

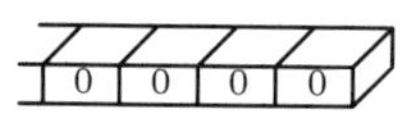

following the pattern given below.

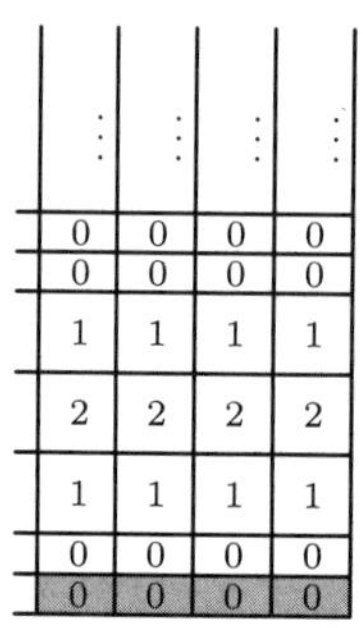

(b) If $\mathfrak{g} = A_5^{(2)}$, we use the colored blocks

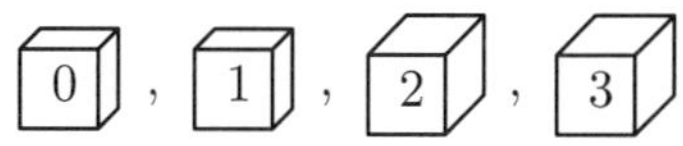

The walls are built on the ground-state wall

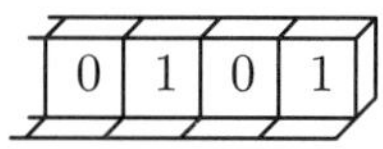

following the pattern given below

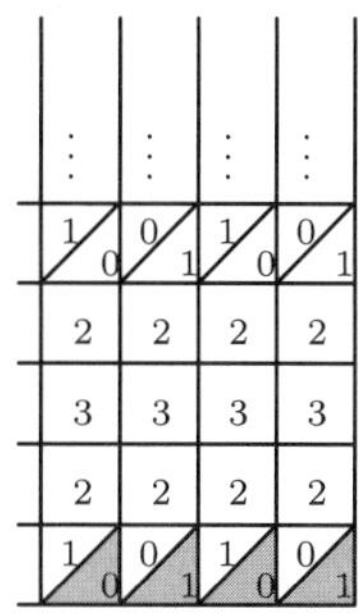

A wall Y built on the ground-state wall Y_Λ following the rules given above is called a *Young wall* on Y_Λ, for the heights of its columns are weakly decreasing as we proceed from right to left. We often write $Y = (y_k)_{k=0}^{\infty} = (\cdots\cdots, y_2, y_1, y_0)$ as an infinite sequence of its columns.

Definition 2.7.4

(a) A column of a Young wall is called a *full column* if its height is a multiple of the unit length and its top is of unit thickness.

(b) A Young wall is said to be *proper* if none of the full columns have the same height.

We denote by $\mathcal{Z}(\Lambda)$ the set of all proper Young walls on Y_Λ. We define the action of Kashiwara operators $\tilde{e}_i$, $\tilde{f}_i$ $(i \in I)$ on $\mathcal{Z}(\Lambda)$ as follows. Let $Y = (y_k)_{k=0}^{\infty}$ be a proper Young wall on Y_Λ. A block of color i in Y is called a *removable i-block* if Y remains a proper Young wall after removing the block. A column in Y is said to be *i-removable* if the top of that column is a removable i-block. Similarly, a place in Y is called an *i-admissible slot* if one may add an i-block to obtain another proper Young wall. A column in Y is said to be *i-admissible* if the top of that column is an i-admissible slot. Fix an index $i \in I$. To each column y_k of Y, we assign its *i-signature* as follows:

(i) we assign $--$ if the column y_k is twice i-removable (the i-block will be of half-unit height in this case);
(ii) we assign $-$ if the column is once i-removable, but not i-admissible (the i-block may be of unit height or of half-unit height);
(iii) we assign $-+$ if the column is once i-removable and once i-admissible (the i-block will be of half-unit height in this case);
(iv) we assign $+$ if the column is once i-admissible, but not i-removable (the i-block may be of unit height or of half-unit height);
(v) we assign $++$ if the column is twice i-admissible (the i-block will be of half-unit height in this case).

Then we get an infinite sequence of $+$'s and $-$'s. From this infinite sequence, we cancel out every $(+,-)$-pair to obtain a finite sequence of $-$'s followed by $+$'s, reading from left to right. This sequence is called the *i-signature* of Y.

(i) We define $\tilde{e}_i Y$ to be the proper Young wall obtained from Y by removing the i-block corresponding to the rightmost $-$ in the i-signature of Y. We define $\tilde{e}_i Y = 0$ if there is no $-$ in the i-signature of Y.

(ii) We define $\tilde{f}_i Y$ to be the proper Young wall obtained from Y by adding an i-block to the column corresponding to the leftmost $+$ in the i-signature of Y. We define $\tilde{f}_i Y = 0$ if there is no $+$ in the i-signature of Y.

We also define the maps

$$\mathrm{wt} : \mathcal{Z}(\Lambda) \longrightarrow P, \quad \varepsilon_i : \mathcal{Z}(\Lambda) \longrightarrow \mathbb{Z}, \quad \varphi_i : \mathcal{Z}(\Lambda) \longrightarrow \mathbb{Z}$$

by

$$\begin{aligned} \mathrm{wt}(Y) &= \Lambda - \sum_{i \in I} k_i \alpha_i, \\ \varepsilon_i(Y) &= \text{the number of } -\text{'s in the } i\text{-signature of } Y, \\ \varphi_i(Y) &= \text{the number of } +\text{'s in the } i\text{-signature of } Y, \end{aligned}$$

where k_i is the number of i-blocks in Y that have been added to the ground-state wall Y_Λ.

Proposition 2.7.5 ([139, Theorem 6.3]) The set $\mathcal{Z}(\Lambda)$ together with the maps $\mathrm{wt} : \mathcal{Z}(\Lambda) \to P$, $\tilde{e}_i, \tilde{f}_i : \mathcal{Z}(\Lambda) \to \mathcal{Z}(\Lambda) \cup \{0\}$ and $\varepsilon_i, \varphi_i : \mathcal{Z}(\Lambda) \to \mathbb{Z}$ becomes a $U_q(\mathfrak{g})$-crystal.

Definition 2.7.6

(a) The part of a column in a proper Young wall is called a *δ-column* if it has the same number of colored blocks as the null root δ in some cyclic order.

(b) A δ-column in a proper Young wall is called *removable* if it can be removed to yield another proper Young wall.

(c) A proper Young wall is said to be *reduced* if none of its columns contain a removable δ-column.

Let $\mathcal{Y}(\Lambda) \subseteq \mathcal{Z}(\Lambda)$ be the set of all reduced proper Young walls on Y_Λ.

Theorem 2.7.7 ([139, Theorem 7.1]) The set $\mathcal{Y}(\Lambda)$ is a connected $U_q(\mathfrak{g})$-crystal. Moreover, there exists a crystal isomorphism

$$B(\lambda) \xrightarrow{\sim} \mathcal{Y}(\Lambda) \quad \text{given by} \quad u_\Lambda \longmapsto Y_\Lambda.$$

Example 2.7.8

(a) If $\mathfrak{g} = A_4^{(2)}$, the top part of the crystal $\mathcal{Y}(\Lambda_0)$ is given as follows.

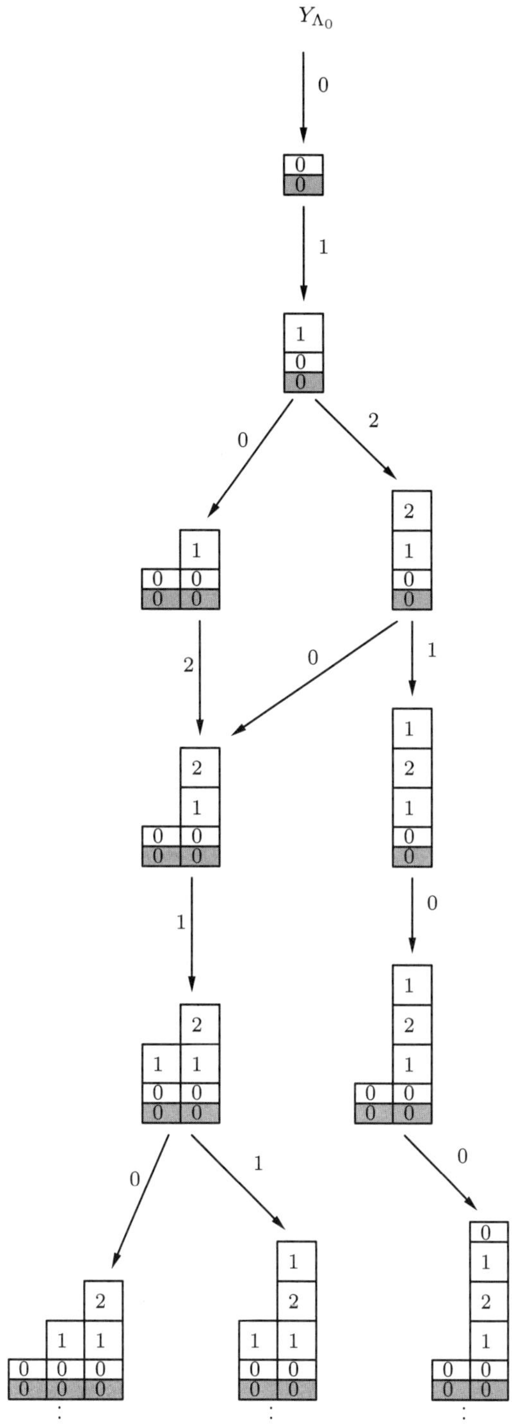
Y_{Λ_0}

(b) If $\mathfrak{g} = A_5^{(2)}$, the top part of the crystal $\mathcal{Y}(\Lambda_0)$ is given as follows.

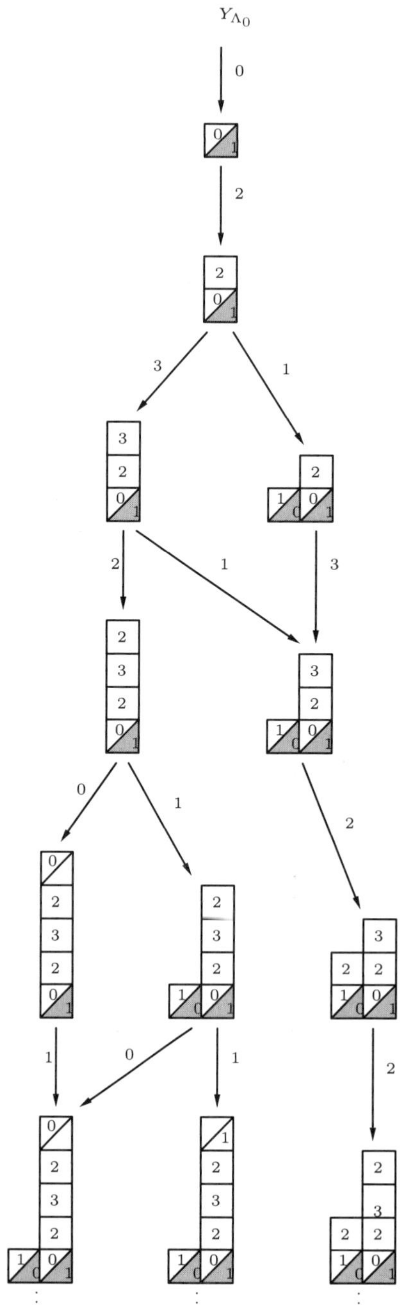

Seok-Jin Kang
Department of Mathematical Sciences
Seoul National University
Seoul, Korea
sjkang@math.snu.ac.kr

Fields Institute Communications
Volume **59**, 2011

CHAPTER 3

Geometric Realizations of Crystals

Alistair Savage

Abstract. We give an introduction to the geometric realization of crystal graphs via the quiver varieties of Lusztig and Nakajima. The emphasis is on motivating the constructions through concrete examples. The relation between the geometric construction of crystals and combinatorial realizations using Young tableaux is also discussed.

Introduction

The goal of these lectures is to give an introduction to the geometric realization of crystal graphs via quiver varieties. The aim throughout is to adequately motivate the definitions so that the reader gains an intuition for the constructions. For this reason, concrete examples are discussed in detail and mathematical rigor is sometimes sacrificed in the name of exposition. The hope is that after studying these lectures, the reader will have an intuitive grasp of the theory and several specific examples at hand that will equip him or her to explore the literature on this subject. References are given for proofs or arguments that have been omitted.

Section 3.1 is dedicated to motivating the definitions to follow in the other four lectures. Several examples are discussed in detail, noting connections to ideas that appeared in the lectures of Kamnitzer (Chapter 1) and Kang (Chapter 2) at the same summer school. Equipped with these examples, the discussion of the general theory begins in Section 3.2 where the notions of quivers and their representations are introduced. In Section 3.3, we define the Lusztig quiver varieties and the crystal structure on their sets of irreducible components. We see how one obtains the crystal corresponding to half of the quantized enveloping algebra of a symmetric Kac-Moody algebra. The lagrangian Nakajima quiver varieties are introduced in Section 3.4. Here we define the crystal structure on the sets of irreducible components and obtain the crystals corresponding to irreducible integrable highest weight representations. Finally, in Section 3.5 we describe the relationship between the geometric realizations of crystals using Nakajima quiver varieties and the well known combinatorial realizations using tableaux. In this final section we restrict our attention to the Lie algebra $\mathfrak{sl}_n$.

Prerequisites. In these notes, we assume a basic knowledge of Kac-Moody Lie algebras. In particular, we assume the reader is familiar with their definitions and the basics of the theory of irreducible integrable highest weight representations. We also presuppose a knowledge of the basic definitions of crystals. Students in the summer school benefited from a course on this subject (Chapter 2). A more comprehensive treatment can be found in the book [122]. Some background in algebraic geometry would be helpful in following these notes, but the reader willing to take some results on faith should be able to follow the presentation.

Acknowledgments. This research is supported by the Natural Sciences and Engineering Research Council (NSERC) of Canada. The author would like to thank all of the participants of the summer school for their enthusiasm and insightful questions and the other speakers of the summer school (Vyjayanthi Chari, Joel Kamnitzer, Seok-Jin Kang, Erhard Neher and Weiqiang Wang) for their interesting lecture series.

3.1 Motivating examples

In this first section, we discuss some motivating examples for the theory that will be introduced in future sections. We will explicitly work out various special cases of the general objects we will introduce later (such as quiver varieties). Our goal is to create a collection of concrete examples that will guide our intuition and serve as motivation for the general definitions to follow. We begin by considering the following table.

Geometry	Algebra (Representation Theory)	Combinatorics (Crystals)
Varieties (components)	Vector space (basis)	Vertex set
Correspondences	Lie algebra action (Chevalley generators)	Crystal operators

In Chapter 1, Kamnitzer explained some relations between the first and second columns in this table. Namely he described certain varieties whose homology yielded the underlying vector space V of a representation of a Lie algebra $\mathfrak{g}$ and correspondences which produced operators realizing the action of Chevalley generators of $\mathfrak{g}$ on V. Then, in Chapter 2, Kang explained the passage from the second column to the third column. More precisely, he described how certain nice bases in representations of $U_q(\mathfrak{g})$ yield the vertices of a crystal graph in the $q \to 0$ limit and how the action of the Chevalley generators is replaced by crystal operators (colored directed edges of this graph).

In the current chapter, we will describe a general process in which one can pass from the first column directly to the third. In particular, for any (symmetric) Kac-Moody algebra, one can define what are called quiver varieties and from these we can obtain the crystal graph directly. The vertex set of the crystal is the set of irreducible components of the varieties and the crystal operators are given by natural geometric operators closely related to correspondences.

We begin by considering a specific example of the construction. Let $\mathfrak{g} = \mathfrak{sl}_n$, the Lie algebra of $n \times n$ traceless matrices. We let E_{ij} be the matrix with (i,j) entry equal to one and all other entries equal to zero. Then

$$\{e_k := E_{k,k+1},\ f_k := E_{k+1,k}\}_{1 \le k \le n-1}$$

are the *Chevalley generators* of $\mathfrak{g}$. We let $\mathfrak{h}$ be the *Cartan subalgebra* consisting of traceless diagonal matrices. Then $\mathfrak{h}$ has basis $\{h_k := E_{k,k} - E_{k+1,k+1}\}_{1 \le k \le n-1}$.

We define $\varepsilon_k \in \mathfrak{h}^*$, by $\varepsilon_k(E_{l,l}) = \delta_{k,l}$ for $1 \le l, k \le n$. Note that $\varepsilon_1 + \cdots + \varepsilon_n = 0$ since we consider traceless matrices. The dual space $\mathfrak{h}^*$ has a basis consisting of *simple roots* $\{\alpha_k := \varepsilon_k - \varepsilon_{k+1}\}_{1\le k\le n-1}$. We also define the *fundamental weights* $\omega_k = \varepsilon_1 + \cdots + \varepsilon_k$, $1 \le k \le n-1$. Then

$$P := \bigoplus_{k=1}^{n-1} \mathbb{Z}\omega_i \quad \text{and} \quad P^+ := \bigoplus_{k=1}^{n-1} \mathbb{N}\omega_i$$

are the *weight lattice* and *dominant weight lattice* respectively. Elements of P^+ are called *dominant integral weights*. Any $w = w_1\omega_1 + \cdots + w_{n-1}\omega_{n-1} \in P^+$ can be written in the form

$$w = \lambda_1\varepsilon_1 + \cdots + \lambda_{n-1}\varepsilon_{n-1}, \quad \lambda_k = w_k + \cdots + w_{n-1}.$$

Thus w corresponds to a partition

$$\lambda(w) = (\lambda_1, \lambda_2, \ldots, \lambda_{n-1}), \quad \lambda_i \le \lambda_{i+1},\ i = 1, \ldots, n-1.$$

In this way, we will often identify the set of partitions of length $n-1$ with the dominant weight lattice.

Finite-dimensional representations V of $\mathfrak{g}$ have a weight space decomposition

$$V = \bigoplus_{\mu\in P} V_\mu, \quad V_\mu = \{v \in V \mid h \cdot v = \mu(h)v\ \forall\ h \in \mathfrak{h}\}.$$

The action of $\mathfrak{g}$ on itself yields the *adjoint representation* and the corresponding weight space decomposition

$$\mathfrak{g} = \mathfrak{h} \oplus \bigoplus_{\alpha\in\Phi} \mathfrak{g}_\alpha, \quad \dim \mathfrak{g}_\alpha = 1\ \forall\ \alpha \in \Phi,\ \dim \mathfrak{h} = n-1,$$

is called the *root space decomposition* of $\mathfrak{g}$. Here $\Phi \subseteq P$ is the *set of roots* of $\mathfrak{g}$.

Irreducible representations of $\mathfrak{g}$ are labeled by their highest weight (an element of P^+). We denote the irreducible representation of $\mathfrak{g}$ of highest weight $w \in P^+$ by $V(w)$, or $V(\lambda(w))$ when we wish to label it by the corresponding partition. The adjoint representation of $\mathfrak{g}$ is isomorphic to $V(\omega_1 + \omega_{n-1})$ as a $\mathfrak{g}$-module.

In Section 2.4, Kang described a realization of the crystals of irreducible representations of $\mathfrak{g} = \mathfrak{sl}_n$ (or $\mathfrak{g} = \mathfrak{gl}_n$) via tableaux (see [122] for a more detailed description). Consider the example of $\mathfrak{g} = \mathfrak{sl}_3$ and $V = V(\omega_1 + \omega_2)$ the adjoint representation. Let $w = \omega_1 + \omega_2$ be the highest weight and let $\lambda = \lambda(w) = (2, 1)$ be the corresponding partition. Then $B(\lambda)$ is the set of semistandard tableaux of shape λ and the crystal is as in Figure 1 (see Example 2.4.7).

Recall that the weight of a tableaux T is equal to $\sum_{i=1}^n (\#\ i\text{'s in } T)\varepsilon_i$. In the above figure, the tableaux are grouped according to weight. The highest weight is $2\varepsilon_1 + \varepsilon_2 = \omega_1 + \omega_{n-1}$ and the lowest weight is $\varepsilon_2 + 2\varepsilon_3 = -\omega_1 - \omega_{n-1}$. All weight spaces are one-dimensional except for the $\varepsilon_1 + \varepsilon_2 + \varepsilon_3 = 0$ weight space $\mathfrak{h}$, which is two dimensional. This corresponds to the fact that there are two tableaux (in the center of the figure) of this weight.

We now recall from Section 1.3, Ginzburg's construction of irreducible representations of $\mathfrak{sl}_n$ (or $\mathfrak{gl}_n$) via Springer fibers. For comparison purposes, let us consider the same example. That is, let $\mathfrak{g} = \mathfrak{sl}_3$ and let $V = V(\omega_1 + \omega_2)$ be the adjoint representation as above. We set $w = \omega_1 + \omega_2$ and $\lambda = \lambda(w) = (2, 1)$. Recall that to construct the representation $V(w_1\omega_1 + \cdots + w_{n-1}\omega_{n-1})$ of $\mathfrak{sl}_n$, we fix a nilpotent endomorphism with w_i Jordan blocks of size i. So we fix a nilpotent $X \in \operatorname{End}(\mathbb{C}^3)$ with Jordan blocks of size 2 and 1.

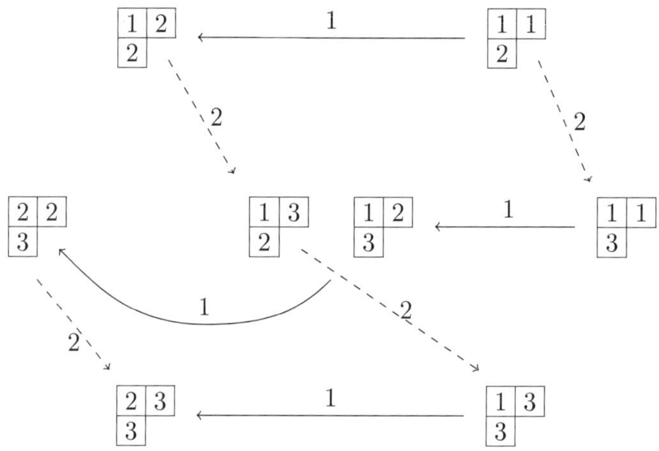

Figure 1 Crystal graph of $B((2,1))$, for $\mathfrak{g} = \mathfrak{sl}_3$.

Choose the standard basis $\{e_1, e_2, e_3\}$ of $\mathbb{C}^3$ and take

$$X = \begin{pmatrix} 0 & 1 & 0 \\ 0 & 0 & 0 \\ 0 & 0 & 0 \end{pmatrix}$$

in this basis. Then the variety we are interested in is

$$\mathrm{Fl}_3(\mathbb{C}^3)^X = \{0 = V_0 \subseteq V_1 \subseteq V_2 \subseteq V_3 = \mathbb{C}^3 \mid X(V_i) \subseteq V_{i-1},\ 1 \le i \le 3\},$$

and we have the decomposition

$$\mathrm{Fl}_3(\mathbb{C}^3)^X = \bigsqcup_{\mu \in \mathbb{N}^3,\, \mu_1+\mu_2+\mu_3=3} \mathrm{Fl}_\mu(\mathbb{C}^3)^X,$$

where $\mathrm{Fl}_\mu(\mathbb{C}^3)^X$ consists of the flags with $\dim V_i/V_{i-1} = \mu_i$.

Let $0 = V_0 \subseteq V_1 \subseteq V_2 \subseteq V_3 = \mathbb{C}^3$ be a flag in $\mathrm{Fl}_\mu(\mathbb{C}^3)^X$. Note that $X(V_3) = X(\mathbb{C}^3) = \langle e_1 \rangle$ and thus we must have $\langle e_1 \rangle \subseteq V_2$. Furthermore $\ker X = X^{-1}(0) = \langle e_1, e_3 \rangle$ and so $V_1 \subseteq \langle e_1, e_3 \rangle$. Now suppose that $\mu = (1,1,1)$. Thus $\dim V_i = i$ for $i = 0,1,2,3$. We see that V_1 can be any one-dimensional subspace of $\langle e_1, e_3 \rangle$ and thus the choice of V_1 yields the projective line $\mathbb{P}^1$. Now, for a fixed V_1, consider the possibilities for V_2. We must have

$$V_1 + \langle e_1 \rangle = V_1 + X(V_3) \subseteq V_2 \subseteq X^{-1}(V_1).$$

Thus, if $V_1 = \langle e_1 \rangle$, the only conditions on V_2 are

$$\langle e_1 \rangle \subseteq V_2 \subseteq \mathbb{C}^3, \quad \dim V_2 = 2,$$

and so the choice of V_2 yields a projective line $\mathbb{P}^1$. On the other hand, if $V_1 \neq \langle e_1 \rangle$, then we must have $V_2 = \langle e_1 \rangle + V_1 = \langle e_1, e_3 \rangle$. Therefore, $\mathrm{Fl}_{(1,1,1)}(\mathbb{C}^3)^X$ consists of two $\mathbb{P}^1$'s meeting at a point.

Exercise 3.1.1 For all $\mu \in \mathbb{N}^3$ with $\mu_1 + \mu_2 + \mu_3 = 3$ and $\mu \neq (1,1,1)$, show that the variety $\mathrm{Fl}_\mu(\mathbb{C}^3)$ is either empty or is a point.

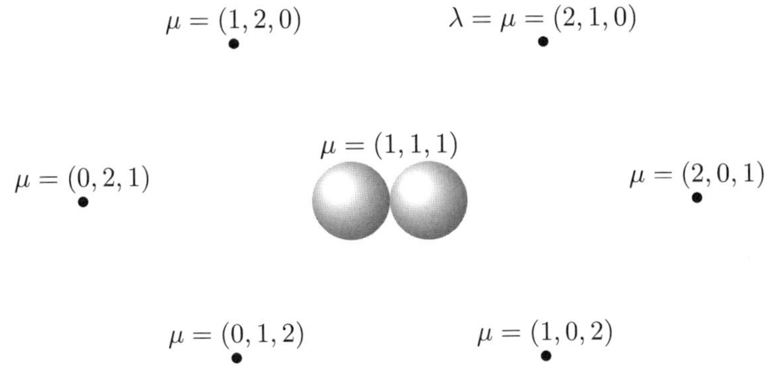

Figure 2 $\mathrm{Fl}_3(\mathbb{C}^3)^X$ where X has Jordan type $(2, 1)$.

Recall that $H_{\mathrm{top}}(\mathrm{Fl}_3(\mathbb{C}^3)^X) \cong V(\lambda)$ and that the decomposition $H_{\mathrm{top}}(\mathrm{Fl}_3(\mathbb{C}^3)^X) = \bigoplus_\mu H_{\mathrm{top}}(\mathrm{Fl}_\mu(\mathbb{C}^3)^X)$ corresponds to the decomposition $V(\lambda) = \bigoplus_\mu V(\lambda)_\mu$. Furthermore, a basis of $H_{\mathrm{top}}(\mathrm{Fl}_\mu(\mathbb{C}^3)^X)$ is given by the fundamental classes of the irreducible components of $\mathrm{Fl}_\mu(\mathbb{C}^3)^X$. Using the above and Exercise 3.1.1, if we draw $\mathrm{Fl}_3(\mathbb{C}^3)^X$ we obtain Figure 2.

Note the similarity between Figures 1 and 2. Our goal is to form a crystal graph whose vertex set is the set of irreducible components of certain varieties (such as the Springer fibers discussed above) and whose crystal operators are defined geometrically. If we do this, we should be able to say which of the two $\mathbb{P}^1$'s in Figure 2 correspond to which of the weight zero tableaux in the center of Figure 1.

Let's look at another example. Consider $\mathfrak{g} = \mathfrak{sl}_n$ and the representation $V = V(N\omega_1)$. This corresponds to $X = 0$. Then

$$\mathrm{Fl}_\mu(\mathbb{C}^N)^X = \mathrm{Fl}_\mu(\mathbb{C}^N) = \{0 = V_0 \subseteq V_1 \subseteq \cdots \subseteq V_n = \mathbb{C}^N \mid \dim V_i/V_{i-1} = \mu_i\}$$

is irreducible for all μ. This corresponds to the fact that all weight spaces of V are one-dimensional. Therefore the set of irreducible components of $\mathrm{Fl}_n(\mathbb{C}^N)$ is precisely

$$\{\mathrm{Fl}_\mu(\mathbb{C}^N) \mid \mu \in \mathbb{N}^n,\ \mu_1 + \cdots + \mu_n = N\}.$$

We wish to construct crystal operators $\tilde{e}_i$ and $\tilde{f}_i$ in a "natural" geometric way. For $i = 1, \ldots, n-1$, define

$$\mu \pm \alpha_i = (\mu_1, \ldots, \mu_{i-1}, \mu_i \pm 1, \mu_{i+1} \mp 1, \mu_{i+1}, \ldots, \mu_n).$$

Define

$$\mathrm{Fl}_{\mu,\mu+\alpha_i}(\mathbb{C}^N) = \{((U_j), (V_j)) \in \mathrm{Fl}_\mu(\mathbb{C}^N) \times \mathrm{Fl}_{\mu+\alpha_i}(\mathbb{C}^N) \mid U_j \subseteq V_j \ \forall\, j\}.$$

Note that the condition $U_j \subseteq V_j$ implies that $U_j = V_j$ for all $j \neq i$. Then we have natural projections

$$\mathrm{Fl}_\mu(\mathbb{C}^N) \xleftarrow{\pi_1} \mathrm{Fl}_{\mu,\mu+\alpha_i}(\mathbb{C}^N) \xrightarrow{\pi_2} \mathrm{Fl}_{\mu+\alpha_i}(\mathbb{C}^N)$$

given by $\pi((U_j), (V_j)) = (U_j)$ and $\pi_2((U_j), (V_j)) = (V_j)$.

What are the fibers of the maps π_1 and π_2? Fix $(U_j) \in \mathrm{Fl}_\mu(\mathbb{C}^N)$. Then

$$\begin{aligned}\pi_1^{-1}((U_j)) &\cong \{V_i \mid U_i \subseteq V_i \subseteq U_{i+1},\ \dim V_i/U_i = 1\} \\ &\cong \{\bar{V} \subseteq U_{i+1}/U_i \mid \dim \bar{V} = 1\} \\ &\cong \mathbb{P}^{\mu_{i+1}-1}.\end{aligned}$$

So π_1 is a fiber bundle with smooth fibers isomorphic to $\mathbb{P}^{\mu_{i+1}-1}$.

Exercise 3.1.2 Show that π_2 is a fiber bundle with smooth fibers isomorphic to $\mathbb{P}^{\mu_i-1}$.

Since π_1 and π_2 are both fiber bundles with smooth fibers, they induce the following bijections (provided $\mu + \alpha_i \in \mathbb{N}^n$).

$$\begin{array}{ccccc}\text{Irreducible} & & \text{Irreducible} & & \text{Irreducible} \\ \text{components of} & \overset{\text{1-1}}{\longleftrightarrow} & \text{components of} & \overset{\text{1-1}}{\longleftrightarrow} & \text{components of} \\ \mathrm{Fl}_\mu(\mathbb{C}^N) & & \mathrm{Fl}_{\mu,\mu+\alpha_i}(\mathbb{C}^N) & & \mathrm{Fl}_{\mu+\alpha_i}(\mathbb{C}^N)\end{array}$$

We can then use these bijections to define the actions of the crystal operators $\tilde{e}_i$ and $\tilde{f}_i$, $i = 1, \dots, n-1$. Namely, $\tilde{e}_i$ sends an irreducible component of $\mathrm{Fl}_\mu(\mathbb{C}^N)$ to the corresponding irreducible component of $\mathrm{Fl}_{\mu+\alpha_i}(\mathbb{C}^N)$ and $\tilde{f}_i$ does the opposite.

This example will motivate the more general construction in Section 3.4. However, we will need to do some extra work because the above example was a bit too simple. In particular, the weight spaces were all one-dimensional and so the varieties involved were all irreducible. We will need to deal with representations whose weight spaces have higher dimension and thus whose corresponding varieties are not irreducible. This will require us to develop slightly more sophisticated operators.

3.2 Quivers

The varieties we will use in our geometric construction of crystal bases are certain varieties attached to quivers. In this section we review some of the basic theory of quivers and their representations.

A *quiver* is simply another name for a directed graph. Thus, a quiver is a quadruple $Q = (Q_0, Q_1, s, t)$ where Q_0 and Q_1 are sets and s and t are maps from Q_1 to Q_0. We call Q_0 and Q_1 the sets of *vertices* and *directed edges* (or *arrows*) respectively. The maps s and t tell us the endpoints of each arrow: for an arrow $a \in Q_1$, we call $s(a)$ the *source* of a and $t(a)$ the *target* of a. Usually we will write $Q = (Q_0, Q_1)$, leaving the maps s and t implied. The quiver Q is said to be *finite* if Q_0 and Q_1 are both finite. A *loop* is an arrow a with $s(a) = t(a)$. In this paper, all quivers will be assumed to be finite and without loops. If we forget the orientation of the edges in a quiver, we obtain a graph, called the *underlying graph* of the quiver. A quiver is said to be of *finite type* its underlying graph is a Dynkin diagram of finite ADE type. Similarly, it is of *affine* (or *tame*) *type* if the underlying graph is a Dynkin diagram of affine type and of *indefinite* (or *wild*) *type* if the underlying graph is a Dynkin diagram of indefinite type.

A *path* in Q is a sequence of arrows lining up tip-to-tail. More precisely, a path is a sequence $\beta = a_l a_{l-1} \cdots a_1$ of arrows such that $t(a_i) = s(a_{i+1})$ for $1 \le i \le l-1$. We call l the *length* of the path. We let $s(\beta) = s(a_1)$ and $t(\beta) = t(a_l)$ denote the initial and final vertices of the path β. For each vertex $i \in Q_0$, we have a trivial path e_i with $s(e_i) = t(e_i) = i$.

The *path algebra* $\mathbb{C}Q$ associated to a quiver Q is the $\mathbb{C}$-algebra whose underlying vector space has basis the set of paths in Q, and with the product of paths given by concatenation. More precisely, if $\beta = a_l \cdots a_1$ and $\beta' = b_m \cdots b_1$ are two paths in Q, then $\beta\beta' = a_l \cdots a_1 b_m \cdots b_1$ if $t(\beta') = s(\beta)$ and $\beta\beta' = 0$ otherwise. This is an associative multiplication. There is a natural grading $\mathbb{C}Q = \bigoplus_{n\in\mathbb{N}}(\mathbb{C}Q)_n$ where $(\mathbb{C}Q)_n$ is the span of the paths of length n. This is a grading as an algebra in the sense that $(\mathbb{C}Q)_n(\mathbb{C}Q)_m \subseteq (\mathbb{C}Q)_{n+m}$ (in fact, we have equality here).

Example 3.2.1 Let Q be the *Jordan quiver*:

Then $\mathbb{C}Q = \mathbb{C}[t]$, the polynomial algebra in one variable. The variable t corresponds to the unique path of length one.

Exercise 3.2.2 Let Q be the following *quiver of type* A_n:

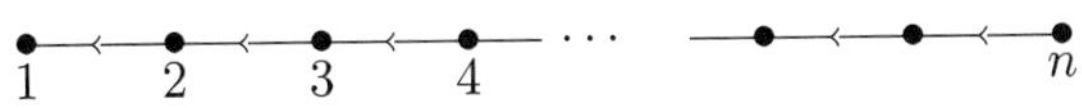

Then for all $1 \le i \le j \le n$, there exists a unique path p_{ij} from j to i. Define a map $f : \mathbb{C}Q \to M_{n\times n}(\mathbb{C})$, the algebra of $n \times n$ matrices with complex entries, by $p_{ij} \mapsto E_{ij}$ and extending by linearity. Show that f is an isomorphism (of algebras) onto the algebra of upper triangular matrices.

A *representation* (V, x) of a quiver Q is a collection

$$\{V_i \mid i \in Q_0\}$$

of finite-dimensional vector spaces, together with a collection

$$\{x_a : V_{s(a)} \to V_{t(a)} \mid a \in Q_1\}$$

of linear maps. Representations of Q are equivalent to representations of the path algebra $\mathbb{C}Q$.

We will often view the collection $\{V_i \mid i \in Q_0\}$ as an Q_0-graded vector space $V = \bigoplus_{i\in Q_0} V_i$. Then the *graded dimension* of V is

$$\mathbf{dim}\, V = (\dim V_i)_{i\in Q_0}.$$

A morphism $(V, x) \to (W, y)$ between two representations of a quiver Q is a collection

$$\{\psi_i : V_i \to W_i \mid i \in Q_0\}$$

of linear maps such that the diagram

$$\begin{array}{ccc} V_{s(a)} & \xrightarrow{x_a} & V_{t(a)} \\ {\scriptstyle \psi_{s(a)}}\downarrow & & \downarrow{\scriptstyle \psi_{t(a)}} \\ W_{s(a)} & \xrightarrow{y_a} & W_{t(a)} \end{array}$$

commutes for all $a \in Q_1$.

Quivers and path algebras play an important role in the representation theory of finite-dimensional associative algebras. In particular, we have the following result.

Proposition 3.2.3 *Let A be a finite-dimensional $\mathbb{C}$-algebra. Then the category of representations of A is equivalent to the category of representations of $\mathbb{C}Q/I$ for some quiver Q and some two-sided ideal I of $\mathbb{C}Q$.*

Proof This follows from results in [105, 106]. The full proof can also be found in [15, Theorem 3.7]. □

Suppose (V, x) and (W, y) are representations of a quiver Q. Then we define the *direct sum* of (V, x) and (W, y) to be the representation $(V \oplus W, x \oplus y)$ where $(V \oplus W)_i = V_i \oplus W_i$ for $i \in Q_0$, and $(x \oplus y)_a = x_a \oplus y_a$ for $a \in Q_1$.

If (V, x) is a representation of a quiver Q and W is an Q_0-graded subspace of V (that is, $W_i \subseteq V_i$ for all $i \in Q_0$) then we say that W is *x-invariant* if $x_a(W_{s(a)}) \subseteq W_{t(a)}$ for all $a \in Q_1$. Then $(W, x|_W)$ is called a *subrepresentation* of (V, x). The representation $(0, 0)$ is called the *trivial representation.*

A quiver representation Q is said to be *irreducible* (or *simple*) if it contains no nontrivial proper subrepresentations. It is called *indecomposable* if cannot be written as a direct sum of two nontrivial subrepresentations.

Example 3.2.4 Fix a vertex $i \in Q_1$ and let S^i be a Q_0-graded vector space with $S^i_i = \mathbb{C}$ and $S^i_j = 0$ for $i \neq j$. Then $(S^i, 0)$ is a simple representation.

Example 3.2.5 Let Q be the quiver of Exercise 3.2.2. For $1 \leq i \leq j \leq n$, define $V^{i,j}$ to be the Q_0-graded vector space defined by

$$V^{i,j}_k = \begin{cases} \mathbb{C} & i \leq k \leq j, \\ 0 & \text{otherwise.} \end{cases}$$

For $a \in Q_1$, define

$$x^{i,j}_a = \begin{cases} \mathrm{id} & i+1 \leq s(a) \leq j, \\ 0 & \text{otherwise.} \end{cases}$$

Then $(V^{i,j}, x^{i,j})$ is indecomposable. However, if $i < j$, then $(V^{i,j}, x^{i,j})$ is not irreducible.

Forgetting the orientation of the edges of a quiver Q, we obtain the *underlying graph* $\Gamma(Q)$ of Q. We then let $\mathfrak{g}(Q)$ be the Kac-Moody algebra whose Dynkin graph is the graph $\Gamma(Q)$. Note that such a Kac-Moody algebra has symmetric Cartan matrix and any Kac-Moody algebra with symmetric Cartan matrix arises as $\mathfrak{g}(Q)$ for some (in general not unique) quiver Q.

Theorem 3.2.6 (Gabriel's Theorem [33, 105]) *A quiver Q has finitely many indecomposable representations (up to isomorphism) if and only if Q is of finite type. Furthermore, if Q is of finite type, there is a one-to-one correspondence between isomorphism classes of indecomposable representations of Q and positive roots of $\mathfrak{g}(Q)$. This correspondence is given by*

$$(V, x) \mapsto \sum_{i \in Q_0} (\dim V_i)\alpha_i,$$

where the α_i, $i \in Q_0$, are the simple roots of $\mathfrak{g}(Q)$.

Proof This result was first proved by Gabriel in [105]. The proof is computational and the relation with Dynkin diagrams is a consequence of this proof and not an important feature of the calculations. A second proof was given by Bernšteĭn,

Gel′fand, and Ponomarev in [33] that involves the machinery of Coxeter functors and Weyl groups. □

Exercise 3.2.7 (Type A_n) Let Q be the quiver of type A_n given in Exercise 3.2.2. Then $\mathfrak{g}(Q) = \mathfrak{sl}_{n+1}$. Show that the representations $(V^{i,j}, x^{i,j})$ are the only indecomposable representations of Q up to isomorphism. The indecomposable representation $(V^{i,j}, x^{i,j})$ corresponds to the positive root

$$\alpha_i + \alpha_{i+1} + \cdots + \alpha_j$$

which is the weight of the root space $\mathbb{C}E_{i,j+1}$ of $\mathfrak{sl}_{n+1}$, where $E_{i,j+1}$ is the elementary matrix with a one in entry $(i, j+1)$ and a zero in all other entries.

Example 3.2.8 (Type $A_n^{(1)}$) Let Q be the quiver

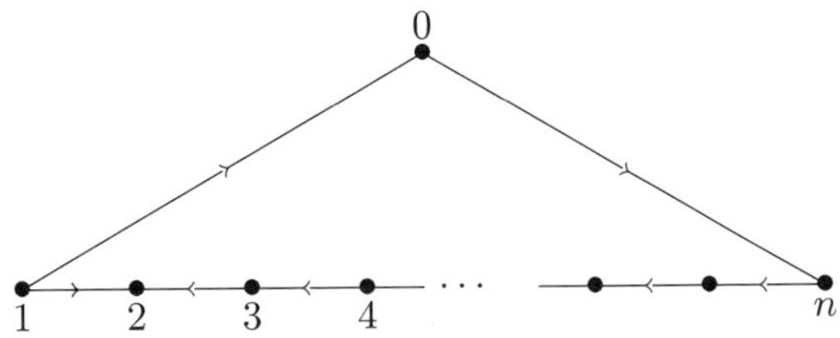

and consider the positive imaginary root $\delta = \sum_{i=0}^{n} \alpha_i$. Then there exists a one parameter family of non-isomorphic representations where $\dim V_i = 1$ for $i = 0, \ldots, n$, and all x_a are non-zero. The parameter is the composition around the cycle.

Theorem 3.2.9 (Kac's Theorem) *Let Q be an arbitrary quiver. Then the graded dimensions of indecomposable representations of Q correspond, via the map*

$$\mathbf{dim}\, V \mapsto \sum_{i \in Q_0} (\dim V_i)\alpha_i,$$

to the positive roots of the root system of $\mathfrak{g}(Q)$.

Remarks 3.2.10 (i) In Kac's Theorem, no assertion is made that positive roots are in one-to-one correspondence with isomorphism classes of indecomposable representations.

(ii) Real roots correspond to graded dimensions with one indecomposable representation of that dimension and imaginary roots correspond to graded dimensions with families of indecomposable representations of that dimension.

Assume for simplicity that Q is of finite type. Then any representation can be written in a unique way (up to isomorphism) as a sum of indecomposable representations labeled by positive roots of $\mathfrak{g} = \mathfrak{g}(Q)$. Let Φ be the set of roots of $\mathfrak{g}$ and $\Phi^{\pm}$ the set of positive/negative roots. Recall that $\mathfrak{g}$ has a *triangular decomposition*

$$\mathfrak{g} = \mathfrak{n}^- \oplus \mathfrak{h} \oplus \mathfrak{n}^+, \quad \mathfrak{n}^- = \bigoplus_{\beta \in \Phi^-} \mathfrak{g}_\beta, \quad \mathfrak{g}_\beta = \mathbb{C} f_\beta, \quad \beta \in \Phi^-.$$

We fix an ordering $\beta_1, \ldots, \beta_m$ of Φ^-. Then, by the PBW Theorem, $U(\mathfrak{n}^-)$ has a basis

$$\{f_{\beta_1}^{l_1} f_{\beta_2}^{l_2} \cdots f_{\beta_m}^{l_m} \mid l_i \in \mathbb{N},\ 1 \le i \le m\}.$$

Thus we have a bijection

$$\{\text{Isomorphism classes of representations of } Q\} \xleftrightarrow{\text{1-1}} \text{Basis of } U(\mathfrak{n}^-).$$

Fix a Q_0-graded vector space $V = \bigoplus_{i \in Q_0} V_i$ and let $\mathbf{v} = \mathbf{dim}\, V$. Then define $A_{\mathbf{v}}$ to be the space of representations of Q on V. More precisely,

$$A_{\mathbf{v}} := \{x = (x_a)_{a \in Q_1} \mid x_a : V_{s(a)} \to V_{t(a)},\ a \in Q_1\}.$$

Let $G = G_V = \prod_{i \in Q_0} GL(V_i)$. Then G acts on $A_{\mathbf{v}}$ by

$$g \cdot x = (g_i)_{i \in Q_0} \cdot (x_a)_{a \in Q_1} = (x'_a)_{a \in Q_1}, \quad x'_a = g_{t(a)} x_a g_{s(a)}^{-1}.$$

In this way, the G-orbits of $A_{\mathbf{v}}$ are precisely the isomorphism classes of representations of Q with graded dimension $\mathbf{v}$. Therefore, we have a bijection

$$\{G\text{-orbits of } A_{\mathbf{v}}\} \overset{1\text{-}1}{\longleftrightarrow} \text{Basis of } U(\mathfrak{n}^-)_{-\sum_i v_i \alpha_i},$$

where $U(\mathfrak{n}^-)_\beta$ denotes the β-weight space of $U(\mathfrak{n}^-)$.

Example 3.2.11 Consider the quiver

Q 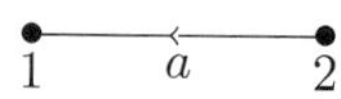

Then $\mathfrak{g}(Q) = \mathfrak{sl}_3$. Let V be the Q_0-graded vector space given by $V_1 = \mathbb{C}$ and $V_2 = \mathbb{C}$ and let $\mathbf{v} = \mathbf{dim}\, V$. Then $A_{\mathbf{v}}$ is the space of linear maps from $\mathbb{C}$ to $\mathbb{C}$, which is isomorphic to $\mathbb{C}$ (since any such map is given by multiplication by some complex number). Then $A_{\mathbf{v}} \cong \mathbb{C}$ decomposes as a disjoint union of two orbits: $\mathbb{C}^*$ and $\{0\}$.

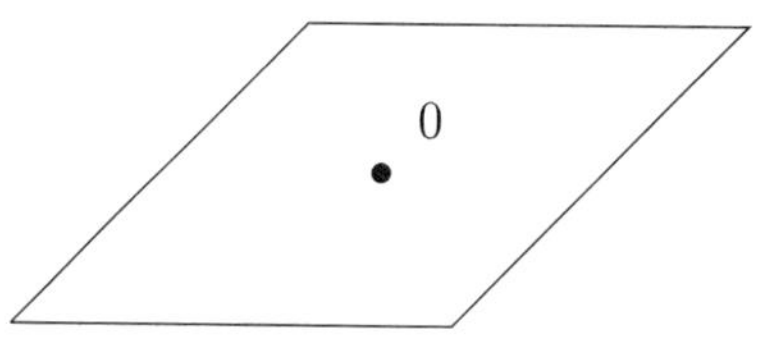

We would like to construct a variety whose irreducible components are in one-to-one correspondence with the elements of a basis of a weight space of $U(\mathfrak{n}^-)$. The variety $A_{\mathbf{v}}$ does not satisfy this property because it has only one irreducible component. The problem is that its orbits fit together in such a way as to form a single irreducible component. We would like each orbit to give rise to its own irreducible component. One way of doing this would be to take the union of the conormal bundles to the orbits in $A_{\mathbf{v}}$.

We recall the definition of the conormal bundle. The conormal bundle to a smooth subvariety S of a smooth variety A is the sub-bundle of the cotangent bundle T^*A whose fiber over any point $x \in S$ consists of those $\phi \in (T^*A)_x$ such that $\phi(v) = 0$ for all $v \in (TS)_x$ and whose fiber over all other points is empty. The dimension of the conormal bundle is equal to the dimension of A (and half the dimension of the cotangent bundle T^*A). The larger the dimension of the orbit, the smaller the dimension of the fibers.

We thus see that the union of the conormal bundles to the orbits in $A_{\mathbf{v}}$ has the property we desire. Namely, its set of irreducible components is in one-to-one correspondence with a basis of the $-\sum_i v_i \alpha_i$ weight space of $U(\mathfrak{n}^-)$. In particular, the conormal bundle to the orbit $\mathbb{C}^*$ is simply the orbit itself while the conormal bundle to the orbit $\{0\}$ is a complex line. We will see in the next section that this observation leads naturally to a "doubling" of the quiver and the definition of the Lusztig quiver variety.

3.3 The Lusztig quiver variety

Let $\mathfrak{g}$ be a Kac-Moody algebra with symmetric Cartan matrix. For example, $\mathfrak{g}$ could be a simple Lie algebra of type A, D or E. We let $Q = (Q_0, Q_1)$ be the *double quiver* associated to the Dynkin graph of $\mathfrak{g}$. That is, Q_0 is the set of vertices of this Dynkin graph and for each (undirected) edge of the Dynkin graph, Q_1 contains two arrows (one in each direction) with the same endpoints. For example, if $\mathfrak{g} = \mathfrak{sl}_{n+1}$ is the Lie algebra of type A_n, then the corresponding double quiver is as follows.

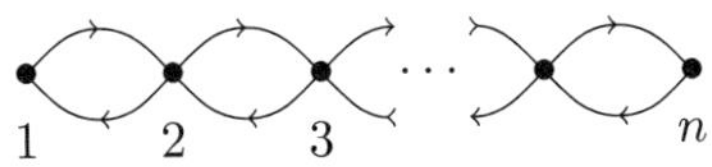

We have a natural involution $a \mapsto \bar{a}$, $a \in Q_1$, which maps the arrow a to the arrow with the same underlying edge but with opposite orientation.

An *orientation* of Q is a choice of subset $\Omega \subseteq Q_1$ such that $\Omega \cup \bar{\Omega} = Q_1$ and $\Omega \cap \bar{\Omega} = \emptyset$. That is, Ω contains exactly one arrow from each pair associated to each edge of the Dynkin graph.

For a Q_0-graded vector space $V = \bigoplus_{i \in Q_0} V_i$, let

$$E_V = \bigoplus_{a \in Q_1} \operatorname{Hom}(V_{s(a)}, V_{t(a)}).$$

Then $E_V = E_{V,\Omega} \oplus E_{V,\bar{\Omega}}$, where

$$E_{V,\Omega} = \bigoplus_{a \in \Omega} \operatorname{Hom}(V_{s(a)}, V_{t(a)}), \quad E_{V,\bar{\Omega}} = \bigoplus_{a \in \bar{\Omega}} \operatorname{Hom}(V_{s(a)}, V_{t(a)}).$$

Recall that $G_V = \prod_{i \in Q_0} GL(V_i)$ acts naturally on E_V, $E_{V,\Omega}$ and $E_{V,\bar{\Omega}}$.

Define

$$\epsilon : Q_1 \to \{\pm 1\}, \quad \epsilon(a) = \begin{cases} +1 & a \in \Omega, \\ -1 & a \in \bar{\Omega}. \end{cases}$$

We then define a G_V-invariant, nondegenerate, symplectic form $\langle \cdot, \cdot \rangle$ on E_V by

$$\langle x, y \rangle = \sum_{a \in Q_1} \epsilon(a) \operatorname{tr}(x_a y_{\bar{a}}).$$

This pairs $E_{V,\Omega}$ with $E_{V,\bar{\Omega}}$ and so we can view $E_{V,\bar{\Omega}}$ as the dual space $(E_{V,\Omega})^*$ and E_V as the tangent space $T^* E_{V,\Omega}$ to $E_{V,\Omega}$. This is simply the observation that $\operatorname{Hom}(V, W)$ is dual to $\operatorname{Hom}(W, V)$ under the trace.

To any Hamiltonian action of a Lie group on a symplectic manifold, there is an associated *moment map*. The G_V-action on E_V is such an action and the corresponding moment map is

$$\psi : E_V \to \mathfrak{gl}_V = \prod_{i \in Q_0} \mathfrak{gl}(V_i) = \prod_{i \in Q_0} \operatorname{End} V_i,$$

with ith component

$$\psi_i(x) = \sum_{a \in Q_1,\, t(a) = i} \epsilon(a) x_a x_{\bar{a}}.$$

Here $\mathfrak{gl}_V$ is the Lie algebra of G_V. Usually the moment map is a map to the dual of the Lie algebra of the group but we have identified $\mathfrak{gl}_V$ with its dual via the trace here.

We say that $x \in E_V$ is *nilpotent* if there exists an $N \geq 1$ such that for any path $\beta = a_N \cdots a_2 a_1$ of length N, we have that

$$x_{a_N} \cdots x_{a_2} x_{a_1} : V_{s(a_1)} \to V_{t(a_N)}$$

is the zero map.

Definition 3.3.1 (Lusztig quiver variety) For a Q_0-graded vector space $V = \bigoplus_{i \in Q_0} V_i$, let

$$\Lambda_V := \{x \in E_V \mid \psi(x) = 0,\ x \text{ is nilpotent}\}.$$

The variety Λ_V is called the *Lusztig quiver variety* associated to Q and V.

For $\mathfrak{g}$ (or Q) of arbitrary type, Λ_V has the following properties:

(i) Λ_V is a closed subvariety of E_V of pure dimension $\frac{1}{2} \dim E_V$. That is, each irreducible component of Λ_V has this dimension.
(ii) Λ_V is a langrangian subvariety of E_V.
(iii) If $x_\Omega \in E_{V,\Omega}$ and $x_{\bar{\Omega}} \in E_{V,\bar{\Omega}}$, then

$$\psi(x_\Omega + x_{\bar{\Omega}}) = 0 \iff \begin{array}{l} x_{\bar{\Omega}} \text{ is orthogonal to the tangent space to the } G_V\text{-} \\ \text{orbit through } x_\Omega \text{ (with respect to the form } \langle \cdot, \cdot \rangle). \end{array}$$

Additionally, if $\mathfrak{g}$ is of finite type, then

(iv) If $x \in E_V$, then $\psi(x) = 0$ implies that x is nilpotent. Thus, the nilpotency condition in the definition of the Lusztig quiver variety is superfluous.
(v) The irreducible components of Λ_V are the closures of the conormal bundles of the G_V-orbits in $E_{V,\Omega}$.

Our goal is to use Lusztig quiver varieties to construct the crystal $B(\infty)$. We want the following relationship between the geometry and the elements of the crystal.

Crystal	Geometry
Vertex set	Irreducible components of $\bigsqcup_V \Lambda_V$
Vertices of weight $-\sum v_i \alpha_i$	Irreducible components of Λ_V, $\mathbf{dim}\, V = \mathbf{v}$
Crystal operators	Natural geometrically defined operators

In the union $\bigsqcup_V \Lambda_V$, we take the union over one Q_0-graded vector space V of each graded dimension.

We now describe this process in more detail. For $\mathbf{v} \in \mathbb{N}^{Q_0}$, set $V^{\mathbf{v}} = \bigoplus_{i \in Q_0} \mathbb{C}^{v_i}$ and $\Lambda(\mathbf{v}) = \Lambda_{V^{\mathbf{v}}}$. For $i \in Q_0$, let $\beta^i \in \mathbb{N}^{Q_0}$ such that $\beta^i_j = \delta_{ij}$ and for $c \in \mathbb{N}$, define $\tilde{\Lambda}(\mathbf{v}, c\beta^i)$ to be the variety of triples $(x, \phi', \bar{\phi})$ where $x \in \Lambda(\mathbf{v})$ and $\phi' = (\phi'_i)_{i \in Q_0}$, $\bar{\phi} = (\bar{\phi}_i)_{i \in Q_0}$ give an exact sequence

$$0 \to V^{\mathbf{v} - c\beta^i} \xrightarrow{\phi'} V^{\mathbf{v}} \xrightarrow{\bar{\phi}} V^{c\beta^i} \to 0$$

such that $\operatorname{im} \phi'$ is x-invariant. Then x induces $x' \in \Lambda(\mathbf{v} - c\beta^i)$ by the restriction to $\operatorname{im} \phi'$ and $\bar{x} \in \Lambda(c\beta^i) = \{0\}$ by passing to the quotient $V^{\mathbf{v}} / \operatorname{im} \phi'$. Note that x is nilpotent if and only if x' is.

Consider the maps

$$\Lambda(\mathbf{v} - c\beta^i) \xleftarrow{p_1} \tilde{\Lambda}(\mathbf{v}, c\beta^i) \xrightarrow{p_2} \Lambda(\mathbf{v})$$

where $p_1(x, \phi', \bar{\phi}) = x'$ and $p_2(x, \phi', \bar{\phi}) = x$. We want to use these maps to identify irreducible components. The problem is that they are not the right type of maps (that is, fiber bundles with smooth fibers). Thus, we need to restrict them.

For $i \in Q_0$, define $\varepsilon_i : \Lambda(\mathbf{v}) \to \mathbb{N}$ by

$$\varepsilon_i(x) = \dim \operatorname{Coker}\left(\bigoplus_{a,\, t(a)=i} V_{s(a)} \xrightarrow{(x_a)} V_i \right).$$

We will see that this map will play the role of the map ε_i in the definition of crystals (see Section 2.5), hence the notation. For $c \in \mathbb{N}$, define

$$\Lambda(\mathbf{v})_{i,c} = \{x \in \Lambda(\mathbf{v}) \mid \varepsilon_i(x) = c\}.$$

This is a locally closed subvariety of $\Lambda(\mathbf{v})$.

If $\Lambda(\mathbf{v})_{i,c} \neq \emptyset$, then

$$p_1^{-1}(\Lambda(\mathbf{v} - c\beta^i)_{i,0}) = p_2^{-1}(\Lambda(\mathbf{v})_{i,c}) := \tilde{\Lambda}(\mathbf{v}, c\beta^i)_{i,0}$$

and we have

$$\Lambda(\mathbf{v} - c\beta^i)_{i,0} \xleftarrow{p_1} \tilde{\Lambda}(\mathbf{v}, c\beta^i)_{i,0} \xrightarrow{p_2} \Lambda(\mathbf{v})_{i,c}. \tag{3.1}$$

Lemma 3.3.2 ([144, Lemma 5.2.3]) *The maps p_1 and p_2 in (3.1) are fiber bundles with smooth fibers.*

Exercise 3.3.3 Describe the fibers (see [144, Lemma 5.2.3]).

Corollary 3.3.4 *If $\Lambda(\mathbf{v})_{i,c} \neq \emptyset$, then we have a bijective correspondence*

$$\textit{Irreducible components of } \Lambda(\mathbf{v} - c\beta^i)_{i,0} \xleftrightarrow{\text{1-1}} \textit{Irreducible components of } \Lambda(\mathbf{v})_{i,c}.$$

Let $B(\mathbf{v}, \infty)$ be the set of irreducible components of $\Lambda(\mathbf{v})$ and $B_g(\infty) = \bigsqcup_{\mathbf{v}} B(\mathbf{v}, \infty)$. Since the sets $\Lambda(\mathbf{v})_{i,c}$ are locally closed, for each $X \in B(\mathbf{v}, \infty)$ there is an open dense subset of X where ε_i takes a fixed value. We define $\varepsilon_i(X)$ to be this value. For $c \in \mathbb{N}$, we let

$$B(\mathbf{v}, \infty)_{i,c} = \{X \in B(\mathbf{v}, \infty) \mid \varepsilon_i(X) = c\}.$$

By Corollary 3.3.4, we have a bijective correspondence

$$B(\mathbf{v} - c\beta^i)_{i,0} \cong B(\mathbf{v})_{i,c}, \quad \bar{X} \leftrightarrow X.$$

Define maps

$$\tilde{f}_i^c : B(\mathbf{v} - c\beta^i, \infty)_{i,0} \to B(\mathbf{v}, \infty)_{i,c}, \quad \tilde{f}_i^c(\bar{X}) = X,$$
$$\tilde{e}_i^c : B(\mathbf{v}, \infty)_{i,c} \to B(\mathbf{v} - c\beta^i, \infty)_{i,0}, \quad \tilde{e}_i^c(X) = \bar{X},$$

and then define

$$\tilde{e}_i, \tilde{f}_i : B_g(\infty) \to B_g(\infty) \sqcup \{0\}$$

by

$$\tilde{e}_i : B(\mathbf{v}, \infty)_{i,c} \xrightarrow{\tilde{e}_i^c} B(\mathbf{v} - c\beta^i, \infty)_{i,0} \xrightarrow{\tilde{f}_i^{c-1}} B(\mathbf{v} - \beta^i, \infty)_{i,c-1}, \quad c > 0,$$
$$\tilde{f}_i : B(\mathbf{v}, \infty)_{i,c} \xrightarrow{\tilde{e}_i^c} B(\mathbf{v} - c\beta^i, \infty)_{i,0} \xrightarrow{\tilde{f}_i^{c+1}} B(\mathbf{v} + \beta^i, \infty)_{i,c+1}.$$

We set $\tilde{e}_i(X) = 0$ for $X \in B(\mathbf{v}, \infty)_{i,0}$. We also define

$$\mathrm{wt} : B_g(\infty) \to P, \quad \mathrm{wt}(X) = -\sum_{i \in Q_0} v_i \alpha_i \text{ for } X \in B(\mathbf{v}, \infty),$$
$$\varphi_i(X) = \varepsilon_i(X) + \langle h_i, \mathrm{wt}(X) \rangle.$$

Here $h_i = [e_i, f_i]$ is the usual element of the Cartan subalgebra of $\mathfrak{g}$ and $\langle \cdot, \cdot \rangle$ denotes the pairing of $\mathfrak{h}$ with $\mathfrak{h}^*$. The following theorem was proved by Kashiwara and Saito.

Theorem 3.3.5 ([144, Theorem 5.3.2]) *The definitions above endow $B_g(\infty)$ with the structure of a $\mathfrak{g}$-crystal and $B_g(\infty)$ is isomorphic to $B(\infty)$, the crystal corresponding to the lower half of the quantized enveloping algebra, $U_q^-(\mathfrak{g})$.*

3.4 The lagrangian Nakajima quiver variety

The goal in this section is to modify the definition of the Lusztig quiver variety to obtain varieties giving a geometric realization of the crystals of irreducible integrable highest weight representations. To motivate the definitions, we first recall the construction of these representations via Verma modules.

As before, we let $\mathfrak{g}$ be a Kac-Moody algebra with symmetric Cartan matrix, consider a triangular decomposition

$$\mathfrak{g} = \mathfrak{n}^- \oplus \mathfrak{h} \oplus \mathfrak{n}^+,$$

and set $\mathfrak{b}^\pm = \mathfrak{h} \oplus \mathfrak{n}^\pm$. We fix a highest weight $\lambda \in P^+$, where P^+ is the dominant weight lattice. Then we define a $U(\mathfrak{b}^+)$-module $\mathbb{C}_\lambda = \mathbb{C}v_\lambda$ by

$$h \cdot v_\lambda = \lambda(h) v_\lambda, \quad h \in \mathfrak{h},$$
$$x \cdot v_\lambda = 0, \quad x \in \mathfrak{n}^+.$$

The *Verma module* of highest weight λ is then defined to be the $U(\mathfrak{g})$-module

$$M(\lambda) := U(\mathfrak{g}) \otimes_{U(\mathfrak{b}^+)} \mathbb{C}_\lambda,$$

with the $U(\mathfrak{g})$-action given by left multiplication on the left factor. It follows that $M(\lambda) \cong U(\mathfrak{n}^-) \otimes_{\mathbb{C}} \mathbb{C}_\lambda$ as vector spaces. There exists a unique maximal submodule $I(\lambda)$ of $M(\lambda)$ and the irreducible integrable highest weight representation of highest weight λ is defined to be the quotient $V(\lambda) := M(\lambda)/I(\lambda)$.

We summarize this construction as follows:

$$U(\mathfrak{n}^-) \xrightarrow{\text{shift weights}} M(\lambda) \xrightarrow{\text{cut}} V(\lambda). \tag{3.2}$$

An analogous phenomenon occurs for crystals. Namely, the crystal $B(\lambda)$ embeds into the crystal $B(\infty)$ in the following manner. For all $\lambda \in P^+$, there exists a map $\psi_\lambda : B(\lambda) \to B(\infty)$ such that

(i) ψ_λ is injective,
(ii) $\psi_\lambda(b_\lambda) = 1$,
(iii) $\psi_\lambda(\tilde{f}_i b) = \tilde{f}_i \psi_\lambda(b)$ when $\tilde{f}_i(b) \neq 0$,
(iv) $\psi_\lambda(\tilde{e}_i b) = \tilde{e}_i \psi_\lambda(b)$ for all $b \in B(\lambda)$, and
(v) $\mathrm{wt}\, \psi_\lambda(b) = \mathrm{wt}\, b - \lambda$, $\varepsilon_i(\psi_\lambda(b)) = \varepsilon_i(b)$ for all $b \in B(\lambda)$.

Property (v) is analogous to the "shifting of weights" in (3.2) and the fact that some $\tilde{f}_i$'s act as zero in $B(\lambda)$ (while this does not occur in $B(\infty)$) corresponds to the "cutting" in (3.2). We now aim to mimic these procedures using quiver varieties.

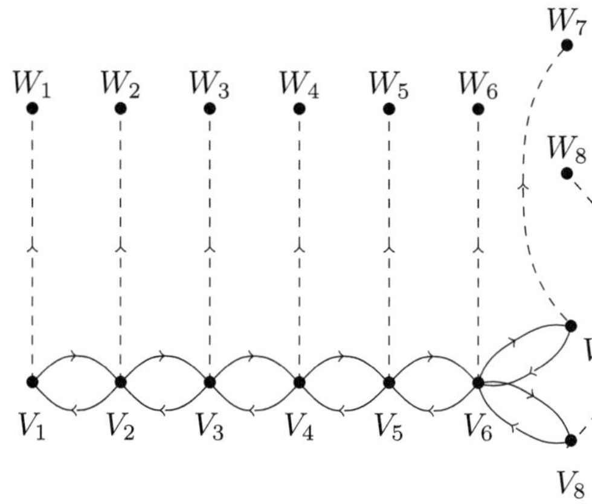

Figure 3 The linear maps involved in the definition of the space $\Lambda(\mathbf{v}, \mathbf{w})$ when $\mathfrak{g} = \mathfrak{so}_{16}$, the simple Lie algebra of type D_8.

Let $Q = (Q_0, Q_1)$ be the double quiver corresponding to our Kac-Moody algebra $\mathfrak{g}$ and choose an orientation Ω. Fix $\mathbf{v}, \mathbf{w} \in \mathbb{N}^{Q_0}$ and let V and W be Q_0-graded vector spaces of graded dimensions $\mathbf{v}$ and $\mathbf{w}$ respectively. The graded dimension $\mathbf{w}$ will correspond to the highest weight

$$\omega_w := \sum_{i \in Q_0} w_i \omega_i \tag{3.3}$$

of the representation whose crystal we would like to construct (here ω_i is the ith fundamental weight of $\mathfrak{g}$). The graded dimension $\mathbf{v}$ corresponds to a weight space in that representation. More specifically, it corresponds to the weight space of weight

$$\omega_w - \alpha_v, \text{ where } \alpha_v = \sum_{i \in Q_0} v_i \alpha_i. \tag{3.4}$$

Here α_i is the ith simple root of $\mathfrak{g}$.

Define

$$\Lambda(\mathbf{v}, \mathbf{w}) := \Lambda(\mathbf{v}) \times \bigoplus_{i \in Q_0} \operatorname{Hom}(V_i, W_i).$$

See Figure 3. The space $\operatorname{Hom}(V_i, W_i)$ is affine and hence we have a bijection

$$\text{Irreducible components of } \Lambda(\mathbf{v}, \mathbf{w}) \overset{\text{1-1}}{\longleftrightarrow} \text{Irreducible components of } \Lambda(\mathbf{v}).$$

We think of the passage from $\Lambda(\mathbf{v})$ to $\Lambda(\mathbf{v}, \mathbf{w})$ as the analogue of the "shifting weights" procedure in (3.2).

Definition 3.4.1 (Stability condition) We say a point $(x, t) \in \Lambda(\mathbf{v}, \mathbf{w})$ is *stable* if the following condition holds: If S is an x-invariant Q_0-graded subspace of V such that $t_i(S_i) = 0$ for all $i \in Q_0$, then $S_i = 0$ for all $i \in Q_0$. We denote the set of stable points in $\Lambda(\mathbf{v}, \mathbf{w})$ by $\Lambda(\mathbf{v}, \mathbf{w})^{\text{st}}$.

The group G_V acts on $\Lambda(\mathbf{v}, \mathbf{w})$ as

$$(g, (x, t)) = (x', t'), \text{ where } x'_a = g_{t(a)} x_a g_{s(a)}^{-1},\ t'_i = t_i g_i^{-1}.$$

Lemma 3.4.2 ([178, Lemma 3.10]) *The stabilizer of any point of $\Lambda(\mathbf{v}, \mathbf{w})^{\text{st}}$ in G_V is trivial.*

Definition 3.4.3 (Lagrangian Nakajima quiver variety) We define

$$\mathfrak{L}(\mathbf{v}, \mathbf{w}) = \Lambda(\mathbf{v}, \mathbf{w})^{\text{st}} / G_V,$$

and we call $\mathfrak{L}(\mathbf{v},\mathbf{w})$ a *lagrangian Nakajima quiver variety.*

Remarks 3.4.4 (i) The name lagrangian Nakajima quiver variety arises from the fact that the varieties defined here are lagrangian subvarieties of what are called (smooth) Nakajima quiver varieties. See [178, 181] for details.

(ii) For a given irreducible component X of $\Lambda(\mathbf{v},\mathbf{w})$, one of the following two conditions holds:

(a) No point of X satisfies the stability condition.

(b) All points in an open dense subset of X satisfy the stability condition.

Thus we speak of each irreducible component as failing or satisfying the stability condition in case the first or second condition (respectively) above holds. Imposing the stability condition corresponds to the "cutting" procedure in (3.2).

(iii) Since G_V acts freely on $\Lambda(\mathbf{v},\mathbf{w})^{\mathrm{st}}$, there is a bijection

$$\text{Irreducible components of } \Lambda(\mathbf{v},\mathbf{w})^{\mathrm{st}} \stackrel{\text{1-1}}{\longleftrightarrow} \text{Irreducible components of } \mathfrak{L}(\mathbf{v},\mathbf{w}).$$

We summarize the geometric construction as follows (compare to (3.2)):

$$\Lambda(\mathbf{v}) \overset{\text{shift weights}}{\rightsquigarrow} \Lambda(\mathbf{v}) \times \bigoplus_{i\in Q_0} \mathrm{Hom}(V_i, W_i) \overset{\text{cut}}{\rightsquigarrow} \Lambda(\mathbf{v},\mathbf{w})^{\mathrm{st}} \rightsquigarrow \mathfrak{L}(\mathbf{v},\mathbf{w}).$$

Example 3.4.5 Let $\mathfrak{g} = \mathfrak{sl}_2$. In this case Q is the quiver consisting of a single vertex and no edges. Then $\mathbf{v} = v$ and $\mathbf{w} = w$ are just natural numbers and

$$\Lambda(v,w)^{\mathrm{st}} = \{t : V \to W \mid t \text{ injective}\},$$

which is empty unless $v \le w$. If this condition is satisfied, then

$$\mathfrak{L}(v,w) = \Lambda(v,w)^{\mathrm{st}}/GL(V) \cong \mathrm{Gr}(v,w),$$

where $\mathrm{Gr}(v,w)$ is the Grassmannian of v-dimensional subspaces of $\mathbb{C}^w$ and the isomorphism is induced by the map $t \mapsto \operatorname{im} t$.

Exercise 3.4.6 Let $\mathfrak{g} = \mathfrak{sl}_{n+1}$ and $\mathbf{w} = N\beta^n$ for some $N \in \mathbb{N}$. Then we can picture the maps involved in the quiver variety as follows:

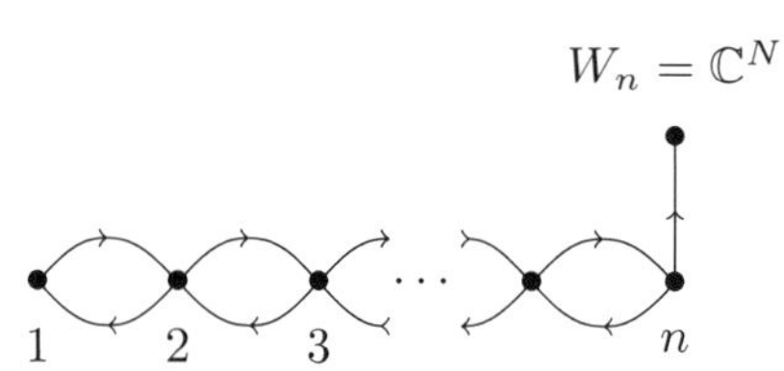

Show that $\mathfrak{L}(v,w)$ is empty unless $v_1 \le v_2 \le \cdots \le v_n \le N$, in which case

$$\mathfrak{L}(\mathbf{v},\mathbf{w}) \cong \{0 \subseteq V_1 \subseteq \cdots \subseteq V_n \subseteq \mathbb{C}^N \mid \dim V_i = v_i\},$$
$$(x,t) \mapsto (V_i), \text{ where } V_i = \operatorname{im} t a_{n-1\to n} a_{n-2\to n-1} \cdots a_{i\to i+1}.$$

Here $a_{j\to j+1}$ is the arrow from vertex j to vertex $j+1$ for $1 \le j \le n-1$. Thus $\mathfrak{L}(\mathbf{v},\mathbf{w})$ is the partial flag variety. See Section 1.3.1.

Example 3.4.7 Let $\mathfrak{g} = \mathfrak{sl}_3$, $\mathbf{w} = \beta^1 + \beta^2$, and $\mathbf{v} = \beta^1 + \beta^2$. This corresponds, under the identifications (3.3) and (3.4), to the weight zero subspace of the adjoint representation of $\mathfrak{sl}_3$ (which is two-dimensional). We can picture the maps involved in our quiver varieties as follows:

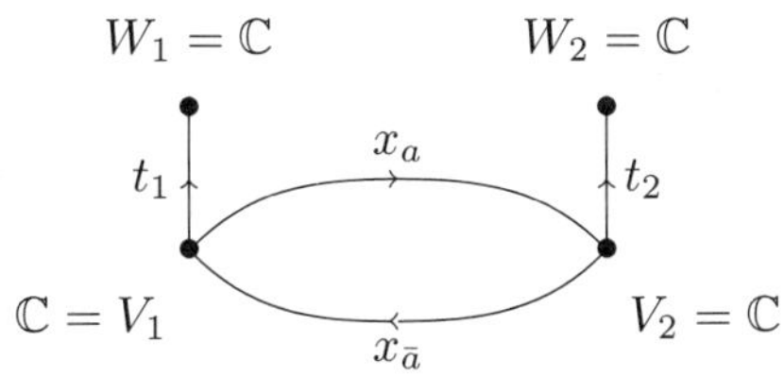

We split the points of the quiver variety into two cases. We first consider the case where $x_a = 0$. Then the stability condition implies that

$$\ker x_{\bar{a}} \cap \ker t_2 \neq 0, \quad t_1 \neq 0.$$

There is a unique element of $GL(V_1)$ whose action changes t_1 to the identity map. Therefore, we can describe the quotient defining the subvariety of $\mathfrak{L}(\mathbf{v}, \mathbf{w})$ given by $x_a = 0$ by fixing $t_1 = 1$ and considering only the quotient by $GL(V_2)$. Thus we can identify this subvariety of the quiver variety with

$$\{(x_{\bar{a}}, t_2) \in \mathbb{C}^2 \mid (x_{\bar{a}}, t_2) \neq (0,0)\}/\mathbb{C}^* \cong \mathbb{P}^1,$$

where the quotient by $\mathbb{C}^*$ is the quotient by $GL(V_2)$. The second case, where $x_{\bar{a}} = 0$ is analogous and also yields $\mathbb{P}^1$. Therefore, the lagrangian Nakajima quiver variety $\mathfrak{L}(\mathbf{v}, \mathbf{w})$ consists of two projective lines meeting at a point (the point where x_a and $x_{\bar{a}}$ are both zero).

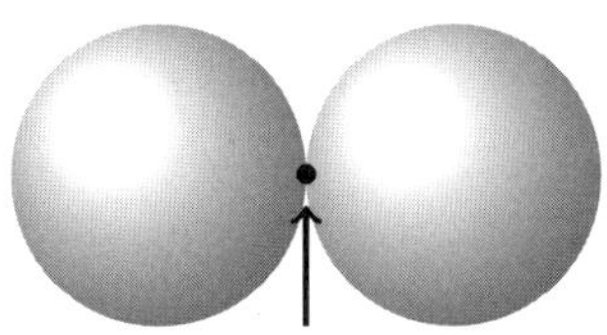

$x_a = 0$, $x_{\bar{a}} = 0$

Exercise 3.4.8 Let $\mathfrak{g} = \mathfrak{sl}_{n+1}$, $\mathbf{w} = \beta^1 + \beta^n$ and $\mathbf{v} = \beta^1 + \beta^2 + \cdots + \beta^n$. Under the identifications (3.3) and (3.4), this corresponds to the n-dimensional weight zero subspace of the adjoint representation of $\mathfrak{sl}_{n+1}$. Show that the lagrangian Nakajima quiver variety is a union of n projective lines as shown below.

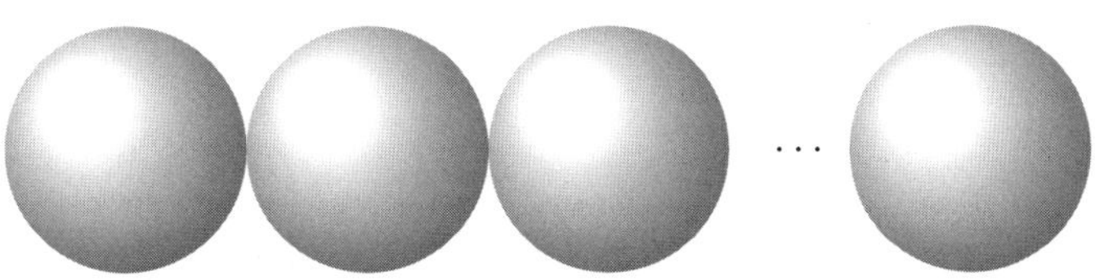

Having defined the lagrangian Nakajima quiver varieties, our goal is to describe the structure of a crystal on the set of irreducible components of $\bigsqcup_{\mathbf{v}} \mathfrak{L}(\mathbf{v}, \mathbf{w})$. For $c \in \mathbb{N}$, define

$$\mathfrak{F}(\mathbf{v}, \mathbf{w}; c\beta^i) = \{(x, t, S) \mid (x, t) \in \Lambda(\mathbf{v}, \mathbf{w})^{\mathrm{st}},\ S \text{ an } x\text{-invariant} \\ Q_0\text{-graded subspace of } V,\ \dim S = \mathbf{v} - c\beta^i\}/GL_V.$$

We then have natural projections

$$\mathfrak{L}(\mathbf{v} - c\beta^i, \mathbf{w}) \xleftarrow{\pi_1} \mathfrak{F}(\mathbf{v}, \mathbf{w}; c\beta^i) \xrightarrow{\pi_2} \mathfrak{L}(\mathbf{v}, \mathbf{w}),$$

where $\pi_1(G_V \cdot (x, t)) = G_S \cdot (x|_S, t|_S)$ and $\pi_2(G_V \cdot (x, t, S)) = G_V \cdot (x, t)$. For $c \in \mathbb{N}$, define

$$\varepsilon_i : \mathfrak{L}(\mathbf{v}, \mathbf{w}) \to \mathbb{N},\ \varepsilon_i(x, t) = \dim \operatorname{Coker}\left(\bigoplus_{a,\, t(a)=i} V_{s(a)} \to V_i\right).$$

For $i \in Q_0$ and $c \in \mathbb{N}$, let

$$\mathfrak{L}(\mathbf{v}, \mathbf{w})_{i,c} = \{G_V \cdot (x, t) \in \mathfrak{L}(\mathbf{v}, \mathbf{w}) \mid \varepsilon_i((x, t)) = c\}.$$

Then $\mathfrak{L}(\mathbf{v}, \mathbf{w})_{i,c}$ is a locally closed subvariety of $\mathfrak{L}(\mathbf{v}, \mathbf{w})$. If $\mathfrak{L}(\mathbf{v}, \mathbf{w})_{i,c} \neq \emptyset$, then

$$\pi_1^{-1}(\mathfrak{L}(\mathbf{v} - c\beta^i, \mathbf{w})_{i,0}) = \pi_2^{-1}(\mathfrak{L}(\mathbf{v}, \mathbf{w})_{i,c}).$$

Setting

$$\mathfrak{F}(\mathbf{v}, \mathbf{w}; c\beta^i)_{i,0} = \pi_1^{-1}(\mathfrak{L}(\mathbf{v} - c\beta^i, \mathbf{w})_{i,0}) = \pi_2^{-1}(\mathfrak{L}(\mathbf{v}, \mathbf{w})_{i,c}),$$

we have the natural projections

$$\mathfrak{L}(\mathbf{v} - c\beta^i, \mathbf{w})_{i,0} \xleftarrow{\pi_1} \mathfrak{F}(\mathbf{v}, \mathbf{w}; c\beta^i)_{i,0} \xrightarrow{\pi_2} \mathfrak{L}(\mathbf{v}, \mathbf{w})_{i,c}.$$

The restriction of π_2 to $\mathfrak{F}(\mathbf{v}, \mathbf{w}; c\beta^i)_{i,0}$ is an isomorphism and $\mathfrak{L}(\mathbf{v} - c\beta^i, \mathbf{w})_{i,0}$ is an open subvariety of $\mathfrak{L}(\mathbf{v} - c\beta^i, \mathbf{w})$.

Lemma 3.4.9 ([202, Lemma 4.2.2]) *(i) For $i \in Q_0$,*

$$\mathfrak{L}(\mathbf{0}, \mathbf{w})_{i,c} = \begin{cases} \mathrm{pt} & \text{if } c = 0, \\ \emptyset & \text{if } c > 0. \end{cases}$$

(ii) If $\mathfrak{L}(\mathbf{v}, \mathbf{w})_{i,c} \neq \emptyset$, then the fiber of the restriction of π_1 to $\mathfrak{F}(\mathbf{v}, \mathbf{w}; c\beta^i)_{i,0}$ is isomorphic to a Grassmannian variety.

Corollary 3.4.10 *If $\mathfrak{L}(\mathbf{v}, \mathbf{w})_{i,c} \neq \emptyset$, then there is a one-to-one correspondence between the set of irreducible components of $\mathfrak{L}(\mathbf{v}-c\beta^i, \mathbf{w})_{i,0}$ and the set of irreducible components of $\mathfrak{L}(\mathbf{v}, \mathbf{w})_{i,c}$.*

Let $B(\mathbf{v}, \mathbf{w})$ be the set of irreducible components of $\mathfrak{L}(\mathbf{v}, \mathbf{w})$ and $B_g(\mathbf{w}) = \bigsqcup_{\mathbf{v}} B(\mathbf{v}, \mathbf{w})$. As in Section 3.3, for $X \in B(\mathbf{v}, \mathbf{w})$, let $\varepsilon_i(X) = \varepsilon_i((x, t))$ for a generic point $G_V \cdot (x, t) \in X$. Then for $c \in \mathbb{N}$, define

$$B(\mathbf{v}, \mathbf{w})_{i,c} = \{X \in B(\mathbf{v}, \mathbf{w}) \mid \varepsilon_i(X) = c\}.$$

Thus, by Corollary 3.4.10 we have a bijective correspondence

$$B(\mathbf{v} - c\beta^i, \mathbf{w})_{i,0} \cong B(\mathbf{v}, \mathbf{w})_{i,c}, \quad \bar{X} \leftrightarrow X.$$

We then define maps

$$\tilde{f}_i^c : B(\mathbf{v} - c\beta^i, \mathbf{w})_{i,0} \to B(\mathbf{v}, \mathbf{w})_{i,c}, \quad \tilde{f}_i^c(\bar{X}) = X,$$
$$\tilde{e}_i^c : B(\mathbf{v}, \mathbf{w})_{i,c} \to B(\mathbf{v} - c\beta^i, \mathbf{w})_{i,0}, \quad \tilde{e}_i^c(X) = \bar{X}.$$

Then we define

$$\tilde{e}_i,\ \tilde{f}_i : B(\mathbf{w}) \to B(\mathbf{w}) \sqcup \{0\}$$

by

$$\tilde{e}_i : B(\mathbf{v},\mathbf{w})_{i,c} \xrightarrow{\tilde{e}_i^c} B(\mathbf{v}-c\beta^i,\mathbf{w})_{i,0} \xrightarrow{\tilde{f}_i^{c-1}} B(\mathbf{v}-\beta^i,\mathbf{w})_{i,c-1},$$

$$\tilde{f}_i : B(\mathbf{v},\mathbf{w})_{i,c} \xrightarrow{\tilde{e}_i^c} B(\mathbf{v}-c\beta^i,\mathbf{w})_{i,0} \xrightarrow{\tilde{f}_i^{c+1}} B(\mathbf{v}+\beta^i,\mathbf{w})_{i,c+1}.$$

We set $\tilde{e}_i(X) = 0$ for $X \in B(\mathbf{v},\mathbf{w})_{i,0}$ and $\tilde{f}_i(X) = 0$ for $X \in B(\mathbf{v},\mathbf{w})_{i,c}$ with $B(\mathbf{v},\mathbf{w})_{i,c+1} = \emptyset$. Furthermore, we define

$$\mathrm{wt} : B_g(\mathbf{w}) \to P, \quad \mathrm{wt}\, X = \omega_{\mathbf{w}} - \alpha_{\mathbf{v}} \text{ for } X \in B(\mathbf{v},\mathbf{w}),$$
$$\varphi_i(X) = \varepsilon_i(X) + \langle h_i, \mathrm{wt}\, X\rangle.$$

The following theorem was proven by Saito.

Theorem 3.4.11 ([202, Theorem 4.6.4]) *The definitions above endow $B_g(\mathbf{w})$ with the structure of a $\mathfrak{g}$-crystal and $B_g(\mathbf{w})$ is isomorphic to the crystal $B(\omega_{\mathbf{w}})$ of the irreducible integrable highest weight representation of highest weight $\omega_{\mathbf{w}}$.*

Remarks 3.4.12 (i) While Theorem 3.4.11 only applies to crystals with symmetric Cartan matrix, one can use a "folding" procedure to produce a geometric construction of the crystals of highest weight representations in arbitrary symmetrizable type (see [203]).

(ii) Nakajima quiver varieties can be used to construct the full structure (as opposed to the crystal structure) of the irreducible integrable highest weight representations of a Kac-Moody algebra $\mathfrak{g}$ with symmetric Cartan matrix. The procedure is similar to the one described in Section 1.3.3. One defines a smooth quiver variety $\mathfrak{M}$ and a singular quiver variety $\mathfrak{M}_0$ and there exists a resolution of singularities $\pi : \mathfrak{M} \to \mathfrak{M}_0$. Then $\mathfrak{L} = \pi^{-1}(0)$ and we define $Z = \mathfrak{M} \times_\pi \mathfrak{M}$. Then convolution gives the homology of Z the structure of an algebra (isomorphic to a quotient of $U(\mathfrak{g})$) and defines an action of this algebra on the homology of $\mathfrak{L}$, realizing the representation in question. For details, see [178, 181].

(iii) In type A (i.e. when $\mathfrak{g} = \mathfrak{sl}_n$), the quiver variety construction and the Ginzburg construction described in Section 1.3 are closely related but slightly different. See [205] for details.

(iv) One can define quiver varieties which yield geometric realizations of tensor products of crystals of highest weight representations (see [167, 168, 180]).

3.5 Connections to combinatorial realizations of crystal graphs

In this section, we will describe the precise relationship between the combinatorial realizations of crystals given by tableaux and the geometric realizations introduced in Section 3.4. For this section, we fix $\mathfrak{g} = \mathfrak{sl}_{n+1}$ to be the simple Lie algebra of type A_n.

Recall that the double quiver Q associated to $\mathfrak{g}$ is as follows.

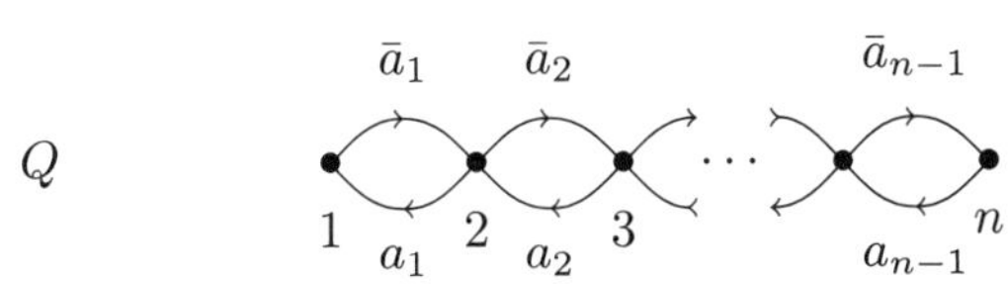

We choose the orientation $\Omega = \{a_1, a_2, \ldots, a_{n-1}\}$ consisting of the left-pointing arrows and let Q_Ω be the corresponding quiver containing only the arrows in Ω:

Q_Ω

Let $V = \bigoplus_{i \in Q_0} V_i$ be a Q_0-graded vector space of graded dimension $\mathbf{dim}\, V = \mathbf{v}$. By Gabriel's Theorem (Theorem 3.2.6) and the discussion in Section 3.2, we have the following bijections.

G_V-orbits in $E_{V,\Omega}$	$\overset{1\text{-}1}{\longleftrightarrow}$	Isom. classes of reps. of Q_Ω	$\overset{1\text{-}1}{\longleftrightarrow}$	Collections of positive roots of $\mathfrak{g}$ adding to $\sum_i v_i \alpha_i$
		Isom. classes of indecomposable reps.	$\overset{1\text{-}1}{\longleftrightarrow}$	Positive roots of $\mathfrak{g}$

Recall that under these bijections, the positive root $\alpha_i + \cdots + \alpha_j$, $i \le j$, corresponds to the isomorphism class of the representation $(V^{i,j}, x^{i,j})$ of Example 3.2.5. We depict this representation by the diagram

$V^{i,j}$

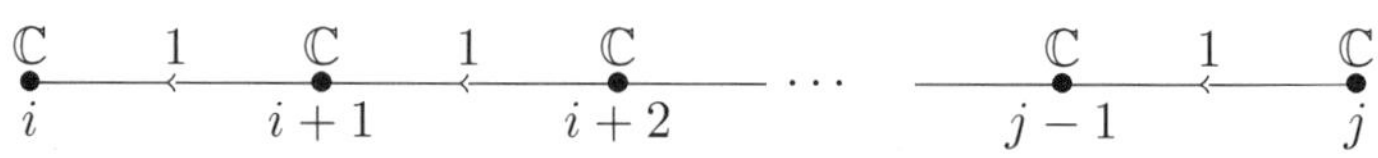

or simply by

$V^{i,j}$

In such pictures, each vertex represents a basis vector in the representation, and we identify the two. Vertex labels (elements of Q_0) indicate the degree in which this vector lives and arrow labels indicate the action of x ($= x^{i,j}$ here). Thus an arrow labeled by $z \in \mathbb{C}$ from a vertex a to a vertex b indicates that the coefficient of b in the expansion of $x(a)$ as a linear combination of the vertices is z. We will always vertically line up vertices of the same degree and degrees will increase from left to right.

Recall that in finite type, the nilpotency condition in the definition of the Lusztig quiver varieties is superfluous and thus

$$\Lambda(\mathbf{v}) = \{x \in E_V = E_{V,\Omega} \oplus E_{V,\bar{\Omega}} \mid \psi(x) = 0\},$$

where $\psi(x) = 0$ is the moment map condition. In our case, the moment map condition is equivalent to the following set of conditions:

$$x_{a_i} x_{\bar{a}_i} = x_{\bar{a}_{i-1}} x_{a_{i-1}}, \quad i \neq 1, n, \tag{3.5}$$

$$x_{a_1} x_{\bar{a}_1} = 0, \tag{3.6}$$

$$x_{\bar{a}_{n-1}} x_{a_{n-1}} = 0. \tag{3.7}$$

We adopt the convention that $x_{a_i} = 0$ and $x_{\bar{a}_i} = 0$ for $i \le 0$ or $i \ge n$.

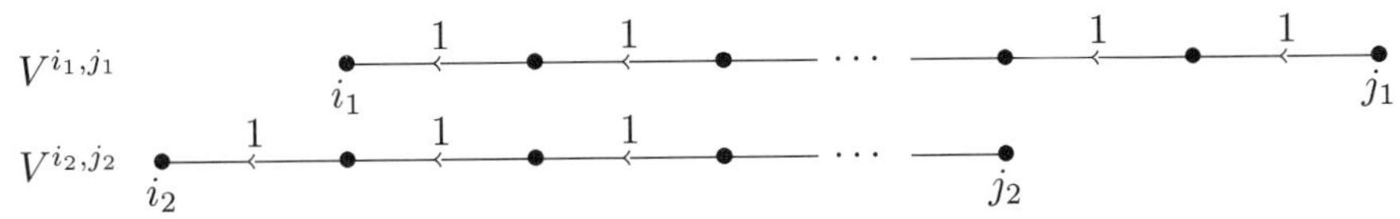

Figure 4 The quiver representation $x^{i_1,j_1} \oplus x^{i_2,j_2}$. Each vertex represents a basis vector in $V^{i_1,j_1} \oplus V^{i_2,j_2}$. Vertex labels indicate the degree of the corresponding basis vector and vertices of the same degree are vertically aligned. Edges indicate the action of $x^{i_1,j_1} \oplus x^{i_2,j_2}$ with coefficients given by edge labels.

We know that the irreducible components of $\Lambda(\mathbf{v})$ are the closures of the conormal bundles to the orbits in $E_{V,\Omega}$ and that $E_{V,\bar{\Omega}}$ corresponds to the cotangent direction. Suppose we consider an orbit corresponding to a single positive root. More precisely, we consider the orbit through the point $(V^{i,j}, x^{i,j})$ for some $i \leq j$. We would like to describe the conormal bundle to this orbit. Because the group G_V acts transitively on the orbit and identifies the fibers over points, it suffices to describe the fiber of the conormal bundle over a particular point, namely $(V^{i,j}, x^{i,j})$. We have seen that this fiber is given by the set

$$\{x = (x^{i,j}, x_{\bar{\Omega}} = (x_{\bar{a}_k})_{1 \leq k \leq n-1}) \in E_V \mid \psi(x) = 0\}.$$

Suppose $x = (x^{i,j}, x_{\bar{\Omega}} = (x_{\bar{a}_k})_{1 \leq k \leq n-1})$ is in the fiber. Since $V_k^{i,j} = 0$ unless $i \leq k \leq j$, we have $x_{\bar{a}_k} = 0$ unless $i \leq k \leq j-1$. Therefore, by (3.5) (if $i > 1$) or (3.6) (if $i = 1$), we have $x_{a_i} x_{\bar{a}_i} = 0$. But $x_{a_i} = 1$ and so $x_{\bar{a}_i} = 0$. Again, one can use (3.5) to show that $x_{\bar{a}_{i+1}} = 0$. Continuing in this manner, we see that in fact $x_{\bar{\Omega}} = 0$. Therefore, the conormal bundle to the orbit is just the orbit itself (i.e. each fiber in the conormal bundle consists of a single point).

Now consider an orbit corresponding to 2 positive roots. More precisely, we consider the orbit through the point $x^{i_1,j_1} \oplus x^{i_2,j_2}$ for some $i_1 \leq j_1$ and $i_2 \leq j_2$. We picture this representation as in Figure 4.

We now wish to describe the conormal bundle to this orbit. Again, it suffices to describe the fiber of the conormal bundle over a particular point, namely $(V^{i_1,j_1} \oplus V^{i_2,j_2}, x^{i_1,j_1} \oplus x^{i_2,j_2})$, and this fiber is given by the set

$$\{x = (x^{i_1,j_1} \oplus x^{i_2,j_2}, x_{\bar{\Omega}}) \in E_V \mid \psi(x) = 0\}.$$

For $l = 1, 2$, let $V_k^{i_l,j_l} = \mathbb{C}v_k^l$, $i_l \leq k \leq j_l$, such that $x_{a_k}^{i,j}(v_{k+1}^l) = v_k^l$ for $i_l \leq k \leq j_l - 1$. Set $v_k^l = 0$ for $k < i_l$ or $k > j_l$. What are the possible values of $x_{\bar{a}_k}(v_k^2)$ for $i_2 \leq k \leq j_2$? In general, $x_{\bar{a}_k}(v_k^2)$ is a linear combination of v_{k+1}^1 and v_{k+1}^2. However, by the same argument as above, one can show that the coefficient of v_{k+1}^2 must in fact be zero. Therefore

$$x_{\bar{a}_k}(v_k^2) = c_k v_{k+1}^1,$$

for some $c_k \in \mathbb{C}$. Suppose that $i_1 \leq k+1 \leq j_1$ and $c_k \neq 0$. Then if $k+1 > j_2$, by (3.5) we have

$$x_{\bar{a}_{k-1}}(v_{k-1}^2) = x_{\bar{a}_{k-1}} x_{a_{k-1}}(v_k^2) = x_{a_k} x_{\bar{a}_k}(v_k^2) = c_k x_{a_k}(v_{k+1}^1) = c_k v_k^1 \neq 0.$$

Continuing in this manner, we see that $x_{\bar{a}_{i_1-1}}(v_{i_1-1}^2) \neq 0$ and thus $i_2 < i_1$.

Now, if $k+1 \leq j_2$, then

$$x_{a_{k+1}} x_{\bar{a}_{k+1}}(v_{k+1}^2) = x_{\bar{a}_k} x_{a_k}(v_{k+1}^2) = x_{\bar{a}_k}(v_k^2) \neq 0.$$

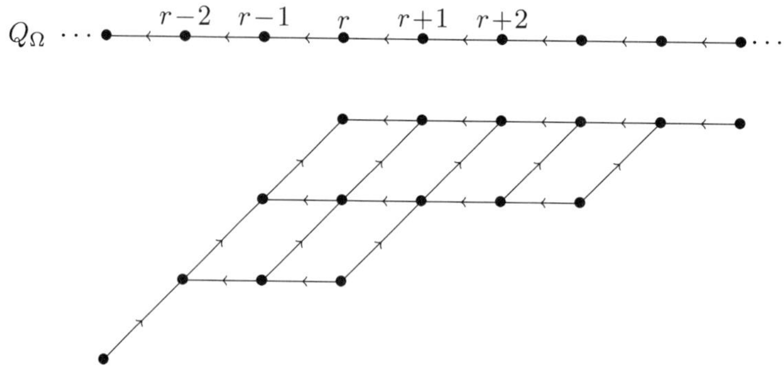

Figure 5 A depiction of an element x of a point $(x,t) \in \Lambda(\mathbf{v},\beta^r)$ satisfying the stability condition.

Therefore $x_{\bar{a}_{k+1}}(v^2_{k+1}) \neq 0$. But $x_{\bar{a}_{k+1}}(v^2_{k+1})$ must be a multiple of v^1_{k+2} as above. Thus, we must have $k+2 \leq j_1$. Continuing in this way, we see that $j_2 < j_1$. Therefore, in order for $x_{\bar{a}_k}(v^2_k)$ to be nonzero for any k, we must have $i_2 < i_1$ and $j_2 < j_1$ (for instance, we could have the situation pictured in Figure 4).

Now, let $x = (x_\Omega, x_{\bar{\Omega}})$ lie in the conormal bundle to the point

$$x_\Omega = \bigoplus_{l=1}^{s} x^{i_l,j_l}.$$

By reordering the indices if necessary, we can assume that $i_1 \geq i_2 \geq \cdots \geq i_s$. As above, let $V_k^{i_l,j_l} = \mathbb{C}v^l_k$, for $1 \leq l \leq s$. By the above arguments, $x_{\bar{a}_k}(v^l_k)$ must be a linear combination of $\{v^a_{k+1}\}_{a<l}$. Thus,

$$v^1_{i_1} \in \ker x_{a_{i_1-1}} \cap \ker x_{\bar{a}_{i_1}}. \tag{3.8}$$

Exercise 3.5.1 Show that a point $(x,t) \in \Lambda(\mathbf{v},\mathbf{w})$ satisfies the stability condition if and only if

$$\ker x_{a_{k-1}} \cap \ker x_{\bar{a}_k} \cap \ker t_k = 0 \quad \forall\, 1 \leq k \leq n.$$

Now consider the lagrangian Nakajima quiver variety $\mathfrak{L}(\mathbf{v},\mathbf{w})$ for $\mathbf{w} = \beta^r$ for some $1 \leq r \leq n$. Suppose a point (x,t) satisfies the stability condition where x is in the conormal bundle to the point $x_\Omega = \bigoplus_{l=1}^{s} x^{i_l,j_l}$. By Exercice 3.5.1 and (3.8), we must have $i_1 = r$ and there can be no other v^l_k in $\ker x_{a_{k-1}} \cap \ker x_{\bar{a}_k}$ for any k. By the above considerations, $v^l_{i_l}$ is in $\ker x_{a_{i_l-1}} \cap \ker x_{\bar{a}_{i_l}}$ unless $i_l + 1 = i_{l-1}$. Thus we have $i_{l+1} = i_l - 1$ and $x_{\bar{a}_{i_{l+1}}}\left(v^{l+1}_{i_{l+1}}\right) = c_l v^l_{i_l} \neq 0$ for $1 \leq l \leq s-1$. Thus, by the above, we must have $j_{l+1} < j_l$ for $1 \leq l \leq s-1$. Thus, the element x can be depicted as in Figure 5.

In the tableaux realization of $B(\Lambda_k)$ (see [122]), the vertices of the crystal graph consist of single column semistandard tableaux with k rows and entries from the set $\{1,2,\ldots,n+1\}$. Note that there is a one-to-one correspondence between this set of tableaux and the set of Young diagrams with at most k rows and whose rows have at most $n+1-k$ boxes. Precisely, the semistandard tableau

$$\begin{array}{|c|}\hline n_1 \\ \hline n_2 \\ \hline \vdots \\ \hline n_k \\ \hline \end{array} \tag{3.9}$$

corresponds to the Young diagram (or partition) $(n_k - k, n_{k-1} - 1, \ldots, n_1 - 1)$. It follows that we have a bijective correspondence between $B(\omega_k)$ and the set of conormal bundles satisfying the stability condition, where the tableaux (3.9) corresponds to the closure of the conormal bundle to the orbit through the representation

$$\left(\bigoplus_{i=1}^k V^{k+1-i,n_i-1}, \bigoplus_{i=1}^k x^{k+1-i,n_i-1}\right).$$

Loosely speaking, we have the correspondence

$\boxed{j}$ in the ith row $\longleftrightarrow$ 

Example 3.5.2 Suppose $\mathfrak{g} = \mathfrak{sl}_{10}$ and $k = 4$. Then we have the following correspondence.

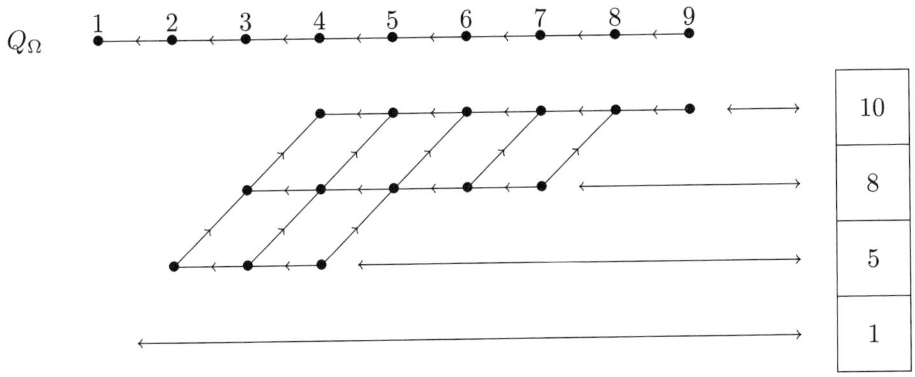

Here we draw the tableau upside down in order to make the correspondence clearer. The tableau

$$T = \begin{array}{|c|}\hline 1 \\ \hline 5 \\ \hline 8 \\ \hline 10 \\ \hline \end{array}$$

then corresponds to the irreducible component of the lagrangian Nakajima quiver variety obtained as follows: Let C_T be the closure of the conormal bundle to the G_V-orbit through

$$x^{4,9} \oplus x^{3,7} \oplus x^{2,4}.$$

Then T corresponds to the irreducible component

$$\left(\left(C_T \times \sum_{i=1}^{n} \mathrm{Hom}(V_i, W_i)\right) \cap \Lambda(\mathbf{v}, \mathbf{w})^{\mathrm{st}}\right) / G_V.$$

In fact, one can show that the Nakajima quiver varieties $\Lambda(\beta^k)$ corresponding to fundamental weights ω_k are all single points (this can be shown directly or by using the dimension formula for Nakaijima quiver varieties and the fact that these varieties are connected).

What about more general highest weights (that is, highest weights that are not fundamental weights)? Consider the highest weight $\omega_{\mathbf{w}} = \sum_{i \in Q_0} w_i \omega_i$. Then an argument similar to the above demonstrates that we can have w_i Young diagrams associated to each vertex i. In order to avoid double counting, we must associate each irreducible component of the lagrangian Nakajima quiver variety to a unique such collection of Young diagrams. We adopt the convention that larger indecomposable representations of Q_Ω (corresponding to the rows of the Young diagrams) are associated to the Young diagrams attached to vertices of higher degree (see [204] for a more precise treatment). Each Young diagram corresponds to a single column tableaux as above and, if we organize these columns into a tableau, our convention corresponds precisely to the condition that this tableau be semistandard. We thus obtain a bijective map from the set $B_g(\mathbf{w})$ of irreducible components of the lagrangian Nakajima quiver variety to the set $B(\omega_w)$ of semistandard tableaux.

Theorem 3.5.3 ([204, Theorem 6.4]) *The identification of $B_g(\mathbf{w})$ with $B(\omega_w)$ described above is an isomorphism of crystals.*

Exercise 3.5.4 If $\mathfrak{g} = \mathfrak{sl}_3$, $\mathbf{w} = e_1 + e_2$ (so $V(\omega_{\mathbf{w}})$ is the adjoint representation) and $\mathbf{v} = e_1 + e_2$ (so $V(\omega_{\mathbf{w}})_{\omega_{\mathbf{w}} - \alpha_{\mathbf{v}}} = \mathfrak{h}$ is the zero weight space), we described the lagrangian Nakajima quiver variety $\mathfrak{L}(\mathbf{v}, \mathbf{w})$ in Example 3.4.7. It consists of two projective lines meeting at a point. Show that the projective line given by $x_a = 0$ corresponds to the tableau

1	2
3	

and the projective line given by $\bar{x}_a = 0$ corresponds to the tableau

1	3
2	

.

Remarks 3.5.5 (i) When $\mathfrak{g} = \mathfrak{so}_{2n}$ is the simple Lie algebra of type D_n, a similar explicit correspondence between the irreducible components of the lagrangian Nakajima quiver varieties and the tableaux appearing in a combinatorial realization of the irreducible highest weight representations of $\mathfrak{g}$ (see [122, § 8.5] for a description of the tableaux appearing in this model) can also be given. However, this description, as well as the tableaux appearing in the combinatorial realization, are slightly more complicated. See [204] for details.

(ii) When $\mathfrak{g} = \widehat{\mathfrak{sl}}_n$ is the Lie algebra of affine type A (see Example 3.2.8), one can give a similar explicit enumeration of the irreducible components of the lagrangian Nakajima quiver variety. However, now the "strings" representing indecomposable representations of the quiver Q_Ω can "wrap around" the quiver and thus have arbitrary length. For level one representations, we get pictures as in Figure 6. If we straighten this picture, flip it (in the line $x = y$ if we view the picture as lying in the standard (x, y)-plane), and replace vertices by boxes, we obtain Figure 7, which is simply one of the Young walls appearing in Section 2.7. Since we are no longer in the finite type case, we must impose the nilpotency condition. This corresponds to the condition that the Young walls be proper.

(iii) In the case where $\mathfrak{g} = \widehat{\mathfrak{sl}}_n$ and $\omega_{\mathbf{w}} = \sum_{i=0}^{n} w_i \omega_i$ is arbitrary (i.e. we consider general dominant integral weights), we get w_i Young walls associated to each vertex i as in the case of $\mathfrak{sl}_n$. Again, choosing a convention to avoid double

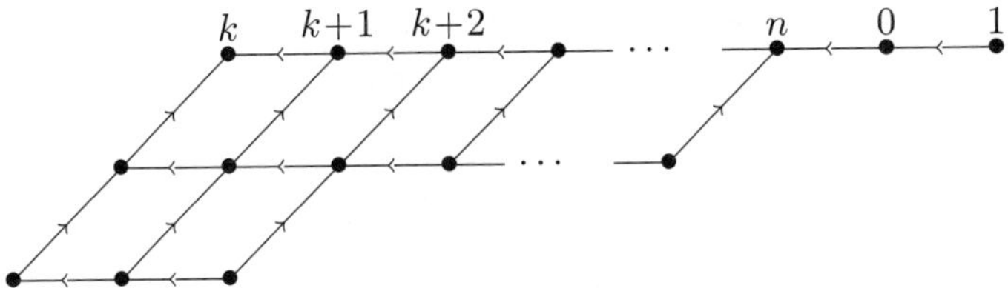

Figure 6 A diagram representing a point in the quiver variety corresponding to a level one representation (of highest weight ω_k) of the Lie algebra $\widehat{\mathfrak{sl}}_n$

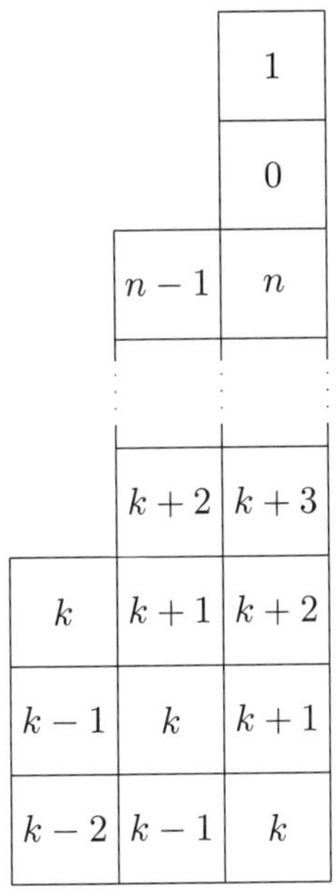

Figure 7 The Young wall corresponding to the quiver representation of Figure 6.

counting becomes a condition on the heights of columns appearing. What remains are combinatorial objects called *Young pyramids*. See Figure 8 for an example of a Young pyramid. Further details on these objects and their connections to quiver varieties can be found in [203].

Figure 8 A Young pyramid corresponding to the highest weight $2\omega_0 + 2\omega_1 + \omega_2$. We think of each vertical wall running diagonally from top-left to bottom-right as a Young diagram or Young wall. The positions of the leftmost columns of these walls are determined by the highest weight. Here the leftmost columns of the back two walls are aligned (corresponding to the summand of $2\omega_0$), the next two walls are aligned one position to the left (corresponding to the summand of $2\omega_1$) and the frontmost wall starts one position further to the left (corresponding to the summand ω_2).

Alistair Savage
Department of Mathematics and Statistics
University of Ottawa
Ottawa, ON, Canada K1N 6N5
alistair.savage@uottawa.ca

Fields Institute Communications
Volume **59**, 2011

CHAPTER 4

Nilpotent Orbits and Finite W-Algebras

WEIQIANG WANG

Abstract. In recent years, the finite W-algebras associated to a semisimple Lie algebra $\mathfrak{g}$ and a nilpotent element of $\mathfrak{g}$ have been studied intensively from different viewpoints. In this chapter, we shall present some basic constructions, connections, and applications of finite W-algebras. The topics include:

- various equivalent definitions of W-algebras
- independence of W-algebras on choices of Lagrangian subspaces
- W-algebras as quantization of Slodowy slices
- classification of good $\mathbb{Z}$-gradings in type A
- independence of W-algebras on choices of good gradings
- category equivalence between W-algebra modules and Whittaker $\mathfrak{g}$-modules
- higher level Schur duality between W-algebras and Hecke algebras
- W-(super)algebras in prime characteristic

Introduction

The finite W-algebras are certain associative algebras associated to a complex semisimple Lie algebra $\mathfrak{g}$ and a nilpotent element $e \in \mathfrak{g}$. In the full generality, they were introduced by Premet [193, 194] into mathematics in a different terminology. Some important special cases go back to Lynch's thesis [166], which is in turn a generalization of Kostant's construction in the case when e is regular nilpotent. For $e = 0$, a finite W-algebra is simply the universal enveloping algebra $U(\mathfrak{g})$. On the other extreme, the finite W-algebra for regular nilpotent e is isomorphic to the center of $U(\mathfrak{g})$ [150]. In the literature of mathematical physics, the finite W-algebras appeared in the work of de Boer and Tjin [82] from the viewpoint of BRST quantum hamiltonian reduction. The history is further complicated as there is an enormous amount of work on affine W-algebras in the 1990's preceding the recent interest of finite W-algebras (and we apologize for inadequate references in this regard). See the book of Bouwknegt and Schoutens [41] and the vast references therein; also see some more precise comments below.

In the last few years, the finite W-algebras and their representations have been extensively studied in mathematics with various new connections developed. The goal of these notes is to provide an introduction to some basic aspects of the fast-growing area of finite W-algebras. We sometimes sketch the main ideas of proofs when it is not practical to reproduce complete proofs. These lecture notes are purely expository in nature and contain no original results.

Premet [193] started out with a characteristic p version of the finite W-algebras, and used it to settle a long-standing conjecture of Kac and Weisfeiler [220] (cf. the review of Jantzen [133] on modular representations of Lie algebras). Subsequently, Premet [194] made a transition to characteristic zero and showed that, with respect to the so-called Kazhdan filtration on the finite W-algebras (cf. [150, 173]), the associated graded algebra is the algebra of functions on the Slodowy slice through e [210]. In Premet's definition, one starts with a $\mathbb{Z}$-grading $\Gamma : \mathfrak{g} = \oplus_{j \in \mathbb{Z}} \mathfrak{g}_j$ which is induced by a standard $\mathfrak{sl}_2$-triple $\{e, h, f\}$ (nowadays such a $\mathbb{Z}$-grading is referred to as a Dynkin $\mathbb{Z}$-grading). The definition of finite W-algebras $\mathcal{W}_{e,\Gamma,\mathfrak{l}}$ depends on the choice of a Dynkin $\mathbb{Z}$-grading Γ and also the choice of a Lagrangian subspace $\mathfrak{l}$ of $\mathfrak{g}_{-1}$ (which is known to be a symplectic vector space).

Gan and Ginzburg [107] developed a purely characteristic zero treatment in a generalized form of Premet's theorem above which allows us to regard finite W-algebras as quantizations of the Slodowy slices. Moreover, Gan and Ginzburg modified Premet's definition by starting with an arbitrary *isotropic* (instead of Lagrangian) subspace $\mathfrak{l}$ of $\mathfrak{g}_{-1}$ and showed that different choices of isotropic (in particular, Lagrangian) subspaces $\mathfrak{l}$ gave rise to isomorphic finite W-algebras $\mathcal{W}_{e,\Gamma}$. The approach of Gan and Ginzburg also provides a new and more transparent proof of Skryabin's equivalence of categories [208] between a module category of $\mathcal{W}_{e,\Gamma}$ and a category of Whittaker modules corresponding to e. The Whittaker $\mathfrak{g}$-module was earlier studied in Kostant's paper [150] when e is regular nilpotent.

Various key ingredients for finite W-algebras actually appeared earlier in the 1990's in the setting of (classical and quantum) affine W-algebras. In particular, Fehér *et al.* [94, Section III] formulated the classical affine W-algebras in the generality of good $\mathbb{Z}$-gradings of $\mathfrak{g}$ (as a natural generalization of the Dynkin $\mathbb{Z}$-gradings) for an arbitrary nilpotent element $e \in \mathfrak{g}$, though the terminology of good $\mathbb{Z}$-gradings was formalized later by Kac, Roan and Wakimoto [138]. Fehér *et al.* [95, Section 3.4] also recognized the key role played by a Lagrangian subspace of $\mathfrak{g}_{-1}$ in classical affine W-algebras. The quantum affine W-algebras associated to a regular nilpotent e from the BRST approach (also called quantized Drinfeld-Sokolov reduction in the affine setup) were studied from the viewpoint of vertex algebras by Frenkel, Kac, and Wakimoto [98]. The finite and affine W-algebras are related via a fundamental construction of Zhu algebra in the theory of vertex algebras [227] (cf. Arakawa [13], De Sole and Kac [84]). The good $\mathbb{Z}$-gradings were subsequently classified by Elashvili and Kac [92]; also see Baur and Wallach [19] for closely related work and further clarification.

Brundan and Goodwin [46] showed that different good $\mathbb{Z}$-gradings Γ on $\mathfrak{g}$ for e led to isomorphic finite W-algebras $\mathcal{W}_e$ (where we drop the dependence on Γ). Actually, they worked and established the results in a somewhat more general and flexible setting of good $\mathbb{R}$-gradings, which allowed them to introduce some suitable geometric object called good grading polytopes.

It is clear from the definition that two nilpotent elements in $\mathfrak{g}$ which are conjugate under the adjoint group G lead to isomorphic finite W-algebras. Hence, the isomorphism classes of finite W-algebras ultimately depend only on the nilpotent orbits in $\mathfrak{g}$.

The representation theory of finite W-algebras has been developed most adequately in type A in a series of papers by Brundan and Kleshchev [48, 49, 50]. In particular, there exists a remarkable higher level duality between finite W-algebras and cyclotomic Hecke algebras, which recovers the classical Schur duality at level one [50].

The constructions of finite W-algebras afford a natural superalgebra generalization. In characteristic p, it has been developed in Lei Zhao's Virginia dissertation (see [216, 217, 226]), in connection with the formulation and proof of a super analogue of the Kac-Weisfeiler conjecture when $\mathfrak{g}$ is a basic classical Lie superalgebra or a queer Lie superalgebra. Many aspects of the representation theory of finite W-algebras over $\mathbb{C}$ afford a superalgebra generalization, and they are currently being investigated.

For lack of expertise and space, we have left out many interesting topics. We refer to Section 4.9 for somewhat more detailed comments on these topics and discussions of some open questions. We also refer to a more recent survey by Losev [161] which covers various complementary topics in detail.

Here is a quick outline on various topics which are (regretfully) *not* to be covered in details in these lecture notes. Losev [157] has developed a powerful deformation quantization approach toward finite W-algebras. Connections between shifted Yangians and finite W-algebras of type A [48] (also cf. [198] in some special cases) allow Brundan and Kleshchev [49] to classify the finite-dimensional irreducible $\mathcal{W}_e$-modules and obtain Kazhdan-Lusztig type character formula in type A (see [85] for formulation of Kazhdan-Lusztig conjecture for finite W-algebras). Deep connections between finite-dimensional modules of finite W-algebras and primitive ideals have been developed [195, 196, 197, 111, 157, 158, 160, 115], and there has also been interesting work on category $\mathcal{O}$ of finite W-algebras, see [195, 47, 159, 114, 218].

The plan of the lecture notes is as follows.

In Section 4.1, a bijection between $\mathfrak{sl}_2$-triples modulo the conjugation by the adjoint group G and nonzero nilpotent orbits in $\mathfrak{g}$ is formulated. We introduce the Dynkin and good $\mathbb{Z}$-gradings, and then explain some of their basic properties.

In Section 4.2, the finite W-algebras are defined in several different ways and these definitions are shown to be equivalent.

In Section 4.3, we formulate the finite W-algebras based on an isotropic subspace $\mathfrak{l}$ of $\mathfrak{g}_{-1}$. Then we introduce the Slodowy slices and the Kazhdan filtration on finite W-algebras, with respect to which the associated graded algebra is isomorphic to the algebra of functions on Slodowy slices. The proof of the independence of the choices of $\mathfrak{l}$ for the isoclasses of finite W-algebras is reduced to some Lie algebra cohomology calculation, which is completed in Section 4.4. The same type of argument is used to derive Skryabin's equivalence in Section 4.4.

In Section 4.5, the classification of good $\mathbb{Z}$-gradings in type A is presented, in terms of some combinatorial objects called pyramids.

In Section 4.6, different good $\mathbb{Z}$-gradings are shown to lead to isomorphic finite W-algebras.

In Section 4.7, we formulate and outline the higher level Schur duality between finite W-algebras of type A and cyclotomic Hecke algebras.

In Section 4.8, we formulate W-superalgebras in characteristic p, and its role in proof of super Kac-Weisfeiler conjecture. This includes the Lie algebra counterpart as its most important special cases.

In Section 4.9, we briefly outline some major topics on finite W-algebras which are not covered in these lecture notes, and then list some open problems.

Acknowledgments. This work was supported in part by NSF Grant DMS–0800280.

4.1 Nilpotent orbits, Dynkin, and good $\mathbb{Z}$-gradings

4.1.1 Basic setups for Lie algebras. For a Lie algebra $\mathfrak{a}$, we will denote by $U(\mathfrak{a})$ the universal enveloping algebra of $\mathfrak{a}$. Let $\mathfrak{a}_e$ denote the centralizer of $e \in \mathfrak{a}$. Sometimes it is convenient to regard $\mathfrak{a}_e = \ker(\operatorname{ad} e)$, the kernel of the adjoint operator $\operatorname{ad} e : \mathfrak{a} \to \mathfrak{a}$.

Let $\mathfrak{g}$ be a finite dimensional semisimple or reductive Lie algebra over $\mathbb{C}$, with a non-degenerate invariant symmetric bilinear form $(\cdot|\cdot)$. Throughout the lectures, we almost always take $\mathfrak{g}$ either to be a simple Lie algebra or the general linear Lie algebra $\mathfrak{gl}_N$. For $\mathfrak{g} = \mathfrak{gl}_N$, we may take the bilinear form by letting $(a|b) = \operatorname{tr}(ab)$ where ab denotes the matrix multiplication of $a, b \in \mathfrak{gl}_N$. An element $e \in \mathfrak{g}$ is *nilpotent* means that $\operatorname{ad} e$ is a nilpotent endomorphism on $\mathfrak{g}$. For example, a nilpotent element in $\mathfrak{gl}_N$ is exactly an $N \times N$ matrix with all eigenvalues 0.

A *$\mathbb{Z}$-grading* of Lie algebra $\mathfrak{g}$ is a decomposition $\mathfrak{g} = \oplus_{j \in \mathbb{Z}} \mathfrak{g}_j$ which is always assumed to satisfy the graded commutation relation: $[\mathfrak{g}_i, \mathfrak{g}_j] \subseteq \mathfrak{g}_{i+j}, \forall i, j$.

4.1.2 Nilpotent orbits. Denote by G the adjoint group associated to the Lie algebra $\mathfrak{g}$, and by $\mathcal{O}_e$ the G-adjoint orbit of e in $\mathfrak{g}$. We refer to Collingwood and McGovern [80] for more on nilpotent orbits.

Denote by $\mathcal{N}$ the null cone which consists of all nilpotent elements in $\mathfrak{g}$. There exists a unique dense open orbit $\mathcal{O}_{reg}$ in $\mathcal{N}$, called the *regular nilpotent orbit*. A nilpotent element $e \in \mathfrak{g}$ is *regular nilpotent* if and only if $\mathfrak{g}_e$ attains the minimal dimension (which equals the rank of $\mathfrak{g}$). The regular nilpotent orbit $\mathcal{O}_{reg}$ consists of all the regular nilpotent elements in $\mathfrak{g}$. There is a unique dense open orbit in $\mathcal{N} \backslash \mathcal{O}_{reg}$, called the *subregular orbit* and denoted by $\mathcal{O}_{sub}$. There exists a unique *minimal orbit* $\mathcal{O}_{min}$ in $\mathcal{N}$ of smallest positive dimension. The dimension of $\mathcal{O}_{min}$ is known [214] to be $2h^\vee - 2$, where $h^\vee$ is the dual Coxeter number of $\mathfrak{g}$. Incidentally, the dual Coxeter number plays a key role in representation theory of affine Lie algebras.

The set of nilpotent orbits in $\mathfrak{g}$ is naturally a poset $\mathcal{P}$ with partial order $\leq$ as follows: $\mathcal{O}' \leq \mathcal{O}$ if and only only if $\mathcal{O}' \subseteq \overline{\mathcal{O}}$ (the closure of $\mathcal{O}$). Then $\mathcal{O}_{reg}$ is the maximum and the zero orbit is the minimum in the poset $\mathcal{P}$. Moreover, $\mathcal{O}_{sub}$ is the maximum in the poset $\mathcal{P} \backslash \{\mathcal{O}_{reg}\}$, while $\mathcal{O}_{min}$ is the minimum in the poset $\mathcal{P} \backslash \{0\}$.

Example 4.1.1 Let $\mathfrak{g} = \mathfrak{gl}_n$. Every nilpotent matrix in $\mathfrak{gl}_n$ is conjugate to a matrix in Jordan canonical form. It follows that the nilpotent orbits in $\mathfrak{g}$ are parameterized by the partitions $\lambda = (\lambda_1, \lambda_2, \ldots)$ of n, where the orbit $\mathcal{O}_\lambda$ contains the matrix in standard Jordan form $\operatorname{diag}(J_{\lambda_1}, J_{\lambda_2}, \ldots)$, and we have denoted by J_k the $k \times k$ Jordan block.

Associated to the four distinguished partitions (n), $(n-1, 1)$, $(2, 1^{n-2})$, (1^n) are the orbits $\mathcal{O}_{reg}$, $\mathcal{O}_{sub}$, $\mathcal{O}_{min}$ and the zero orbit $\{0\}$, respectively.

4.1.3 Jacobson-Morozov Theorem. Associated to a nonzero nilpotent element $e \in \mathfrak{g}$, there always exists $\{e, h, f\} \subseteq \mathfrak{g}$ which satisfy the $\mathfrak{sl}_2$ commutation relation:

$$[e, f] = h, \quad [h, e] = 2e, \quad [h, f] = -2f.$$

We will refer to this as a *standard* $\mathfrak{sl}_2$*-triple* or simply an $\mathfrak{sl}_2$*-triple*.

This statement is known as the *Jacobson-Morozov Theorem*, and it can be proved by induction on the dimension of $\mathfrak{g}$. See Carter [53, 5.3] or [80, 3.3].

Example 4.1.2 Let $\mathfrak{g} = \mathfrak{gl}_n$, $e = J_n$, $h = \mathrm{diag}\,(n-1, n-3, \ldots, 3-n, 1-n)$, and let

$$f = \begin{pmatrix} 0 & & & 0 \\ a_1 & 0 & & \\ & \ddots & \ddots & \\ 0 & & a_{n-1} & 0 \end{pmatrix}$$

with exactly one nonzero (sub)diagonal with entries $a_i = i(n-i)$ for $1 \le i \le n-1$. Then $\{e, h, f\}$ form a standard $\mathfrak{sl}_2$-triple.

4.1.4 The Dynkin gradings. By representation theory of $\mathfrak{sl}_2 = \mathbb{C}\{e, h, f\}$, the eigenspace decomposition of the adjoint action $\mathrm{ad}\, h : \mathfrak{g} \to \mathfrak{g}$ provides a $\mathbb{Z}$-grading

$$\mathfrak{g} = \bigoplus_{j \in \mathbb{Z}} \mathfrak{g}_j,$$

that is, $\mathfrak{g}_j = \{x \in \mathfrak{g} \mid \mathrm{ad}\, h(x) = jx\}$. A $\mathbb{Z}$-grading of $\mathfrak{g}$ arising from $\mathfrak{sl}_2$ this way will be referred to as a *Dynkin $\mathbb{Z}$-grading*. A Dynkin grading affords the following favorable properties:

$$e \in \mathfrak{g}_2, \tag{4.1}$$
$$\mathrm{ad}\, e : \mathfrak{g}_j \to \mathfrak{g}_{j+2} \text{ is injective for } j \le -1, \tag{4.2}$$
$$\mathrm{ad}\, e : \mathfrak{g}_j \to \mathfrak{g}_{j+2} \text{ is surjective for } j \ge -1, \tag{4.3}$$
$$\mathfrak{g}_e \subseteq \oplus_{j \ge 0} \mathfrak{g}_j, \tag{4.4}$$
$$(\mathfrak{g}_i | \mathfrak{g}_j) = 0 \text{ unless } i + j = 0, \tag{4.5}$$
$$\dim \mathfrak{g}_e = \dim \mathfrak{g}_0 + \dim \mathfrak{g}_1. \tag{4.6}$$

Note that (4.1) follows from the $\mathfrak{sl}_2$-triple definition. Claims (4.2) and (4.3) are immediate consequences of the highest weight theory of $\mathfrak{sl}_2$ when applied to the the adjoint module $\mathfrak{g}$. Claim (4.4) follows from (4.2) and (4.3). Formulas (4.5) and (4.6) are also easy, and they will be proved in a generalized setting in Proposition 4.1.5 below using only (4.1), (4.2), and (4.3).

For $k \in \mathbb{Z}$, we shall denote $\mathfrak{g}_{>k} = \oplus_{j>k} \mathfrak{g}_j$. Similarly, we define $\mathfrak{g}_{\ge k}, \mathfrak{g}_{<k}, \mathfrak{g}_{\le k}$.

4.1.5 The good $\mathbb{Z}$-gradings. In this subsection, we recall the notion of good $\mathbb{Z}$-gradings (see [94, (3.3)] and [138]) and its basic properties [92].

Definition 4.1.3 A $\mathbb{Z}$-grading $\Gamma : \mathfrak{g} = \oplus_{j \in \mathbb{Z}} \mathfrak{g}_j$ for $\mathfrak{g}$ semisimple is called a *good* $\mathbb{Z}$-grading for e if it satisfies the conditions (4.1)-(4.3) above.

For $\mathfrak{g}$ reductive, a good $\mathbb{Z}$-grading $\mathfrak{g} = \oplus_{j \in \mathbb{Z}} \mathfrak{g}_j$ for e is required to satisfy (4.1)-(4.3) and an additional condition that the center $\mathfrak{z}(\mathfrak{g})$ of $\mathfrak{g}$ is contained in $\mathfrak{g}_0$.

In particular, $\mathrm{ad}\, e : \mathfrak{g}_{-1} \to \mathfrak{g}_1$ is always a bijection.

The following simple lemma will be repeatedly used.

Lemma 4.1.4 *For any $\mathbb{Z}$-grading $\Gamma : \mathfrak{g} = \oplus_{j \in \mathbb{Z}} \mathfrak{g}_j$, there exists a semisimple element $h_\Gamma \in [\mathfrak{g}, \mathfrak{g}]$ such that Γ coincides with the eigenspace decomposition of* $\mathrm{ad}\, h_\Gamma$, *i.e.* $\mathfrak{g}_j = \{x \in \mathfrak{g} \mid [h_\Gamma, x] = jx\}$.

Proof The degree operator $\partial : \mathfrak{g} \to \mathfrak{g}$ which sends $x \mapsto jx$ for $x \in \mathfrak{g}_j$ is a derivation on the semisimple Lie algebra $[\mathfrak{g}, \mathfrak{g}]$, hence an inner derivation on $[\mathfrak{g}, \mathfrak{g}]$, given by $\operatorname{ad} h_\Gamma$ for some semisimple element $h_\Gamma \in [\mathfrak{g}, \mathfrak{g}]$. Then $\partial = \operatorname{ad} h_\Gamma$ as derivations on $\mathfrak{g} = [\mathfrak{g}, \mathfrak{g}] \oplus \mathfrak{z}(\mathfrak{g})$ since the equality holds on $\mathfrak{z}(\mathfrak{g})$ too. □

Proposition 4.1.5 *Properties (4.4)-(4.6) remain to be valid for every good $\mathbb{Z}$-grading $\Gamma : \mathfrak{g} = \oplus_{j \in \mathbb{Z}} \mathfrak{g}_j$.*

Proof Clearly $\mathfrak{g}_e$ is a $\mathbb{Z}$-graded Lie subalgebra of $\mathfrak{g}$, and (4.4) follows from (4.2).

For $x \in \mathfrak{g}_i, y \in \mathfrak{g}_j$, we have $-i(x|y) = ([x, h_\Gamma]|y) = (x|[h_\Gamma, y]) = j(x|y)$ by Lemma 4.1.4. This implies (4.5).

Finally, (4.6) follows by an exact sequence of vector spaces:

$$0 \longrightarrow \mathfrak{g}_e \longrightarrow \mathfrak{g}_{-1} + \mathfrak{g}_0 + \mathfrak{g}_{>0} \stackrel{\operatorname{ad} e}{\longrightarrow} \mathfrak{g}_{>0} \longrightarrow 0$$

which is well-defined by (4.2) and (4.3). □

Example 4.1.6 Associated to $e = 0$, we have a good $\mathbb{Z}$-grading with the whole $\mathfrak{g}$ concentrated on degree zero, i.e. $\mathfrak{g}_0 = \mathfrak{g}$.

Example 4.1.7 Let $\mathfrak{g} = \mathfrak{gl}_3$ and $e = E_{13}$.

(i) Let $h = \operatorname{diag}(1, 0, -1)$ and $f = E_{31}$. Then $\{e, h, f\}$ forms an $\mathfrak{sl}_2$-triple, and it gives rise to a Dynkin $\mathbb{Z}$-grading of $\mathfrak{gl}_3$ for $e \in \mathfrak{g}_2$ whose degrees on the matrix units E_{ij} are listed in the matrix below:

$$\begin{pmatrix} 0 & 1 & 2 \\ -1 & 0 & 1 \\ -2 & -1 & 0 \end{pmatrix}$$

A basis of the (5-dimensional) centralizer $\mathfrak{g}_e$ consists of 2 degree-zero elements $E_{11} + E_{33}, E_{22}$, 2 degree-one elements E_{12}, E_{23}, and a degree-two element E_{13}.

(ii) The eigenspace decomposition of the diagonal element diag $(1, 1, -1)$ gives rises to a good (but non-Dynkin) $\mathbb{Z}$-grading for $e = E_{13}$ whose degrees on the matrix units E_{ij} are listed in the matrix below:

$$\begin{pmatrix} 0 & 0 & 2 \\ 0 & 0 & 2 \\ -2 & -2 & 0 \end{pmatrix}$$

Note now $\mathfrak{g}_{-1} = 0$.

In short, *Dynkin is good*, but *being good is not good enough for being Dynkin.*

Lemma 4.1.8 *Properties* (4.2) *and* (4.3) *are equivalent for any $\mathbb{Z}$-grading $\mathfrak{g} = \oplus_{j \in \mathbb{Z}} \mathfrak{g}_j$.*

Proof Note the proof of (4.5) in Proposition 4.1.5 only uses the $\mathbb{Z}$-grading of $\mathfrak{g}$, but does not use (4.2) and (4.3). Then we have a non-degenerate pairing between $\mathfrak{g}_i$ and $\mathfrak{g}_{-i}$ induced from the non-degenerate form $(\cdot|\cdot)$ on $\mathfrak{g}$. The lemma follows by this non-degeneracy and the equation $(x_i|[e, y_j]) = ([x_i, e]|y_j)$ for $i + j = -2$ and $i \leq -1 \leq j$. □

Let $\Gamma : \mathfrak{g} = \oplus_{j\in\mathbb{Z}}\mathfrak{g}_j$ be a good $\mathbb{Z}$-grading of $\mathfrak{g}$. Note that h_Γ defined in Lemma 4.1.4 actually lies in $\mathfrak{g}_0$. Then $\mathfrak{g}_0 = \mathfrak{g}_{h_\Gamma}$ is a Levi (reductive) subalgebra of $\mathfrak{g}$. Choose $\mathfrak{h}$ to be a Cartan subalgebra of $\mathfrak{g}_0$, and hence also a Cartan subalgebra of $\mathfrak{g}$. Let $\mathfrak{g} = \mathfrak{h} \oplus \oplus_{\alpha\in\Phi}\mathfrak{g}^\alpha$ be a root space decomposition. Fix a system of positive roots Δ_0^+ of $(\mathfrak{g}_0, \mathfrak{h})$.

Lemma 4.1.9 *Let $\Gamma : \mathfrak{g} = \oplus_{j\in\mathbb{Z}}\mathfrak{g}_j$ be a good $\mathbb{Z}$-grading of $\mathfrak{g}$. Then,*

(i) each root subspace $\mathfrak{g}^\alpha$ lies in $\mathfrak{g}_j$ for some $j \in \mathbb{Z}$;
(ii) $\Delta^+ := \Delta_0^+ \bigcup\{\alpha \mid \mathfrak{g}^\alpha \subseteq \mathfrak{g}_{>0}\}$ is a system of positive roots of $(\mathfrak{g}, \mathfrak{h})$.

Proof (1) Take $h_\Gamma \in \mathfrak{g}$ as in Lemma 4.1.4. Then $[h_\Gamma, x] = \alpha(h_\Gamma)x$ for $x \in \mathfrak{g}^\alpha$ implies that $\mathfrak{g}^\alpha \subseteq \mathfrak{g}_j$ for $j = \alpha(h_\Gamma) \in \mathbb{Z}$.

(2) Choose a regular element $r \in \mathfrak{h}$. Then $\epsilon r + h_\Gamma$ for small $\epsilon > 0$ is also regular, and so $\Delta^+ = \{\alpha \in \Phi \mid \alpha(\epsilon r + h_\Gamma) > 0\}$ is a system of positive roots of $\mathfrak{g}$. □

Let Π be the set of simple roots for Δ^+, and denote $\mathbb{N} = \{0, 1, 2, \ldots\}$. Then,

$$\Pi = \bigcup_{j\in\mathbb{N}} \Pi_j, \qquad \text{where } \Pi_j = \{\alpha \in \Pi \mid \mathfrak{g}^\alpha \subseteq \mathfrak{g}_j\}.$$

Lemma 4.1.10 *For a good $\mathbb{Z}$-grading Γ, we have $\Pi = \Pi_0 \cup \Pi_1 \cup \Pi_2$.*

Proof Assume to the contrary that there exists $\beta \in \Pi_s$ for $s > 2$. Since $e \in \mathfrak{g}_2$ and $\mathfrak{g}_2$ is contained in the subalgebra generated by $\mathfrak{g}^\alpha$ with $\alpha \in \Pi_0 \cup \Pi_1 \cup \Pi_2$, we have $[e, \mathfrak{g}^{-\beta}] = 0$, or in other words, $\mathfrak{g}^{-\beta} \subseteq \mathfrak{g}_e$. This contradicts with the fact (4.4) that $\mathfrak{g}_e$ has $\mathbb{N}$-grading as $\mathfrak{g}^{-\beta} \subseteq \mathfrak{g}_{-s}$. □

4.1.6 A bijective map Ω. Clearly, the group G acts on the collection of $\mathfrak{sl}_2$-triples in $\mathfrak{g}$ by conjugation. Define a map

$$\Omega : \{\mathfrak{sl}_2\text{-triples}\}/G \longrightarrow \{\text{nonzero nilpotent orbits}\},$$
$$\Omega(\{e, h, f\}) = \mathcal{O}_e.$$

Theorem 4.1.11 *The map Ω is a bijection.*

Sketch of a proof According to Jacobson-Morozov Theorem in 4.1.3, the map Ω is surjective. A theorem of Kostant asserts that the map Ω is also injective, as we sketch below (see [80, 3.4] for details).

To that end, take two $\mathfrak{sl}_2$-triples which are mapped by Ω to the same nilpotent orbit $\mathcal{O}_e$. Applying a G-conjugation, we may assume without loss of generality that the two triples are $\{e, h, f\}$ and $\{e, h', f'\}$, that is, they share the same e.

Let $\mathfrak{u}_e := \mathfrak{g}_e \cap [\mathfrak{g}, e]$. Note that $h' - h \in \mathfrak{u}_e$ and $\mathfrak{u}_e$ is an $\operatorname{ad} h$-invariant, nilpotent ideal of $\mathfrak{g}_e$. Actually $\mathfrak{u}_e = \oplus_{j>0}\mathfrak{g}_e \cap \mathfrak{g}_j$. Denote by U_e the subgroup of G with Lie algebra $\mathfrak{u}_e$.

By using the fact that $\mathfrak{u}_e$ is a $\mathbb{N}$-graded nilpotent ideal, one shows easily that $\operatorname{Ad} U_e(h) = h + \mathfrak{u}_e$. Hence, there exists $x \in U_e$ such that $\operatorname{Ad} x(h) = h'$.

Note that we automatically have $\operatorname{Ad} x(e) = e$ for $x \in U_e$. Now the injectivity of Ω follows from the *rigidity of* $\mathfrak{sl}_2$ below.

Lemma 4.1.12 *Let $\{e, h, f\}$ and $\{e, h, f'\}$ be standard $\mathfrak{sl}_2$-triples of $\mathfrak{g}$. Then $f = f'$.*

□

Exercise 4.1.13 Prove Lemma 4.1.12.

Remark 4.1.14 If one works a bit harder (cf. [80, 3.4]), one recovers a theorem of Molcev which asserts the injectivity of the map

$$\Theta : \{\mathfrak{sl}_2\text{-triples}\}/G \longrightarrow \{\text{nonzero semisimple orbits}\},$$
$$\Theta(\{e,h,f\}) = \mathcal{O}_h.$$

Given an $\mathfrak{sl}_2$-triple $\{e,h,f\}$, by Lemma 4.1.10 (which applies to the special case of Dynkin grading $\operatorname{ad} h$) we have a system of simple roots of $\mathfrak{g}$ whose weights on h only take values from $\{0,1,2\}$. Therefore the number of semisimple orbits with such weight restriction is $\leq 3^{\operatorname{rank}(\mathfrak{g})}$. It now follows by the properties of Ω and Θ above that the number of nilpotent orbits in $\mathfrak{g}$ is finite.

An alternative way of obtaining the finiteness of nilpotent orbits of $\mathfrak{g}$ of classical type is to use the parametrization by partitions (cf. [134, Chapter 1]). But such an alternative ceases to work well for Lie algebras $\mathfrak{g}$ of exceptional type.

4.2 Definitions of W-algebras

4.2.1 The endomorphism algebra definition. Assume we are given a reductive Lie algebra $\mathfrak{g}$ with a non-degenerate invariant bilinear form $(\cdot|\cdot)$, an nilpotent element $e \in \mathfrak{g}$, and a good $\mathbb{Z}$-grading $\Gamma : \mathfrak{g} = \oplus_{j\in\mathbb{Z}}\mathfrak{g}_j$ for e.

There exists a unique $\chi \in \mathfrak{g}^*$ such that $\chi(x) = (x|e)$. Define a bilinear form $\langle .,. \rangle$ on $\mathfrak{g}_{-1}$ as follows:

$$\langle \cdot,\cdot \rangle : \mathfrak{g}_{-1} \times \mathfrak{g}_{-1} \to \mathbb{C}, \qquad \langle x,y \rangle := ([x,y]|e) = \chi([x,y]).$$

The following lemma was noticed earlier [95, 3.4] in the setting for classical affine W-algebras.

Lemma 4.2.1 *The bilinear form $\langle \cdot,\cdot \rangle$ on $\mathfrak{g}_{-1}$ is skew-symmetric and non-degenerate.*

Proof The skew-symmetry follows by definition. The non degeneracy follows by the bijection $\operatorname{ad} e : \mathfrak{g}_{-1} \to \mathfrak{g}_1$ and the identity $\langle x,y \rangle = (x|[y,e])$. □

It follows that $\mathfrak{g}_{-1}$ is even-dimensional. Let us choose a Lagrangian (= maximal isotropic) subspace $\mathfrak{l}$ of $\mathfrak{g}_{-1}$ with respect to the form $\langle \cdot,\cdot \rangle$. Then $\dim \mathfrak{l} = \frac{1}{2}\dim \mathfrak{g}_{-1}$. Introduce the following important nilpotent subalgebra of $\mathfrak{g}$:

$$\mathfrak{m} := \mathfrak{l} \bigoplus \Big(\bigoplus_{j \leq -2} \mathfrak{g}_j \Big).$$

Exercise 4.2.2 Prove that

$$\dim \mathfrak{m} = \frac{1}{2}\dim \mathcal{O}_e.$$

The restriction of χ to $\mathfrak{m}$, denoted by $\chi : \mathfrak{m} \to \mathbb{C}$, defines a one-dimensional representation $\mathbb{C}_\chi$ of $\mathfrak{m}$, thanks to the Lagraingian condition on $\mathfrak{l}$. Let $I_\chi^{\mathfrak{m}}$ denote the kernel of the corresponding associative algebra homomorphism $U(\mathfrak{m}) \to \mathbb{C}$, i.e. the two-sided ideal of $U(\mathfrak{m})$ generated by $a - \chi(a)$ for $a \in \mathfrak{m}$. Let I_χ denote the left ideal of $U(\mathfrak{g})$ generated by $a - \chi(a)$ for $a \in \mathfrak{m}$. Define the induced $\mathfrak{g}$-module (called *generalized Gelfand-Graev module*)

$$Q_\chi := U(\mathfrak{g}) \otimes_{U(\mathfrak{m})} \mathbb{C}_\chi \cong U(\mathfrak{g})/I_\chi.$$

The *finite W-algebra* (or simply W-algebra in this paper) $\mathcal{W}_\chi$ is defined to be the endomorphism algebra

$$\mathcal{W}_\chi := \mathrm{End}_{U(\mathfrak{g})}(Q_\chi)^{\mathrm{op}}. \tag{4.7}$$

This is the original definition of Premet [194].

Actually, the above notions depend in addition on $\Gamma, \mathfrak{l}$, and it would be more precise to write $Q_{\chi,\Gamma,\mathfrak{l}}$, $I_{\chi,\Gamma,\mathfrak{l}}$, $\mathcal{W}_{\chi,\Gamma,\mathfrak{l}}$, etc. But we choose to use the simplified notations, partly because we will eventually show that the isoclasses of finite W-algebras do not depend on Γ and $\mathfrak{l}$. At different occasions later on, we may use some other indices among $\chi, \Gamma, \mathfrak{l}$ to put an emphasis on the dependence of those indices.

Example 4.2.3 Let $e = 0$. Then $\chi = 0$, $\mathfrak{g}_0 = \mathfrak{g}$, $\mathfrak{m} = 0$, $Q_\chi = U(\mathfrak{g})$, and $\mathcal{W}_\chi = U(\mathfrak{g})$.

4.2.2 The Whittaker model definition. Since $Q_\chi = U(g)/I_\chi$ is a cyclic module, any endomorphism of the $\mathfrak{g}$-module Q_χ is determined by the image of $\bar{1}$, where $\bar{y}$ denotes the coset $y+I_\chi$ of $y \in U(\mathfrak{g})$. The image of $\bar{1}$ must be annihilated by I_χ. Hence we obtain the following identification of $\mathcal{W}_\chi$ (as the space of Whittaker vectors in $U(\mathfrak{g})/I_\chi$ in the terminology introduced in a later Section 4.4.2):

$$\mathcal{W}_\chi = \{\bar{y} \in U(\mathfrak{g})/I_\chi \mid (a - \chi(a))y \in I_\chi, \forall a \in \mathfrak{m}\}. \tag{4.8}$$

By definition of I_χ, $\mathcal{W}_\chi$ can be further identified with the subspace of $\mathrm{ad}\,\mathfrak{m}$-invariants in Q_χ:

$$\mathcal{W}_\chi = (Q_\chi)^{\mathrm{ad}\,\mathfrak{m}} := \{\bar{y} \in U(\mathfrak{g})/I_\chi \mid [a, y] \in I_\chi, \forall a \in \mathfrak{m}\}. \tag{4.9}$$

Transferred via the above identification, the algebra structure on $\mathcal{W}_\chi$ is given by

$$\bar{y}_1\bar{y}_2 = \overline{y_1y_2}$$

for $y_i \in U(\mathfrak{g})$ such that $[a, y_i] \in I_\chi$ for all $a \in \mathfrak{m}$ and $i = 1, 2$.

Exercise 4.2.4 Without using the identification with the definition (4.7), check directly that

(i) the ideal I_χ is $\mathrm{ad}\,\mathfrak{m}$-invariant, hence $(Q_\chi)^{\mathrm{ad}\,\mathfrak{m}}$ in (4.9) as a vector space is well-defined;

(ii) for y_i satisfying $[a, y_i] \in I_\chi$ for all $a \in \mathfrak{m}$ and $i = 1, 2$, we have $[a, y_1y_2] \in I_\chi$ for all $a \in \mathfrak{m}$. Hence, $(Q_\chi)^{\mathrm{ad}\,\mathfrak{m}}$ in (4.9) as an algebra is well-defined.

Hence, we may regard (4.9), or (4.8), as a second definition of W-algebras.

4.2.3 A simplified definition for even good gradings. In this subsection, we will assume $\mathfrak{g}$ is equipped with an *even* good $\mathbb{Z}$-grading. A good $\mathbb{Z}$-grading $\mathfrak{g} = \oplus_{j\in\mathbb{Z}}\mathfrak{g}_j$ is called *even*, if $\mathfrak{g}_j = 0$ unless j is an even integer.

All the complications of choice of a Lagrangian/isotropic subspace $\mathfrak{l}$ disappear for even $\mathbb{Z}$-gradings, since $\mathfrak{g}_{-1} = 0$. Now we have $\mathfrak{m} = \oplus_{j\le -2}\mathfrak{g}_j$, and a parabolic subalgebra

$$\mathfrak{p} := \bigoplus_{j\ge 0} \mathfrak{g}_j. \tag{4.10}$$

It follows by the PBW theorem that

$$U(\mathfrak{g}) = U(\mathfrak{p}) \bigoplus I_\chi.$$

The projection $\mathrm{pr}_\chi : U(\mathfrak{g}) \to U(\mathfrak{p})$ along this direct sum decomposition induces an isomorphism

$$\overline{\mathrm{pr}}_\chi : U(\mathfrak{g})/I_\chi \xrightarrow{\sim} U(\mathfrak{p}).$$

By the above observation, the algebra $\mathcal{W}_\chi$ can be regarded as a *subalgebra* of $U(\mathfrak{p})$. Consider a χ-twisted adjoint action of $\mathfrak{m}$ on $U(\mathfrak{p})$ by $a.y := \mathrm{pr}_\chi([a,y])$ for $a \in \mathfrak{m}$ and $y \in U(\mathfrak{p})$. We identify $\mathcal{W}_\chi$ as

$$\mathcal{W}_\chi = U(\mathfrak{p})^{\mathrm{ad}\,\mathfrak{m}} := \{y \in U(\mathfrak{p}) \mid [a,y] \in I_\chi, \forall a \in \mathfrak{m}\}. \tag{4.11}$$

Since $\overline{\mathrm{pr}}_\chi : U(\mathfrak{g})/I_\chi \xrightarrow{\sim} U(\mathfrak{p})$ is an isomorphism of $\mathfrak{m}$-modules, the equivalence of (4.11) with the second definition (for even $\mathbb{Z}$-gradings) is clear.

Hence, we may take (4.11) as a third definition of the W-algebras. This is the original definition used by Kostant and Lynch [150, 166] (who only considered even Dynkin $\mathbb{Z}$-gradings). As observed by Brundan-Kleshchev [48], it works equally well in the generality of even good gradings by dropping "Dynkin". Brundan-Goodwin-Kleshchev [47, Section 2] also has a (necessarily) more complicated version of this definition without assuming the grading to be even.

Example 4.2.5 According to a theorem of Kostant (see Theorem 4.3.8 below), for a regular nilpotent element e in $\mathfrak{g}$, $\mathcal{W}_\chi$ is isomorphic to $\mathcal{Z}(\mathfrak{g})$, the center of $U(\mathfrak{g})$.

Let $e = E_{12} \in \mathfrak{gl}_2$, which is regular nilpotent. Then $\mathfrak{m} = \mathbb{C}f$ with $f = E_{21}$, and $\mathfrak{p} = \mathbb{C}e + \mathbb{C}E_{11} + \mathbb{C}E_{22}$. A direct computation shows that $E_{11}+E_{22}$ and $e+\frac{1}{4}h^2-\frac{1}{2}h$ lie in $U(\mathfrak{p})^{\mathrm{ad}\,\mathfrak{m}}$. (If we combine with Kostant's theorem, then $\mathcal{W}_\chi = U(\mathfrak{p})^{\mathrm{ad}\,\mathfrak{m}}$ is the polynomial algebra generated by these two elements.)

Let $\mathfrak{g} = \mathfrak{gl}_n$ and $e = J_n$ as in Example 4.1.2. Let

$$\Omega(u) := \begin{pmatrix} E_{11}+u+1 & E_{12} & & \cdots & E_{1n} \\ 1 & E_{22}+u+2 & E_{23} & & \vdots \\ 0 & & \ddots & \ddots & \vdots \\ \vdots & & \ddots & \ddots & E_{n-1,n} \\ 0 & \cdots & 0 & 1 & E_{nn}+u+n \end{pmatrix}$$

The row determinant of a matrix $A = (a_{i,j})_{n\times n}$ with non-commutative entries is defined to be

$$\mathrm{rdet}\, A = \sum_{\pi\in S_n} \mathrm{sgn}(\pi) a_{1,\pi 1} \cdots a_{n,\pi n}. \tag{4.12}$$

Write $\mathrm{rdet}\,\Omega(u) = u^n + \sum_{i=1}^n w_i u^{n-i}$ for $w_i \in U(\mathfrak{p})$. As shown in [48, Section 12], $w_i, 1 \le i \le n$ are commuting elements which lie in $U(\mathfrak{p})^{\mathrm{ad}\,\mathfrak{m}}$, and they freely generate $\mathcal{W}_\chi = U(\mathfrak{p})^{\mathrm{ad}\,\mathfrak{m}}$.

4.2.4 The BRST definition. There is yet another definition of W-algebras using the BRST complex.

For the sake of simplicity, we will work under the assumption of an *even* good $\mathbb{Z}$-grading $\Gamma : \mathfrak{g} = \oplus_{j\in\mathbb{Z}}\mathfrak{g}_j$ for $e \in \mathfrak{g}_2$. In this case, we have $\mathfrak{m} = \oplus_{j\le -2}\mathfrak{g}_j$.

Let us take another copy of $\mathfrak{m}$, which will be denoted by $\hat{\mathfrak{m}}$.

Endow the vector space $\mathfrak{m}^* \oplus \hat{\mathfrak{m}}$ with the symmetric bilinear form induced by the pairing between $\mathfrak{m}^*$ and $\hat{\mathfrak{m}}$. The corresponding Clifford algebra on $\mathfrak{m}^* \oplus \hat{\mathfrak{m}}$

can be naturally identified with $\wedge(\mathfrak{m}^*) \otimes \wedge(\hat{\mathfrak{m}})$. Consider the tensor algebra of this Clifford algebra with $U(\mathfrak{g})$

$$\mathcal{B}^\bullet := \wedge(\mathfrak{m}^*) \otimes U(\mathfrak{g}) \otimes \wedge(\hat{\mathfrak{m}}).$$

It admits a BRST (cohomological) $\mathbb{Z}$-grading with the assignment of degrees to generators in $\mathfrak{m}^*, \mathfrak{g}, \hat{\mathfrak{m}}$ to be $1, 0, -1$, respectively. This $\mathbb{Z}$-grading is compatible with the superalgebra structure on $\mathcal{B}^\bullet$ by declaring generators in $\mathfrak{g}$ to be even and generators in $\mathfrak{m}^* \oplus \hat{\mathfrak{m}}$ odd.

Take a basis $\{b_i\}$ of $\mathfrak{m}$ and let $\{f^i\}$ be its dual basis for $\mathfrak{m}^*$. Let $d = [\phi, -]$ be the super-derivation of $\mathcal{B}^\bullet$ (of BRST degree 1), which is defined to be the super-commutator with the following odd element

$$\phi = f^i(b_i - \chi(b_i)) - \frac{1}{2} f^i f^j [b_i, b_j]^\wedge$$

where $a^\wedge$ denotes the corresponding element in $\hat{\mathfrak{m}}$ for $a \in \mathfrak{m}$. Here and below we have adopted the convention of summation over repeated indices. It is easy to check that ϕ is independent of the choices of the dual bases.

In more concrete terms, one finds that, for $x \in \mathfrak{g}, f \in \mathfrak{m}^*$ and $a^\wedge \in \hat{\mathfrak{m}}$ corresponding to $a \in \mathfrak{m} \subseteq \mathfrak{g}$,

$$\begin{aligned} d(x) = f^i[b_i, x], \qquad & d(f) = \tfrac{1}{2} f^i \operatorname{ad}^* b_i(f), \\ & d(a^\wedge) = a - \chi(a) + f^i[b_i, a]^\wedge, \end{aligned} \tag{4.13}$$

where we have denoted by ad^* the coadjoint action.

Exercise 4.2.6 Verify that $d^2 = 0$.

So we have defined the so-called BRST complex $(\mathcal{B}^\bullet, d)$ and we can define as usual its cohomology $H^*(\mathcal{B}^\bullet)$. The BRST definition of the W-algebra (cf. de Boer and Tjin [82]) is

$$\mathcal{W}_\chi = H^0(\mathcal{B}^\bullet, d).$$

This is justified by the following isomorphism theorem, which confirms a conjecture of Premet [194, 1.10].

Theorem 4.2.7 ([81]) *The BRST cohomology $H^*(\mathcal{B}^\bullet)$ is concentrated in cohomological degree 0 and there is an algebra isomorphism between $H^0(\mathcal{B}^\bullet, d)$ and $(Q_\chi)^{\operatorname{ad} \mathfrak{m}}$.*

Sketch of a proof Note that in the definition of the BRST complex $(\mathcal{B}^\bullet, d)$, $\mathcal{B}^\bullet$ does not depend on χ, but d does. We introduce a complex $(\mathcal{B}^\bullet_\chi, d')$ which is isomorphic to the BRST complex $(\mathcal{B}^\bullet, d)$ but with a shift of the dependence of χ from the differential to the complex.

The complex $\mathcal{B}^\bullet_\chi := \wedge(\mathfrak{m}^*) \otimes (U(\mathfrak{g}) \otimes \mathbb{C}_\chi) \otimes \wedge(\hat{\mathfrak{m}})$ is obtained from twisting $\mathcal{B}^\bullet$ by χ, and $U(\mathfrak{g}) \otimes \mathbb{C}_\chi$ is naturally an $\mathfrak{m}$-bimodule; the differential d' equals the sum of two (anti)commuting differentials $d_{\mathfrak{m}}$ and $d^{\mathfrak{m}}$, where $d_{\mathfrak{m}}$ (and respectively, $d^{\mathfrak{m}}$) is a $\mathfrak{m}$-homology (respectively, $\mathfrak{m}$-cohomology) differential. Then $(\mathcal{B}^\bullet_\chi, d)$ can be regarded as a double complex, and the first term of a spectral sequence computation gives us $H^{\bullet,0}(\mathcal{B}^\bullet_\chi, d_{\mathfrak{m}}) \cong \wedge^\bullet(\mathfrak{m}^*) \otimes Q_\chi$ (the $\mathfrak{m}$-cohomology complex with coefficient in Q_χ); moreover, $H^{p,q}(\mathcal{B}^\bullet_\chi, d^{\mathfrak{m}}) = 0$ whenever $q \neq 0$. The spectral sequence stabilizes at the second term, which shows $H^i(\mathcal{B}^\bullet_\chi, d) \cong \operatorname{gr} H^i(\mathcal{B}^\bullet_\chi, d) = H^i(\mathfrak{m}, Q_\chi)$, for all i.

We shall see from Theorem 4.4.1 later on that $H^0(\mathfrak{m}, Q_\chi) = Q_\chi^{\operatorname{ad} \mathfrak{m}}$ and that $H^i(\mathfrak{m}, Q_\chi) = 0$ for $i > 0$.

A quasi-isomorphism $\wedge^\bullet(\mathfrak{m}^*) \otimes Q_\chi \to \mathcal{B}^\bullet_\chi$ can then be constructed explicitly, and it induces the algebra isomorphism on 0th cohomology. Putting all things together, we have obtained the theorem. $\square$

We shall describe such an isomorphism in a more concrete fashion. Note by the PBW theorem that $\mathcal{B}^0 = U(\mathfrak{g}) \oplus \mathfrak{m}^*\mathcal{B}^0\hat{\mathfrak{m}}$, with $\mathfrak{m}^*\mathcal{B}^0\hat{\mathfrak{m}}$ being a two-sided ideal. So we can consider the composition q of the two natural maps $\mathcal{B}^0 \twoheadrightarrow U(\mathfrak{g}) \twoheadrightarrow Q_\chi$, which fits into the following commutative diagram:

$$\begin{array}{ccccccccc}
0 & \longrightarrow & I_\chi \oplus \mathfrak{m}^*\mathcal{B}^0\hat{\mathfrak{m}} & \longrightarrow & \mathcal{B}^0 & \xrightarrow{q} & Q_\chi & \longrightarrow & 0 \\
 & & \downarrow & & \downarrow{\scriptstyle \mathrm{pr}} & & \| & & \\
0 & \longrightarrow & I_\chi & \longrightarrow & U(\mathfrak{g}) & \longrightarrow & Q_\chi & \longrightarrow & 0
\end{array}$$

By (4.13), we check that d maps $\mathcal{B}^{-1}$ into $I_\chi \oplus \mathfrak{m}^*\mathcal{B}^0\hat{\mathfrak{m}}$. So q induces a well-defined linear map $q : H^0(\mathcal{B}^\bullet, d) \to Q_\chi$. Then, q is the algebra isomorphism between $H^0(\mathcal{B}^\bullet, d)$ and $Q_\chi^{\mathfrak{m}}$ claimed in Theorem 4.2.7.

Remark 4.2.8 The BRST formulation of W-algebras works for an arbitrary good $\mathbb{Z}$-gradings without the assumption of "even". One needs to add an additional "neutral fermion" tensor factor to $\mathcal{B}^\bullet$ which corresponds to the presence of nonzero $\mathfrak{g}_{-1}$ (cf. [138, 81]). Again here, one has the flexibility of choosing a Lagrangian/isotropic subspace $\mathfrak{l}$ of $\mathfrak{g}_{-1}$ in defining the modified BRST complex.

Remark 4.2.9 There is yet another equivalent definition of finite W-algebras, due to Losev [157], via deformation quantization.

4.3 Quantization of the Slodowy slices

In this section, we will explain the independence of $\mathfrak{l}$ in the definition of finite W-algebras, denoted now by $\mathcal{W}_\mathfrak{l}$, associated to an isotropic subspace $\mathfrak{l} \subseteq \mathfrak{g}_{-1}$, following Gan and Ginzburg [107]. To that end, we explain the geometric picture behind the W-algebras in terms of Slodowy slices, due to Premet [194] (and generalized in [107]). Everything works in a general good $\mathbb{Z}$-grading setting, as remarked in the Introduction of [48].

4.3.1 The isotropic subspace definition of W-algebras. Fix an isotropic subspace $\mathfrak{l}$ of $\mathfrak{g}_{-1}$ with respect to $\langle\cdot,\cdot\rangle$, which means that $\langle \mathfrak{l}, \mathfrak{l}\rangle = 0$, and let

$$\mathfrak{l}' = \{x \in \mathfrak{g}_{-1} \mid \langle x, \mathfrak{l}\rangle = 0\}.$$

Clearly we have $\mathfrak{l} \subseteq \mathfrak{l}'$. In this section, we will introduce and study the generalized Gelfand-Graev modules and W-algebras (using notations $Q_\mathfrak{l}$ and $\mathcal{W}_\mathfrak{l}$ to emphasis the dependence on $\mathfrak{l}$) in such a setting.

Define the following nilpotent subalgebras of $\mathfrak{g}$ (where $\mathfrak{m} \subseteq \mathfrak{m}'$):

$$\mathfrak{m} = \mathfrak{l} \bigoplus \Big(\bigoplus_{i \le -2} \mathfrak{g}_i\Big) \qquad \text{and} \qquad \mathfrak{n} = \mathfrak{l}' \bigoplus \Big(\bigoplus_{i \le -2} \mathfrak{g}_i\Big).$$

Note that χ restricts to a character on $\mathfrak{m}$ (actually this is equivalent to the requirement that $\mathfrak{l}$ is isotropic in $\mathfrak{g}_{-1}$). Denote by $\mathbb{C}_\chi$ the corresponding 1-dimensional $U(\mathfrak{m})$-module, and let

$$Q_\mathfrak{l} = U(\mathfrak{g}) \bigotimes_{U(\mathfrak{m})} \mathbb{C}_\chi = U(\mathfrak{g})/I_\mathfrak{l}$$

be the induced $U(\mathfrak{g})$-module, where $I_{\mathfrak{l}}$ denotes the left ideal of $U(\mathfrak{g})$ generated by $a - \chi(a), \forall a \in \mathfrak{m}$. The same proof for Exercise 4.2.4 shows that $I_{\mathfrak{l}}$ is $\operatorname{ad}\mathfrak{n}$-invariant. Thus, there is an induced $\operatorname{ad}\mathfrak{n}$-action on $Q_{\mathfrak{l}}$.

Let

$$\mathcal{W}_{\mathfrak{l}} := (U(\mathfrak{g})/I_{\mathfrak{l}})^{\operatorname{ad}\mathfrak{n}} \equiv \{\bar{y} \in U(\mathfrak{g})/I_{\mathfrak{l}} \mid [a,y] \in I_{\mathfrak{l}}, \forall a \in \mathfrak{n}\}.$$

In the same spirit of Exercise 4.2.4, letting $\bar{y}_1\bar{y}_2 = \overline{y_1y_2}$, for $\bar{y}_1, \bar{y}_2 \in \mathcal{W}_{\mathfrak{l}}$, provides a well-defined algebra structure on $\mathcal{W}_{\mathfrak{l}}$.

Clearly, the above definition of $Q_{\mathfrak{l}}$ and $\mathcal{W}_{\mathfrak{l}}$ reduces to the earlier one when $\mathfrak{l}$ is Lagrangian.

Theorem 4.3.1 ([107]) *The algebras $\mathcal{W}_{\mathfrak{l}}$ are all isomorphic for different choices of isotropic (in particular, Lagrangian) subspaces $\mathfrak{l} \subseteq \mathfrak{g}_{-1}$.*

The proof of this theorem requires some preparations.

4.3.2 Γ-graded $\mathfrak{sl}_2$-triples.

Lemma 4.3.2 *Let $\Gamma : \mathfrak{g} = \oplus_{j\in\mathbb{Z}}\mathfrak{g}_j$ be a good $\mathbb{Z}$-grading for $0 \neq e \in \mathfrak{g}_2$. Then there exists $h \in \mathfrak{g}_0$ and $f \in \mathfrak{g}_{-2}$ such that $\{e,h,f\}$ form an $\mathfrak{sl}_2$-triple.*

This will be referred to as a Γ-*graded* $\mathfrak{sl}_2$-*triple* associated to e. This Γ-graded $\mathfrak{sl}_2$-triple will be used often.

Proof By Jacobson–Morozov Theorem in Section 4.1.3, there exists an $\mathfrak{sl}_2$-triple $\{e,h',f'\}$ in $\mathfrak{g}$. Denote by $h' = \sum_{j\in\mathbb{Z}} h_j$, $f' = \sum_{j\in\mathbb{Z}} f'_j$ the decompositions with respect to the $\mathbb{Z}$-grading Γ, and take $h = h_0 \in \mathfrak{g}_0$. Then $[h,e] = 2e$ and $h = [e, f'_{-2}] \in [e,\mathfrak{g}]$. Taking $\tilde{f}$ to be the degree -2 component of f'_{-2} with respect to $\operatorname{ad}h$, we obtain a new $\mathfrak{sl}_2$-triple $\{e,h,\tilde{f}\}$. Finally, taking f to be the degree -2 component of $\tilde{f}$ with respect to Γ, we obtain the desired $\mathfrak{sl}_2$-triple $\{e,h,f\}$. (Indeed, $\tilde{f} = f$ by Lemma 4.1.12; in other word, $\tilde{f}$ is already Γ-homogeneous.) □

Given a subspace V of $\mathfrak{g}$, we let

$$V^{\perp} = \{x \in \mathfrak{g} \mid (x|v) = 0, \forall v \in V\}, \quad V^{*,\perp} = \{\xi \in \mathfrak{g}^* \mid \xi(v) = 0, \forall v \in V\}.$$

Lemma 4.3.3 *Let $\{e,h,f\}$ be a Γ-graded $\mathfrak{sl}_2$-triple. Then, we have $\mathfrak{m}^{\perp} = [\mathfrak{n},e] \oplus \mathfrak{g}_f$.*

Proof We have properties (4.1)-(4.6) associated to e and their f-counterparts available. Now the lemma follows from the four facts below:

(i) $\mathfrak{m}^{\perp} \supseteq \mathfrak{g}_f$. This follows from $\mathfrak{m}^{\perp} \supseteq \mathfrak{g}_{\leq 0}$ by (4.5) and the f-counterpart to (4.4) which says that $\mathfrak{g}_f \subseteq \mathfrak{g}_{\leq 0}$.
(ii) $\mathfrak{m}^{\perp} \supseteq [\mathfrak{n},e]$. This can be seen by a direct computation: $(m|[m',e]) = ([m,m']|e) = 0$ for $m \in \mathfrak{m}$ and $m' \in \mathfrak{m}'$.
(iii) $[\mathfrak{n},e] \cap \mathfrak{g}_f = 0$. This follows by the $\mathfrak{sl}_2$ representation theory.
(iv) $\dim \mathfrak{m}^{\perp} = \dim \mathfrak{n} + \dim \mathfrak{g}_0 + \dim \mathfrak{g}_{-1} = \dim[\mathfrak{n},e] + \dim \mathfrak{g}_f$. This follows by the bijection $\mathfrak{n} \to [\mathfrak{n},e]$, $x \mapsto [x,e]$, by (4.2), and the f-counterpart to (4.6).

□

4.3.3 Some $\mathbb{C}^*$-actions. The embedding of the Γ-graded $\mathfrak{sl}_2$-triple in $\mathfrak{g}$ exponentiates to a rational homomorphism $\tilde{\gamma} : \mathrm{SL}_2(\mathbb{C}) \to G$. We put

$$\gamma : \mathbb{C}^* \to G, \quad t \mapsto \tilde{\gamma}(\mathrm{diag}\,(t, t^{-1})).$$

Note that $(\mathrm{Ad}\,\gamma(t))(e) = t^2 e$. The desired action of $\mathbb{C}^*$ on $\mathfrak{g}$, to be denoted by ρ, is defined by

$$\rho(t)(x) = t^2 \cdot \big(\mathrm{Ad}\,\gamma(t^{-1})\big)(x), \quad \forall t \in \mathbb{C}^*, x \in \mathfrak{g}.$$

Note that $\rho(t)(e + x) = e + \rho(t)(x)$. Thus, since $\rho(t)$ stabilizes $\mathfrak{g}_f$ and $\mathfrak{m}^\perp$, it also stabilizes $e + \mathfrak{g}_f$ and $e + \mathfrak{m}^\perp$, respectively. Note that $\lim_{t\to 0} \rho(t)(x) = e, \forall x \in e + \mathfrak{m}^\perp$, i.e. the $\mathbb{C}^*$-action on $e + \mathfrak{m}^\perp$ is contracting.

Let $\mathtt{M}'$ denote the closed subgroup of G whose Lie algebra is $\mathfrak{m}'$. Define a $\mathbb{C}^*$-action on $\mathtt{M}' \times (e + \mathfrak{g}_f)$ by

$$t \cdot (g, x) = (\gamma(t^{-1}) g \gamma(t), \rho(t)(x)).$$

Note that for any $(g, x) \in \mathtt{M}' \times (e + \mathfrak{g}_f)$, we have $\lim_{t\to 0} t \cdot (g, x) = (1, e)$.

4.3.4 Denote by $\kappa : \mathfrak{g} \to \mathfrak{g}^*$ the isomorphism induced by the non-degenerate bilinear form $(\cdot|\cdot)$. Following the terminology of Gan and Ginzburg, we will call

$$\mathcal{S} := \chi + \ker \mathrm{ad}^* f \equiv \kappa(e + \mathfrak{g}_f)$$

the *Slodowy slice* (through χ), or $e + \mathfrak{g}_f$ the *Slodowy slice* (through e). Note that $e + \mathfrak{g}_f$ is a transversal slice to the adjoint orbit through e.

Lemma 4.3.4 *The adjoint action map*

$$\alpha : \mathtt{M}' \times (e + \mathfrak{g}_f) \longrightarrow e + \mathfrak{m}^\perp$$

is an isomorphism of affine varieties.

Sketch of a proof The action map α is $\mathbb{C}^*$-equivariant.

By Lemma 4.3.3, the differential map of α is an isomorphism between the tangent spaces at the $\mathbb{C}^*$-fixed points $(1, e)$ and e.

Now, Lemma 4.3.4 follows from the following general nonsense: an equivariant morphism $\alpha : X_1 \to X_2$ of smooth affine $\mathbb{C}^*$-varieties with contracting $\mathbb{C}^*$-actions which induces an isomorphism between the tangent spaces of the $\mathbb{C}^*$-fixed points must be an isomorphism. □

4.3.5 The remainder of this section is to make sense the following commutative diagram:

$$\begin{array}{ccccc}
\mathrm{gr}\, U(\mathfrak{g}) & = & \mathbb{C}[\mathfrak{g}^*] & \xrightarrow{\sim} & \mathbb{C}[\mathfrak{g}] \\
\downarrow & & \downarrow & & \downarrow \\
\mathrm{gr}\, Q_{\mathfrak{l}} & = & \mathbb{C}[\chi + \mathfrak{m}^{*,\perp}] & \xrightarrow{\sim} & \mathbb{C}[e + \mathfrak{m}^\perp] \\
\uparrow & & \downarrow & & \downarrow \\
\mathrm{gr}\, \mathcal{W}_{\mathfrak{l}} & \xrightarrow[\nu]{\sim} & \mathbb{C}[\mathcal{S}] & \xrightarrow{\sim} & \mathbb{C}[e + \mathfrak{g}_f]
\end{array} \tag{4.14}$$

Remark 4.3.5 When reading (4.14), it is instructive to keep in mind the isomorphism $\mathbb{C}[e + \mathfrak{g}_f] \cong \mathbb{C}[e + \mathfrak{m}^\perp]^{\mathrm{Ad}\,\mathtt{M}'}$ by Lemma 4.3.4 and the identity $\mathcal{W}_{\mathfrak{l}} = Q_{\mathfrak{l}}^{\mathrm{ad}\,\mathfrak{m}'}$ by definition. Another way to rephrase this isomorphism is that $\mathcal{S}$ is obtained by the symplectic reduction for the coadjoint action of $\mathtt{M}'$ on $\mathfrak{g}^*$ at the point χ.

Some parts of the diagram (4.14) are easy to explain. The vertical arrows in the third column are restriction maps (e.g., $\mathfrak{g}_f \subseteq \mathfrak{m}^\perp$ by Lemma 4.3.3). The second column (which is more conceptual from the coadjoint orbit philosophy) is transferred from the (simpler) third one by the isomorphism $\kappa : \mathfrak{g} \to \mathfrak{g}^*$.

It remains to explain the first column (where $\operatorname{gr} A$ stands for the associated graded space/algebra for a Kazhdan-filtered space/algebra A) and its identification with the second one. See Section 4.3.6 below.

Granting the diagram (4.14), we can complete the proof of Theorem 4.3.1.

Proof of Theorem 4.3.1 Given two isotropic subspaces $\mathfrak{l}_1, \mathfrak{l}$ of $\mathfrak{g}_{-1}$ such that $\mathfrak{l}_1 \subseteq \mathfrak{l}$, we have a natural $U(\mathfrak{g})$-homomorphism $Q_{\mathfrak{l}_1} \to Q_{\mathfrak{l}}$, which gives rise to an algebra homomorphism $\mathcal{W}_{\mathfrak{l}_1} \to \mathcal{W}_{\mathfrak{l}}$. The associated graded map $\operatorname{gr} \mathcal{W}_{\mathfrak{l}_1} \to \operatorname{gr} \mathcal{W}_{\mathfrak{l}}$ is an algebra isomorphism, by Theorem 4.3.7. Hence the canonical map $\mathcal{W}_{\mathfrak{l}_1} \to \mathcal{W}_{\mathfrak{l}}$ is indeed an isomorphism.

Taking $\mathfrak{l}_1 = 0$ implies that the isoclass of the algebra $\mathcal{W}_{\mathfrak{l}}$ is independent of the choice of an isotropic subspace $\mathfrak{l}$. This completes the proof of the theorem. □

4.3.6 Kazhdan grading and filtration. Let $\{U_j(\mathfrak{g})\}_{j\leq 0}$ be the standard PBW filtration on $U(\mathfrak{g})$. Recall that h_Γ is introduced in Lemma 4.1.4. The action of $\operatorname{ad} h_\Gamma$ induces a grading on each $U_j(\mathfrak{g})$ by

$$U_j(\mathfrak{g})_i = \{x \in U_j(\mathfrak{g}) \mid \operatorname{ad} h_\Gamma(x) = ix\}.$$

The *Kazhdan filtration* on $U(\mathfrak{g})$ is defined by letting

$$F_n U(\mathfrak{g}) = \sum_{i+2k\leq n} U_k(\mathfrak{g})_i,$$

and it enjoys the following favorable properties:

(a) The canonical map $\operatorname{gr} U(\mathfrak{g}) \to S[\mathfrak{g}] = \mathbb{C}[\mathfrak{g}^*]$ is an isomorphism of graded commutative algebras. Indeed, for $x \in \mathfrak{g}_i, y \in \mathfrak{g}_j$, we have $x \in F_{i+2}U(\mathfrak{g}), y \in F_{j+2}U(\mathfrak{g})$, and $[x, y] \in F_{i+j+2}U(\mathfrak{g})$.

The Kazhdan grading on $\mathfrak{g}$ and so on $S(\mathfrak{g})$ (which is compatible with the one on $\operatorname{gr} U(\mathfrak{g})$) can be described directly.

(b) There is a Kazhdan filtration $\{F_n Q_{\mathfrak{l}}\}$ on $Q_{\mathfrak{l}} = U(\mathfrak{g})/I_{\mathfrak{l}}$ induced from $U(\mathfrak{g})$.
- $F_n Q_{\mathfrak{l}} = 0$ unless $n \geq 0$. Indeed, the generators $\{a - \chi(a) \mid \forall a \in \mathfrak{m}\}$ of the ideal $I_{\mathfrak{l}}$ contains all the negative-degree generators of $U(\mathfrak{g})$.
- $\operatorname{gr} Q_{\mathfrak{l}} = \operatorname{gr} U(\mathfrak{g})/ \operatorname{gr} I_{\mathfrak{l}}$ is a commutative $\mathbb{N}$-graded algebra.
- The ideal $\operatorname{gr} I_{\mathfrak{l}}$ in $\operatorname{gr} U(\mathfrak{g}) = \mathbb{C}[\mathfrak{g}^*]$ can be identified with the ideal of polynomial functions on $\mathfrak{g}^*$ which vanish on $\chi + \mathfrak{m}^{*,\perp}$.
- The canonical map $\operatorname{gr} Q_{\mathfrak{l}} \to \mathbb{C}[\chi + \mathfrak{m}^{*,\perp}]$ is an algebra isomorphism.

(c) There is an induced Kazhdan filtration on the subspace $\mathcal{W}_{\mathfrak{l}}$ of $Q_{\mathfrak{l}}$ such that $F_n \mathcal{W}_{\mathfrak{l}} = 0$ unless $n \geq 0$.

Remark 4.3.6 One could define the graded algebra structures on the second and third columns directly by some canonical $\mathbb{C}^*$-action on $\mathfrak{g}$ which preserves $e+\mathfrak{m}^\perp$ and $e + \mathfrak{g}_f$ (similar to Section 4.3.3), and then show that the horizontal maps between the last two columns in (4.14) are isomorphisms of graded algebras.

We have chosen not to discuss the canonical Poisson algebra structures for various graded algebras above (see [107] for details).

4.3.7 To complete the commutative diagram (4.14), we define the map $\nu : \operatorname{gr} \mathcal{W}_{\mathfrak{l}} \to \mathbb{C}[\mathcal{S}]$ to be the composite of the three natural maps

$$\operatorname{gr} \mathcal{W}_{\mathfrak{l}} \longrightarrow \operatorname{gr} Q_{\mathfrak{l}} \longrightarrow \mathbb{C}[\chi + \mathfrak{m}^{*,\perp}] \longrightarrow \mathbb{C}[\mathcal{S}].$$

The last piece for the completion of the diagram (4.14) is the following result (due to Premet [194] for Lagrangian $\mathfrak{l}$ and Gan-Ginzburg [107] for isotropic $\mathfrak{l}$), whose proof will be postponed to the next section.

Theorem 4.3.7 *The map $\nu : \operatorname{gr} \mathcal{W}_{\mathfrak{l}} \to \mathbb{C}[\mathcal{S}]$ is an isomorphism of [graded] algebras.*

4.3.8 We have the following fundamental theorem of Kostant [150]. Here we follow the approach of Premet [194].

Theorem 4.3.8 *Let e be a regular nilpotent element in $\mathfrak{g}$. Then $\mathcal{W}_{\mathfrak{l}} \cong \mathcal{Z}(\mathfrak{g})$, the center of $U(\mathfrak{g})$.*

Sketch of a proof The algebra of invariants $S(\mathfrak{g}^*)^G$ is known to be a polynomial algebra in $r = \operatorname{rank}(\mathfrak{g})$ variables. Let $f_1, \ldots, f_r$ denote the algebraically independent homogenous generators of $S(\mathfrak{g}^*)^G$, with $\deg f_i = m_i + 1$, where m_i denotes the exponents of $\mathfrak{g}$. Since the associated graded algebra of $\mathcal{Z}(\mathfrak{g})$ with respect to the PBW filtration $\{U_j(\mathfrak{g})\}_{j \geq 0}$ is isomorphic to $S(\mathfrak{g}^*)^G$ (where we have identified $\mathfrak{g} \cong \mathfrak{g}^*$), we can choose lifts $\tilde{f}_i \in U_{m_i+1}(\mathfrak{g}) \cap \mathcal{Z}(\mathfrak{g})$ of each f_i.

Restricting the adjoint quotient map $\mathfrak{g} \to \mathfrak{g}//G = \mathbb{C}^r$, $x \mapsto (f_1(x), \ldots, f_r(x))$ to the Slodowy slice gives rise to a $\mathbb{C}^*$-equivariant isomorphism of affine varieties from $e + \mathfrak{g}_f$ to $\mathbb{C}^r$ [150, 210], where the $\mathbb{C}^*$-action $\tilde{\rho}$ on $e + \mathfrak{g}_f$ is such that $\tilde{\rho}(t^2) = \rho(t)$ for $t \in \mathbb{C}^*$. Hence, $\operatorname{gr} \mathcal{W}_{\mathfrak{l}}$ is generated by (the restrictions of) f_i for $1 \leq i \leq r$.

One can show that the restriction of $U(\mathfrak{g}) \to \operatorname{End}_{\mathbb{C}}(Q_{\mathfrak{l}})$ to the center $\mathcal{Z}(\mathfrak{g})$ is always injective, hence we obtain a monomorphism $\mathcal{Z}(\mathfrak{g}) \to \mathcal{W}_{\mathfrak{l}} = \operatorname{End}_{\mathfrak{g}}(Q_{\mathfrak{l}})$ by (4.7), whose associated graded map is an isomorphism. If follows from a filtered algebra argument that this has to be an isomorphism. □

Remark 4.3.9 The W-algebras associated to subregular nilpotent elements are particularly interesting, as it can be viewed as a noncommutative deformation of the simple singularities. See Premet [194] and Gordon-Rumynin [116] for details.

4.4 An equivalence of categories

In this section, we formulate an equivalence of categories due to Skryabin. The proof of this category equivalence here, due to [107], uses similar ideas of the proof of Theorem 4.3.7, which we will first complete.

4.4.1 Proof of Theorem 4.3.7. Recall that $\mathfrak{n}$ is graded with respect to the grading Γ. We view $U(\mathfrak{g})$ and $Q_{\mathfrak{l}}$ as a $\mathfrak{n}$-modules via the adjoint $\mathfrak{n}$-action. Then, $U(\mathfrak{g})$ and $Q_{\mathfrak{l}}$ are Kazhdan-filtered $\mathfrak{n}$-modules and the canonical map $p : U(\mathfrak{g}) \to Q_{\mathfrak{l}}$ is $\mathfrak{n}$-module homomorphism. Thus, $\operatorname{gr} U(\mathfrak{g})$ and $\operatorname{gr} Q_{\mathfrak{l}}$ are Kazhdan-graded $\mathfrak{n}$-modules, and $\operatorname{gr}\ p : \operatorname{gr} U(\mathfrak{g}) \to \operatorname{gr} Q_{\mathfrak{l}}$ is an $\mathfrak{n}$-module homomorphism.

Note that $\mathcal{W}_{\mathfrak{l}} = H^0(\mathfrak{n}, Q_{\mathfrak{l}})$, the 0th Lie algebra cohomology of $\mathfrak{n}$ with coefficient in $Q_{\mathfrak{l}}$. We reformulate and prove Theorem 4.3.7 as follows.

Theorem 4.4.1 ([107]) *The map $\nu : \operatorname{gr} \mathcal{W}_{\mathfrak{l}} \to \mathbb{C}[\mathcal{S}]$ is equal to the composite $\nu_2 \nu_1$:*

$$\operatorname{gr} H^0(\mathfrak{n}, Q_{\mathfrak{l}}) \xrightarrow{\nu_1} H^0(\mathfrak{n}, \operatorname{gr} Q_{\mathfrak{l}}) \xrightarrow{\nu_2} \mathbb{C}[\mathcal{S}]$$

where ν_1 *and* ν_2 *are isomorphisms. Moreover,* $H^i(\mathfrak{n}, Q_\mathfrak{l}) = H^i(\mathfrak{n}, \mathrm{gr}\, Q_\mathfrak{l}) = 0$ *for* $i > 0$.

Proof By (b) of 4.3.6 and Lemma 4.3.4, we obtain isomorphisms of vector spaces

$$\mathrm{gr}\, Q_\mathfrak{l} \cong \mathbb{C}[\chi + \mathfrak{m}^{*,\perp}] \cong \mathbb{C}[\mathtt{M}'] \otimes \mathbb{C}[\mathcal{S}].$$

These isomorphisms are actually on the level of $\mathfrak{n}$-modules, where the $\mathfrak{n}$-module structure on the third space comes from the $\mathfrak{n}$–adjoint action on its first tensor factor $\mathbb{C}[\mathtt{M}']$. Now

$$H^i(\mathfrak{n}, \mathbb{C}[\mathtt{M}']) = \delta_{i,0}\mathbb{C}$$

since the standard cochain complex for Lie algebra cohomology with coefficients in $\mathbb{C}[\mathtt{M}']$ is just the algebraic de Rham complex for $\mathtt{M}'$ which admits trivial cohomology for an affine space such as $\mathtt{M}'$. Putting all things together, we have proved that ν_2 is an isomorphism and that $H^i(\mathfrak{n}, \mathrm{gr}\, Q_\mathfrak{l}) = 0$ for $i > 0$.

Recall from (b) in 4.3.6 that the Kazhdan-filtration on $Q_\mathfrak{l}$ has no negative-degree component. Note in addition that $\mathfrak{n}$ is a negatively graded subalgebra of $\mathfrak{g}$ with respect to the grading Γ, and so its dual $\mathfrak{n}^*$ is positively graded (with respect to Γ). We write this graded decomposition as $\mathfrak{n}^* = \bigoplus_{i \geq 1} \mathfrak{n}_i^*$.

Consider the standard cochain complex for computing the $\mathfrak{n}$-cohomology of $Q_\mathfrak{l}$:

$$0 \longrightarrow Q_\mathfrak{l} \longrightarrow \mathfrak{n}^* \otimes Q_\mathfrak{l} \longrightarrow \ldots \longrightarrow \wedge^k \mathfrak{n}^* \otimes Q_\mathfrak{l} \longrightarrow \ldots. \tag{4.15}$$

A filtration on $\wedge^k \mathfrak{n}^* \otimes Q_\mathfrak{l}$ is defined by letting $F_p(\wedge^k \mathfrak{n}^* \otimes Q_\mathfrak{l})$ be the subspace of $\wedge^k \mathfrak{n}^* \otimes Q_\mathfrak{l}$ spanned by $(x_1 \wedge \ldots \wedge x_k) \otimes v$, for all $x_1 \in \mathfrak{n}_{i_1}^*, \ldots, x_k \in \mathfrak{n}_{i_k}^*$ and $v \in F_j Q_\mathfrak{l}$ such that $i_1 + \ldots + i_n + j \leq p$. This defines a filtered complex structure on (4.15).

Taking the associated graded complex of (4.15) gives the standard cochain complex for computing the $\mathfrak{n}$-cohomology of $\mathrm{gr}\, Q_\mathfrak{l}$.

Now consider the spectral sequence with

$$E_0^{p,q} = F_p(\wedge^{p+q}\mathfrak{n}^* \otimes Q_\mathfrak{l})/F_{p-1}(\wedge^{p+q}\mathfrak{n}^* \otimes Q_\mathfrak{l}).$$

Then $E_1^{p,q} = H^{p+q}(\mathfrak{n}, \mathrm{gr}_p\, Q_\mathfrak{l})$, and the spectral sequence converges to $E_\infty^{p,q} = F_p H^{p+q}(\mathfrak{n}, Q_\mathfrak{l})/F_{p-1}H^{p+q}(\mathfrak{n}, Q_\mathfrak{l})$. The remaining parts of Theorem 4.4.1 follow from this and the parts about $\mathrm{gr}\, Q_\mathfrak{l}$ established above. □

4.4.2 The Whittaker functor. As we have established the independence of the W-algebras from the choice of isotropic subspaces $\mathfrak{l}$, we will switch the notations for the generalized Gelfand-Graev module and the W-algebra back to $Q_\chi, \mathcal{W}_\chi$, to emphasize the crucial dependence on χ.

In the remainder of this section, we will set up the connections between W-algebras and the category of Whittaker modules. To that end, we shall fix an nilpotent element e (and hence χ) and a *Lagrangian* subspace $\mathfrak{l}$ of $\mathfrak{g}_{-1}$ once for all.

Definition 4.4.2 A $\mathfrak{g}$-module E is called a *Whittaker module* if $a - \chi(a)$, $\forall a \in \mathfrak{m}$, acts on E locally nilpotently. A *Whittaker vector* in a Whittaker $\mathfrak{g}$-module E is a vector $x \in E$ which satisfies $(a - \chi(a))x = 0, \forall a \in \mathfrak{m}$.

Let $\mathfrak{g}\text{-}\mathtt{Wmod}^\chi$ be the category of finitely generated Whittaker $\mathfrak{g}$-modules.

Denote the subspace of all Whittaker vectors in E by

$$\mathrm{Wh}(E) = \{v \in E \mid (a - \chi(a))v = 0, \forall a \in \mathfrak{m}\}.$$

Recall the second definition $\mathcal{W}_\chi = (U(\mathfrak{g})/I_\chi)^{\mathrm{ad}\,\mathfrak{m}}$, and we denote by $\bar{y} \in U(\mathfrak{g})/I_\chi$ the coset associated to $y \in U(\mathfrak{g})$.

Lemma 4.4.3 *(i) Given a Whittaker $\mathfrak{g}$-module E with an action map ϱ, $Wh(E)$ is naturally a $\mathcal{W}_\chi$-module by letting $\bar{y}.v = \varrho(y)v$ for $v \in Wh(E)$ and $\bar{y} \in \mathcal{W}_\chi$.*

(ii) For $V \in \mathcal{W}_\chi$-$\mathtt{mod}$, $Q_\chi \otimes_{\mathcal{W}_\chi} V$ is a Whittaker $\mathfrak{g}$-module by letting

$$y.(q \otimes v) = (y.q) \otimes v, \qquad \text{for } y \in U(\mathfrak{g}), q \in Q_\chi = U(\mathfrak{g})/I_\chi, v \in V.$$

Proof (1) Let $v \in \mathrm{Wh}(E)$. The formula $\bar{y}.v = \varrho(y)v$ is well-defined since $\varrho(y)v = 0$ for all $y \in I_\chi$ and $v \in \mathrm{Wh}(E)$. Being $\mathrm{ad}\,\mathfrak{m}$-invariants, $\bar{y} \in \mathcal{W}_\chi$ satisfies $[\varrho(y), a] \in I_\chi$ and so $\bar{y}.v \in \mathrm{Wh}(E)$. It follows from the $\mathfrak{g}$-module homomorphism ϱ that the formula defines an action of $\mathcal{W}_\chi$.

(2) When we use the first definition $\mathcal{W}_\chi = \mathrm{End}_{U(\mathfrak{g})}(Q_\chi)^{\mathrm{op}}$, it is trivial to see that Q_χ is a $(\mathfrak{g}, \mathcal{W}_\chi)$-bimodule and then $Q_\chi \otimes_{\mathcal{W}_\chi} V$ is a $\mathfrak{g}$-module.

To check that $Q_\chi \otimes_{\mathcal{W}_\chi} V$ lies in $\mathfrak{g}$-$\mathtt{Wmod}^\chi$, it suffices to check that $a - \chi(a)$, $\forall a \in \mathfrak{m}$ acts locally nilpotently on Q_χ by the definition of the $U(\mathfrak{g})$-action on $Q_\chi \otimes_{\mathcal{W}_\chi} V$. Since $\mathfrak{m}$ is negatively graded with respect to the grading Γ, $\mathrm{ad}\, a = \mathrm{ad}(a - \chi(a))$ for any $a \in \mathfrak{m}$ acts locally nilpotently on $\mathfrak{g}$ and so locally nilpotently on $U(\mathfrak{g})$ (and also on $Q_\chi = U(\mathfrak{g})/I_\chi$) by induction on the PBW filtration length for $U(\mathfrak{g})$. □

Let $\mathcal{W}_\chi$-$\mathtt{mod}$ be the category of finitely generated $\mathcal{W}_\chi$-modules. We define the *Whittaker functor*

$$\mathrm{Wh} : \mathfrak{g}\text{-}\mathtt{Wmod}^\chi \longrightarrow \mathcal{W}_\chi\text{-}\mathtt{mod}, \qquad E \mapsto \mathrm{Wh}(E).$$

We define another functor

$$Q_\chi \otimes_{\mathcal{W}_\chi} - : \mathcal{W}_\chi\text{-}\mathtt{mod} \longrightarrow \mathfrak{g}\text{-}\mathtt{Wmod}^\chi, \qquad V \mapsto Q_\chi \otimes_{\mathcal{W}_\chi} V.$$

4.4.3 The Skryabin equivalence. The following theorem is due to Skryabin [208], and here we follow the new proof of [107].

Theorem 4.4.4 *The functor $Q_\chi \otimes_{\mathcal{W}_\chi} - : \mathcal{W}_\chi$-$\mathtt{mod} \longrightarrow \mathfrak{g}$-$\mathtt{Wmod}^\chi$ is an equivalence of categories, with $Wh : \mathfrak{g}$-$\mathtt{Wmod}^\chi \longrightarrow \mathcal{W}_\chi$-$\mathtt{mod}$ as its quasi-inverse.*

Proof (1) We first prove that $\mathrm{Wh}(Q_\chi \otimes_{\mathcal{W}_\chi} V) = V$.

Assume V is generated as a $\mathcal{W}_\chi$-module by a finite-dimensional subspace V_0. This gives rise to a $\mathcal{W}_\chi$-filtered module structure on V by letting $F_n V = (F_n \mathcal{W}_\chi).V_0$. Note that $H^0(\mathfrak{m}, Q_\chi \otimes_{\mathcal{W}_\chi} V) = \mathrm{Wh}(Q_\chi \otimes_{\mathcal{W}_\chi} V)$, where we regard $Q_\chi \otimes_{\mathcal{W}_\chi} V$ as an $\mathfrak{m}$-module with χ-twisted action.

We shall establish the following Claim in Lie algebra cohomology; this readily implies that $H^0(\mathfrak{m}, Q_\chi \otimes_{\mathcal{W}_\chi} V) \cong V$, which is equivalent to $\mathrm{Wh}(Q_\chi \otimes_{\mathcal{W}_\chi} V) = V$.

Claim 1. $\mathrm{gr}\, H^0(\mathfrak{m}, (Q_\chi \otimes_{\mathcal{W}_\chi} V)) \overset{\mu_1}{\simeq} H^0(\mathfrak{m}, \mathrm{gr}(Q_\chi \otimes_{\mathcal{W}_\chi} V)) \overset{\mu_2}{\simeq} \mathrm{gr}\, V$; moreover, $H^i(\mathfrak{m}, (Q_\chi \otimes_{\mathcal{W}_\chi} V)) = H^i(\mathfrak{m}, \mathrm{gr}(Q_\chi \otimes_{\mathcal{W}_\chi} V)) = 0$ for $i > 0$.

The two isomorphisms μ_1, μ_2 in the Claim are parallel to the two isomorphisms ν_1, ν_2 in Theorem 4.4.1, and they will be proved in the same strategy as in the proof of Theorem 4.4.1.

By Lemma 4.3.4 and the diagram (4.14), we have

$$\begin{aligned}\mathrm{gr}(Q_\chi \otimes_{\mathcal{W}_\chi} V) &\cong \mathrm{gr}\, Q_\chi \otimes_{\mathrm{gr}\,\mathcal{W}_\chi} \mathrm{gr}\, V \\ &\cong (\mathbb{C}[\mathtt{M}'] \otimes \mathrm{gr}\, \mathcal{W}_\chi) \otimes_{\mathrm{gr}\,\mathcal{W}_\chi} \mathrm{gr}\, V \cong \mathbb{C}[\mathtt{M}'] \otimes \mathrm{gr}\, V.\end{aligned}$$

The isomorphism μ_2 follows quickly from this.

The isomorphism μ_1 follows exactly as the argument for the isomorphism ν_1 in the proof of Theorem 4.4.1, once we note that $Q_\chi \otimes_{\mathcal{W}_\chi} V$ is filtered with no negative-degree components and $\mathfrak{m}^*$ is $\mathbb{N}$-graded, and then apply a spectral sequence argument.

(2) We shall show that, for any $E \in \mathfrak{g}\text{-}\mathtt{Wmod}^{\chi}$, the canonical map

$$\gamma : Q_\chi \otimes_{\mathcal{W}_\chi} \mathrm{Wh}(E) \longrightarrow E, \quad \bar{y} \otimes v \mapsto y.v$$

is an isomorphism, where $\bar{y} \in Q_\chi = U(\mathfrak{g})/I_\chi$ is associated to $y \in U(\mathfrak{g})$.

We first note that γ is well-defined since $Q_{\mathfrak{l}} = U(\mathfrak{g})/I_\chi$ and I_χ is generated by $a - \chi(a), \forall a \in \mathfrak{m}$. Also observe that $\mathrm{Wh}(E) = 0$ implies that $E = 0$.

Take an exact sequence of $\mathfrak{g}$-modules:

$$0 \longrightarrow E' \longrightarrow Q_\chi \otimes_{\mathcal{W}_\chi} \mathrm{Wh}(E) \xrightarrow{\gamma} E \longrightarrow E'' \longrightarrow 0. \tag{4.16}$$

Let us show that γ is injective (i.e. $E' = 0$). Indeed,

$$\mathrm{Wh}(E') = E' \cap \mathrm{Wh}(Q_\chi \otimes_{\mathcal{W}_\chi} \mathrm{Wh}(E)) = E' \cap \mathrm{Wh}(E) = 0$$

where the second equality follows from Part (1) above and the third equality follows by definition of γ. Hence $E' = 0$.

Now we prove that γ is surjective (i.e. $E'' = 0$). (4.16) reduces now to a short exact sequence

$$0 \longrightarrow Q_\chi \otimes_{\mathcal{W}_\chi} \mathrm{Wh}(E) \xrightarrow{\gamma} E \longrightarrow E'' \longrightarrow 0$$

which gives rise to a long exact sequence

$$0 \to H^0(\mathfrak{m}, Q_\chi \otimes_{\mathcal{W}_\chi} \mathrm{Wh}(E)) \xrightarrow{\gamma^*} H^0(\mathfrak{m}, E) \to H^0(\mathfrak{m}, E'') \to H^1(\mathfrak{m}, Q_\chi \otimes_{\mathcal{W}_\chi} \mathrm{Wh}(E))$$

where $H^1\big(\mathfrak{m}, Q_\chi \otimes_{\mathcal{W}_\chi} \mathrm{Wh}(E)\big) = 0$ by Claim 1. Then, this exact sequence can be rewritten as

$$0 \longrightarrow \mathrm{Wh}(Q_\chi \otimes_{\mathcal{W}_\chi} \mathrm{Wh}(E)) \xrightarrow{\gamma^*} \mathrm{Wh}(E) \longrightarrow \mathrm{Wh}(E'') \longrightarrow 0$$

By (1), $\mathrm{Wh}(Q_\chi \otimes_{\mathcal{W}_\chi} \mathrm{Wh}(E)) = \mathrm{Wh}(E)$ and γ^* is an isomorphism. So, $\mathrm{Wh}(E'') = 0$, whence $E'' = 0$. □

4.5 Good $\mathbb{Z}$-gradings in type A

In this section, we will describe the classification of good $\mathbb{Z}$-gradings on $\mathfrak{g} = \mathfrak{gl}_N$ or $\mathfrak{sl}_N$, due to Elashvili and Kac [92].

4.5.1 Pyramids of shape λ. Given a partition $\lambda = (\lambda_1, \lambda_2, \ldots)$ of N, we construct a combinatorial object, called pyramids (of shape λ). We will illustrate the process by building pyramids of shape $\lambda = (3, 2, 2)$.

We start with a (first) row of $\lambda_1 = 3$ boxes of size 2 units by 2 units, with column numbers $1 - \lambda_1, 3 - \lambda_1, \cdots \lambda_1 - 1$ (which is $-2, 0, 2$ in this example). We mark by $\bullet$ to indicate the column number 0.

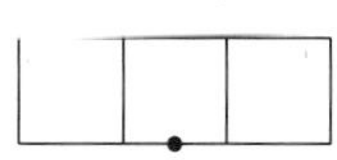

Then, we add a (second) row of $\lambda_2 = 2$ boxes on top of the row 1. The rule is: keep the stair shape with permissible shifts by integer units. In this example, we have three possibilities, and this gives rise to 3 pyramids of shape $(3, 2)$.

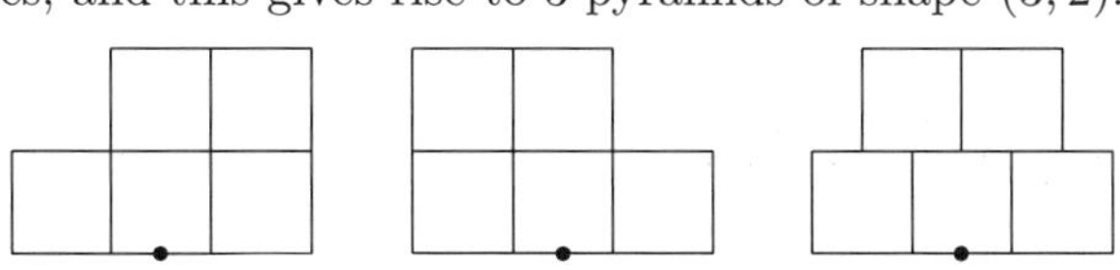

Then we repeat the process with the same rule by adding now a (third) row of $\lambda_3 = 2$ boxes on top of row 2. In this example, we have only one permissible way of doing so, and so obtain three pyramids of shape $(3, 2, 2)$ below (where the column numbers are also indicated).

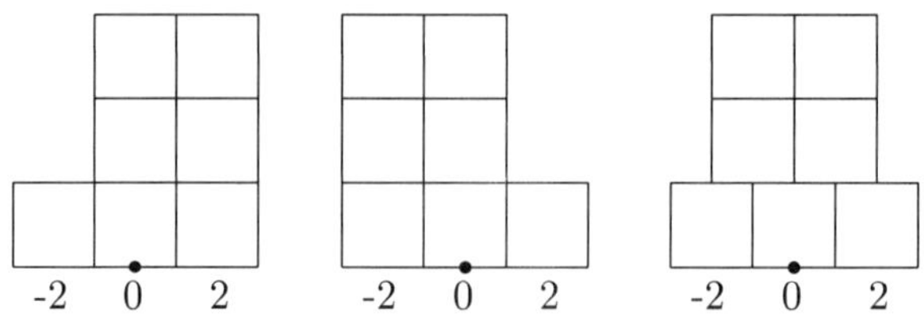

Exercise 4.5.1 There are 7 pyramids of shape $(4, 1)$ as there are 7 permissible way of putting one box on top of

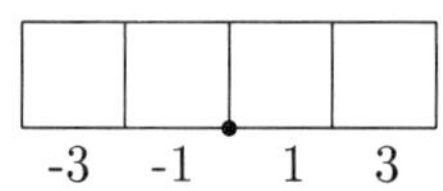

4.5.2 Given a pyramid P of shape λ, let us fix a labeling by $\{1, 2, \ldots, N\}$ of the N boxes in P. A convenient choice is to label downward from left to right in an increasing order.

Let us take the second pyramid of shape $(3, 2, 2)$ as an example with $N = 7$. Our labeled pyramid reads

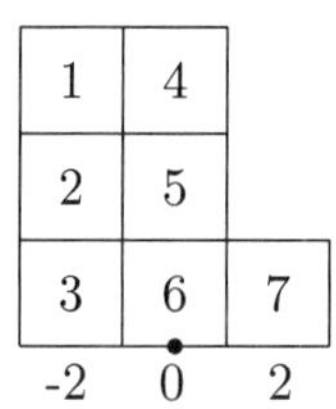

We fix a standard basis $v_i (1 \leq i \leq N)$ of $\mathbb{C}^N$ associated to the above labelling. Let

$$e = e^P = E_{14} + E_{25} + E_{36} + E_{67}$$

be the nilpotent element in $\mathfrak{g}$ which sends a vector v_i to $v_{\mathtt{R}(i)}$ where $\mathtt{R}(i)$ denotes the label to the right of i in the labelled pyramid P (by convention $v_{\mathtt{R}(i)} = 0$ whenever $\mathtt{R}(i)$ is not defined). Note that e has J_λ as its Jordan form.

A $\mathbb{Z}$-grading Γ^P of $\mathfrak{g}$ is determined by letting the degree of a root vector for a simple root $\alpha_i = \varepsilon_i - \varepsilon_{i+1}$ to be $\mathrm{col}_{i+1} - \mathrm{col}_i$, where col_i denotes the column number of the box labelled by i in P. In other words, the grading operator is $h^P = h^P_\partial = -\mathrm{diag}\,(\mathrm{col}_1, \mathrm{col}_2, \ldots, \mathrm{col}_N)$ (note that h^P is unique up to a shift of a scalar multiple of the identity matrix I_N, and there is a unique shift which results to a traceless matrix).

The grading Γ is even if and only if a pyramid is *even* in the sense that every box in P lies immediately on top of at most one box. For example, for $\lambda = (3, 2, 2)$ above, the first and second pyramids are even while the third one is not.

Example 4.5.2 Let $\mathfrak{g} = \mathfrak{gl}_3$ and $e = E_{13}$. Then the good grading for e in Example 4.1.7 (2) corresponds to the first pyramid and the Dynkin grading for e corresponds to the third pyramid below:

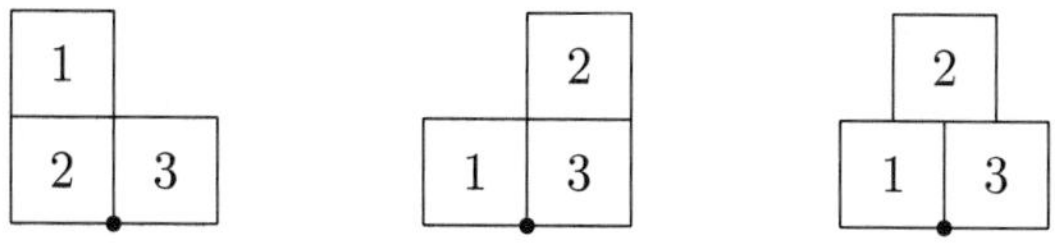

Exercise 4.5.3 Describe explicitly the good $\mathbb{Z}$-grading corresponding to the second pyramid.

Theorem 4.5.4 ([92]) *Let $\mathfrak{g} = \mathfrak{gl}_N$ or $\mathfrak{sl}_N$, and let λ be a partition of N. There exists a one-to-one correspondence between the set of pyramids of shape λ and the set of good $\mathbb{Z}$-gradings for a nilpotent matrix of Jordan shape λ up to GL_N-conjugation,*

$$\{\text{Pyramids of shape } \lambda\} \longrightarrow \{\text{good } \mathbb{Z}\text{-gradings for a nilpotent matrix in } \mathcal{O}_\lambda\}/GL_N$$

by sending P to Γ^P.

Proof It is easy to see the map is well-defined: different choices of labeling of P leads to another nilpotent matrix in the same nilpotent orbit.

To see that Γ^P is a good $\mathbb{Z}$-grading for $e = e^P$, we will actually construct a $\mathbb{N}$-graded homogeneous basis for the centralizer $\mathfrak{g}_e$ explicitly. Let $\ell = \lambda'_1$. We will choose to label the first box of row $1, \dots, \ell$ of P upward by $1, \dots, \ell$. Note that $\{e^{k_i} v_i \mid 1 \le i \le \ell, 0 \le k_i \le \lambda_i - 1\}$ is a linear basis of $\mathbb{C}^N$, and indeed each such $e^{k_i} v_i$ is equal to v_j for some $1 \le j \le N$ which appears in the same row as i in P.

An element z in $\mathfrak{g}_e$ is determined by the values $z(v_i)$ for $1 \le i \le \ell$, since $z(e^k v_i) = e^k z(v_i)$. Consider for now $z_{j,i;k_j} \in \mathfrak{g}_e (1 \le i, j \le \ell)$ such that $z_{j,i;k_j}(v_i) = e^{k_j} v_j$ and $z_{j,i;k_j}(v_{i'}) = 0$ for $1 \le i' \le \ell, i' \ne i$. Then since $e^{\lambda_i + k_j} v_j = z_{j,i;k_j}(e^{\lambda_i} v_i) = 0$ and recall $0 \le k_j < \lambda_j$, we see that the k_j's have to satisfy the inequality:

$$\lambda_j > k_j \ge \max(0, \lambda_j - \lambda_i), \quad \forall i, j$$

and this condition is sufficient for $z_{j,i;k_j} \in \mathfrak{g}_e$ to be well-defined.

Since $\min(\lambda_i, \lambda_j) = \lambda_j - \max(0, \lambda_j - \lambda_i)$, there are in total $\sum_{1 \le i,j \le \ell} \min(\lambda_i, \lambda_j)$ of such $z_{j,i;k_j} \in \mathfrak{g}_e$. These elements $z_{j,i;k_j}$'s in $\mathfrak{g}_e$ are manifestly homogeneous, $\mathbb{N}$-graded and linearly independent. Hence, they must form a basis for $\mathfrak{g}_e$, thanks to the well-known identity $\dim \mathfrak{g}_e = \sum_{1 \le i,j \le \ell} \min(\lambda_i, \lambda_j)$ (which is also equal to $\sum_{i \ge 1} (\lambda'_i)^2$ in terms of the conjugate partition λ').

Now we shall construct an inverse map. Let e be a nilpotent element in the nilpotent orbit $\mathcal{O}_\lambda$ and let $\Gamma = \operatorname{ad} h$ be a good $\mathbb{Z}$-grading for e. Up to GL_N-conjugation, we can assume that $e = J_\lambda$, the canonical Jordan canonical form, and that h is a diagonal matrix. We can suitably relabel the standard basis vectors v_i's of $\mathbb{C}^N$ (or by a conjugation by some permutation matrix in GL_N), so that the standard basis of $\mathbb{C}^N$ consists of $\{e^{k_i} v_i \mid 1 \le i \le \ell, 0 \le k_i \le \lambda_i - 1\}$. In particular, E_{ji} for all i, j are homogeneous with respect to the grading Γ. Now we visualize a labeled "generalized" pyramid P of shape λ as follows (here generalized means that the staircase conditions on both sides of P are not verified at the moment): the row one of P is fixed with column numbers $1 - \lambda_1, 3 - \lambda_1, \cdots \lambda_1 - 1$ with the leftmost box labelled by 1, and the remaining rows are fixed by letting the leftmost box of row j to have a column number $\deg E_{j1} + 1 - \lambda_1$.

Take the basis $z_{j,i;k_j}$ for $\mathfrak{g}_e$ constructed earlier. The good grading Γ implies that $z_{j,i;0} = E_{ji} + \dots$ have non-negative integral degrees, and so do E_{ji}, for $1 \le j < i \le \ell$. This ensures that the left-hand side of the generalized pyramid P is of staircase shape. Similarly, the consideration of $z_{j,i;\lambda_j-\lambda_i}$ for $1 \leq j < i \leq \ell$ shows the right-hand side of P is of staircase shape.

Hence, P is indeed a pyramid. It follows by the construction of P that $e = e^P$. □

Corollary 4.5.5 *There exists a bijection between even pyramids of size N and even good $\mathbb{Z}$-gradings of $\mathfrak{g}$ up to GL_N-conjugation (by sending P to Γ^P).*

Proof Follows from the way a grading is defined using a pyramid. □

Corollary 4.5.6 *Given a nilpotent element $e \in \mathfrak{g}$, there exists an even good $\mathbb{Z}$-grading of $\mathfrak{g}$ for e.*

Proof Assume the Jordan form of e corresponds to a partition λ. The Young diagram λ in French fashion is an even pyramid of shape λ. □

A pyramid P is call *symmetric* if P is self-dual with respect to the reflection along the vertical line passing through the mid-point of row one of P. For example, the third pyramid of shape (3,2,2) in Section 4.5.1 is symmetric.

Corollary 4.5.7 *There exists a bijection from the set of symmetric pyramids of size N to the set of Dynkin gradings of $\mathfrak{g}$ up to GL_N-conjugation (by sending P to Γ^P).*

Proof Let P be a symmetric pyramid of size N. Then the good $\mathbb{Z}$-grading Γ^P of $\mathfrak{g}$ is Dynkin, since we can take $\mathfrak{sl}_2$ to be the diagonal subalgebra of the direct sum of the $\mathfrak{sl}_2$'s associated to each row of P (which corresponds to a Jordan block) as constructed in Example 4.1.2.

Any two Dynkin gradings for a given nilpotent element e are conjugated by GL_N, as we have seen in the course of establishing the bijection Ω in 4.1.6. □

Remark 4.5.8 The classification of good $\mathbb{Z}$-gradings for other types was also carried out in [92]. See Baur and Wallach [19] for closely related classification of nice parabolic subalgebras and further clarification.

4.6 W-algebras and independence of good gradings

In this section, we shall prove that the isoclasses of finite W-algebras are independent of the choices of the good $\mathbb{R}$-gradings Γ, following Brundan and Goodwin [46].

4.6.1 In this subsection, we fix notations which will be used throughout Section 4.6.

Let $e \in \mathfrak{g}$ be a nilpotent element. Fix an $\mathfrak{sl}_2$-triple $\{e, h, f\}$ in $\mathfrak{g}$, and denote this copy of $\mathfrak{sl}_2$ by $\mathfrak{s}$.

Recall that the ad h-eigenspace decomposition of $\mathfrak{g}$ defines the Dynkin $\mathbb{Z}$-grading

$$\Gamma^{\emptyset} : \mathfrak{g} = \bigoplus_{j \in \mathbb{Z}} \mathfrak{g}_j, \qquad \text{where } \mathfrak{g}_0 = \mathfrak{g}_h.$$

Let $\mathfrak{c} = \mathfrak{g}_h$ and let C be the corresponding closed connected subgroup of the adjoint group G. Also let $\mathfrak{r} = \bigoplus_{j>0} \mathfrak{g}_j$ (associated to $\Gamma^{\emptyset}$) and let R be the corresponding closed connected subgroup of G. It is well known that C_e is a maximal reductive subgroup of G_e, with Lie algebra $\mathfrak{c}_e$, and that R_e is the unipotent radical of G_e, with Lie algebra $\mathfrak{r}_e$.

Fix a maximal torus T of G contained in C and containing a maximal torus of C_e. An important role is played by the centralizer $\mathfrak{t}_e$ of e in the Lie algebra $\mathfrak{t}$ of T. It is a Cartan subalgebra of the reductive part $\mathfrak{c}_e$ of the centralizer $\mathfrak{g}_e$.

4.6.2 An $\mathbb{R}$-grading

$$\Gamma : \mathfrak{g} = \bigoplus_{j \in \mathbb{R}} \mathfrak{g}_j$$

of $\mathfrak{g}$ is called a *good grading* (or *good $\mathbb{R}$-grading*) for e if the conditions (4.1)-(4.3) in 4.1.4 holds (where the indices are taken in $\mathbb{R}$). Lemma 4.1.8 (i.e. (4.2) $\Leftrightarrow$ (4.3)) remains to be valid for $\mathbb{R}$-gradings.

We can associate to any $\mathbb{R}$-grading a semisimple element h_Γ, as in Lemma 4.1.4.

Definition 4.6.1 We say that a grading $\Gamma : \mathfrak{g} = \bigoplus_{j \in \mathbb{R}} \mathfrak{g}_j$ is *compatible with* $\mathfrak{t}$ if $\mathfrak{t} \subseteq \mathfrak{g}_0$, or equivalently, if $h_\Gamma \in \mathfrak{t}$.

Remark 4.6.2 We have $h = h_{\Gamma^\emptyset} \in \mathfrak{t}$ since $\mathfrak{t} \subseteq \mathfrak{c}$ implies $[h, \mathfrak{t}] = 0$ and $\mathfrak{t}$ is maximal abelian. Hence, the Dynkin grading $\Gamma^\emptyset$ is compatible with $\mathfrak{t}$.

Lemma 4.6.3 *Every $\mathbb{R}$-grading Γ is G-conjugate to a $\mathbb{R}$-grading that is compatible with* $\mathfrak{t}$.

Proof Every semisimple element of $\mathfrak{g}$, in particular h_Γ, is G-conjugate to an element of $\mathfrak{t}$. The lemma follows. □

Let E_e be the $\mathbb{R}$-form for $\mathfrak{t}_e$ consisting of all $p \in \mathfrak{t}_e$ such that the eigenvalues of $\operatorname{ad} p$ on $\mathfrak{g}$ are real.

Lemma 4.6.4 *(i) Any good $\mathbb{R}$-grading for e is G_e-conjugate to a good $\mathbb{R}$-grading Γ for e compatible with* $\mathfrak{t}$.

(ii) For such a Γ, we have that $h \in \mathfrak{g}_0, f \in \mathfrak{g}_{-2}$, and $h_\Gamma = h + p$ for some point $p \in E_e$.

Proof (1) Recall that any two $\mathfrak{sl}_2$-triples in $\mathfrak{g}$ sharing the same e are conjugate by G_e, as seen from the proof of Theorem 4.1.11. By Lemma 4.3.2, we may assume that the $\mathfrak{sl}_2$-triple $\mathfrak{s} = \{e, h, f\}$ we start with is $\Gamma^\emptyset$-graded, that is, $h \in \mathfrak{g}_0$ and $f \in \mathfrak{g}_{-2}$.

Take the semisimple element $h_{\Gamma^\emptyset} \in \mathfrak{g}_0 = \mathfrak{g}_h$ as in Lemma 4.1.4. Let $p' = h_{\Gamma^\emptyset} - h$, which is a semisimple element lying in $\mathfrak{c}_e$ (which is the centralizer of h and e), since $[p', e] = [h_{\Gamma^\emptyset}, e] - [h, e] = 2e - 2e = 0$. Then p' is C_e-conjugate to an element in $\mathfrak{t}_e$, a Cartan subalgebra of $\mathfrak{c}_e$. Write $\operatorname{Ad} g(p') = p \in \mathfrak{t}_e$ for $g \in C_e \subseteq G_e$. Recall $h \in \mathfrak{t}$ from Remark 4.6.2. Hence, $\operatorname{Ad} g(h_{\Gamma^\emptyset}) = \operatorname{Ad} g(h + p') = h + p \in \mathfrak{t}$, and the good grading $\Gamma := \operatorname{Ad} g(\Gamma^\emptyset)$ for e is compatible with $\mathfrak{t}$.

(2) For such a Γ as in (1), we have $h \in \mathfrak{g}_0$ since $\mathfrak{t} \subseteq \mathfrak{g}_0$ and $h \in \mathfrak{t}$. Then $p = h_\Gamma - h$ centralizes e and h, hence also f, so $[h_\Gamma, f] = [h, f] = -2f$ and $f \in \mathfrak{g}_{-2}$. Finally, p belongs to $\mathfrak{t}_e$, hence to E_e, since Γ is an $\mathbb{R}$-grading. □

4.6.3 For $p \in E_e$, we let Γ^p denote the $\mathbb{R}$-grading of $\mathfrak{g}$ defined by $\operatorname{ad}(h + p)$ (which is consistent with the notation for the Dynkin grading $\Gamma^\emptyset$). Such a grading Γ^p is automatically compatible with $\mathfrak{t}$, but it may not be a good grading.

Note that $\mathfrak{t}_e$ commutes with h and e, and hence with $\mathfrak{s}$. For $\alpha \in \mathfrak{t}_e^*$ and $i \geq 0$, let $L(\alpha, i)$ denote the irreducible $\mathfrak{t}_e \oplus \mathfrak{s}$-module of dimension $(i+1)$ on which $\mathfrak{t}_e$ acts by weight α. Decompose $\mathfrak{g}$ as an $\mathfrak{t}_e \oplus \mathfrak{s}$-module

$$\mathfrak{g} \cong \bigoplus_{\alpha \in \Phi_e \cup \{0\}} \bigoplus_{i \geq 0} m(\alpha, i) L(\alpha, i)$$

where Φ_e denotes the set of nonzero weights of $\mathfrak{t}_e$ (or E_e) of $\mathfrak{g}$. One can show that Φ_e is a so-called restricted root system on E_e (cf. [46]), but for our purpose we only need to record two simple facts: $\alpha \in \Phi_e$ implies $\pm\alpha \in \Phi_e$ and Φ_e spans E_e^*. For $\alpha \in \Phi_e$, let

$$d(\alpha) = 1 + \min\{i \geq 0 \mid m(\alpha, i) \neq 0\}$$

be the minimal dimension of a simple $\mathfrak{s}$-submodule of the α-weight space of $\mathfrak{g}$.

Proposition 4.6.5 *Let $p \in E_e$. The grading Γ^p is a good grading for e if and only if $|\alpha(p)| < d(\alpha)$ for all $\alpha \in \Phi_e$.*

Proof Recall that $\Gamma^p : \mathfrak{g} = \bigoplus_{j \in \mathbb{Z}} \mathfrak{g}_j$ is a good grading for e if and only if $\mathfrak{g}_e \subseteq \bigoplus_{j > -1} \mathfrak{g}_j$. Note that $h + p$ acts on the highest weight vector of $L(\alpha, i)$ as the scalar $\alpha(p) + i$ for $\alpha \in \Phi_e$. So, Γ^p is a good grading for e if and only if $\alpha(p) + i > -1$ for all $\alpha \in \Phi_e$; the latter (by considering $\pm\alpha \in \Phi_e$ simultaneously) is equivalent to $|\alpha(p)| < 1 + i$ for all $\alpha \in \Phi_e$. □

The *good grading polytope* for e is defined to be

$$\mathscr{P}_e = \{p \in E_e \mid \Gamma^p \text{ is a good } \mathbb{R}\text{-grading for } e\}.$$

Since Φ_e spans E_e^*, $\mathscr{P}_e$ is an open convex polytope in E_e by Proposition 4.6.5: indeed, $|\alpha(tp_1 + (1-t)p_2)| \leq t|\alpha(p_1)| + (1-t)|\alpha(p_2)| < d(\alpha)$, for $p_1, p_2 \in \mathscr{P}_e$ and $0 \leq t \leq 1$.

4.6.4 Two good $\mathbb{R}$-gradings $\Gamma : \mathfrak{g} = \bigoplus_{i \in \mathbb{R}} \mathfrak{g}_i$ and $\Gamma' : \mathfrak{g} = \bigoplus_{j \in \mathbb{R}} \mathfrak{g}'_j$ are *adjacent* if

$$\mathfrak{g} = \bigoplus_{i^- \leq j \leq i^+} \mathfrak{g}_i \cap \mathfrak{g}'_j,$$

where i^- denotes the largest integer strictly smaller than i and i^+ denotes the smallest integer strictly greater than i.

Lemma 4.6.6 *The following conditions are equivalent for $i, j \in \mathbb{R}$:*
(i) $i^- \leq j \leq i^+$;
(ii) i and j lie in the same unit interval $[a, a+1]$ for some $a \in \mathbb{Z}$;
(iii) $j^- \leq i \leq j^+$.

Exercise 4.6.7 Prove Lemma 4.6.6.

The next lemma allows us to reduce the study of adjacent good $\mathbb{R}$-gradings to the good grading polytope $\mathscr{P}_e$.

Lemma 4.6.8 *Every pair of adjacent good $\mathbb{R}$-gradings for e is G_e-conjugate to a pair of adjacent good $\mathbb{R}$-gradings for e compatible with $\mathfrak{t}$.*

Sketch of a proof For adjacent gradings Γ and Γ', $\operatorname{ad} h_\Gamma$ and $\operatorname{ad} h_{\Gamma'}$ are by definition simultaneously diagonalizable, and hence $[h_\Gamma, h_{\Gamma'}] = 0$.

We can refine Lemma 4.3.2 with essentially the same proof to find an $\mathfrak{sl}_2$-triple $\mathfrak{s} = \{e, h, f\}$ which is doubly Γ-graded and Γ'-graded, that is, $h \in \mathfrak{g}_0 \cap \mathfrak{g}'_0$ and $f \in \mathfrak{g}_{-2} \cap \mathfrak{g}'_{-2}$.

The lemma can now be viewed as a "double" analogue of Lemma 4.6.4 (1), and it can be proved by refining the arguments therein with the help of the doubly graded $\mathfrak{sl}_2$-triple above. $\square$

Exercise 4.6.9 Fill in the details of the proof of Lemma 4.6.8.

Consider the hyperplanes

$$H_{\alpha,k} = \{p \in E_e \mid \alpha(p) = k\}, \quad \alpha \in \Phi_e, k \in \mathbb{Z}.$$

The connected components of $E_e \backslash \bigcup_{\alpha,k} H_{\alpha,k}$ will be referred to as (open) *alcoves*.

Lemma 4.6.10 *Let $p, p' \in \mathscr{P}_e$. Then, Γ^p and $\Gamma^{p'}$ are adjacent if and only if p and p' belong to the closure of the same alcove.*

Proof By definition, $\mathfrak{g}_\alpha (\forall \alpha \in \Phi)$ lies in the degree $\alpha(h+p)$ subspace of $\mathfrak{g}$ with respect to Γ^p, and a similar remark applies to $\Gamma^{p'}$. Note that $\alpha(h) \in \mathbb{Z}$. It follows that Γ^p and $\Gamma^{p'}$ are adjacent if and only if $\alpha(p)^- \leq \alpha(p') \leq \alpha(p)^+, \forall \alpha \in \Phi$, and by Lemma 4.6.6 (2), if and only if p and p' belong to the closure of the same alcove. $\square$

Proposition 4.6.11 *Given any two good gradings Γ and Γ' for e, there exists a chain $\Gamma = \Gamma_0, \Gamma_1, \ldots, \Gamma_n = \Gamma'$ of good gradings for e such that Γ_i is adjacent to Γ_{i+1} for each $i = 0, \ldots, n-1$.*

Proof A simultaneous conjugation preserves pairs of adjacent good gradings. So by Lemma 4.6.8, we may assume that $\Gamma = \Gamma^p$ and $\Gamma' = \Gamma^q$ are good gradings for e compatible with $\mathfrak{t}$, for some $p, q \in \mathscr{P}_e$. As a convex polytope, $\overline{\mathscr{P}_e}$ is connected, and it follows from Proposition 4.6.5 that the closure $\overline{\mathscr{P}_e}$ is a union of the closures of finitely many alcoves. Now the proposition follows from Lemma 4.6.10. $\square$

Lemma 4.6.12 *Let $\Gamma : \mathfrak{g} = \bigoplus_{i \in \mathbb{R}} \mathfrak{g}_i$ and $\Gamma' : \mathfrak{g} = \bigoplus_{j \in \mathbb{R}} \mathfrak{g}'_j$ be adjacent good gradings for e. Then there exist Lagrangian subspaces $\mathfrak{l}$ of $\mathfrak{g}_{-1}$ and $\mathfrak{l}'$ of $\mathfrak{g}'_{-1}$ such that*

$$\mathfrak{l} \bigoplus (\bigoplus_{i<-1} \mathfrak{g}_i) = \mathfrak{l}' \bigoplus (\bigoplus_{j<-1} \mathfrak{g}'_j).$$

Proof We have that

$$\mathfrak{g}_{-1} = ((\mathfrak{g}_{-1} \cap \mathfrak{g}'_{<-1}) \bigoplus (\mathfrak{g}_{-1} \cap \mathfrak{g}'_{>-1})) \bigoplus (\mathfrak{g}_{-1} \cap \mathfrak{g}'_{-1}),$$

$$\mathfrak{g}'_{-1} = ((\mathfrak{g}_{<-1} \cap \mathfrak{g}'_{-1}) \bigoplus (\mathfrak{g}_{>-1} \cap \mathfrak{g}'_{-1})) \bigoplus (\mathfrak{g}_{-1} \cap \mathfrak{g}'_{-1}).$$

Note the last summand $\mathfrak{g}_{-1} \cap \mathfrak{g}'_{-1}$ is orthogonal to the first two summands in either identity above with respect to $\langle \cdot, \cdot \rangle = (e|[\cdot, \cdot])$. Hence $\langle \cdot, \cdot \rangle$ regarded as a (skew-symmetric) bilinear form on $\mathfrak{g}_{-1} \cap \mathfrak{g}'_{-1}$ remains to be non-degenerate. Take a Lagrangian subspace $\mathfrak{k}$ of $\mathfrak{g}_{-1} \cap \mathfrak{g}'_{-1}$, and set $\mathfrak{l} = \mathfrak{k} \oplus (\mathfrak{g}_{-1} \cap \mathfrak{g}'_{<-1})$ and $\mathfrak{l}' = \mathfrak{k} \oplus (\mathfrak{g}_{<-1} \cap \mathfrak{g}'_{-1})$. The lemma follows by using the assumption of adjacency to check the inclusion of subspaces $\mathfrak{l} \oplus (\oplus_{i<-1} \mathfrak{g}_i) \subseteq \mathfrak{l}' \oplus (\oplus_{j<-1} \mathfrak{g}'_j)$ and the opposite inclusion by symmetry. $\square$

Theorem 4.6.13 ([46]) *The finite W-algebras associated to any two good gradings Γ and Γ' for e are isomorphic.*

Proof In this proof, we shall write a finite W-algebra as $\mathcal{W}_{\mathfrak{l},\Gamma}$, indicating its dependence on a Lagrangian $\mathfrak{l}$ and a good $\mathbb{R}$-grading Γ.

Fix a chain $\Gamma = \Gamma_0, \Gamma_1, \ldots, \Gamma_n = \Gamma'$ of good gradings for e such that Γ_i is adjacent to Γ_{i+1} for each i, as in Proposition 4.6.11.

By Lemma 4.6.12, we can suitably choose Lagrangians $\mathfrak{l}_i$ and $\mathfrak{l}'_i$ with respect to the gradings Γ_i and Γ_{i+1}, respectively, which result an equality of the corresponding Lie subalgebras $\mathfrak{m}_i = \mathfrak{m}_{i+1}$. So by the Whittaker model definition of W-algebras, we have $\mathcal{W}_{\mathfrak{l}_i,\Gamma_i} = \mathcal{W}_{\mathfrak{l}'_i,\Gamma_{i+1}}$. Accordingly to Gan-Gunzburg's Theorem 4.3.1 (or rather its generalization in the setting of $\mathbb{R}$-gradings), the algebras $\mathcal{W}_{\mathfrak{l},\Gamma}$ are isomorphic for a fixed grading Γ and different choices of isotropic subspaces $\mathfrak{l} \subseteq \mathfrak{g}_{-1}$. Hence, the theorem follows by composing a sequence of algebra isomorphisms:

$$\mathcal{W}_{\mathfrak{l}_1,\Gamma_1} = \mathcal{W}_{\mathfrak{l}'_1,\Gamma_2} \cong \mathcal{W}_{\mathfrak{l}_2,\Gamma_2} = \mathcal{W}_{\mathfrak{l}'_2,\Gamma_3} \cong \cdots \cong \mathcal{W}_{\mathfrak{l}_{n-1},\Gamma_{n-1}} = \mathcal{W}_{\mathfrak{l}'_{n-1},\Gamma'}.$$

□

4.7 Higher level Schur duality

In this section, we let $\mathfrak{g} = \mathfrak{gl}_N$. We present a duality between finite W-algebras and degenerate cyclotomic Hecke algebras, due to Brundan and Kleshchev [50].

4.7.1 Schur duality. Let $V = \mathbb{C}^N$. Then the tensor space $V^{\otimes d}$ is naturally a $(U(\mathfrak{g}), S_d)$-bimodule, which will be expressed as

$$U(\mathfrak{g}) \curvearrowright V^{\otimes d} \curvearrowleft S_d.$$

The celebrated Schur duality states that the images of $U(\mathfrak{g})$ and $\mathbb{C}S_d$ form mutual centralizers in $\mathrm{End}(V^{\otimes d})$. Since the S_d-module $V^{\otimes d}$ is semisimple, one further obtains a multiplicity-free decomposition of $V^{\otimes d}$ as a $(U(\mathfrak{g}), S_d)$-bimodule.

4.7.2 A duality of graded algebras. Let $e \in \mathfrak{g}$ be nilpotent, and let $\chi \in \mathfrak{g}^*$ be associated to e as before by $(\cdot|\cdot)$. Assume the Jordan canonical form of e is given by a partition λ of N. We choose the (even) pyramid to be the Young diagram λ in the French fashion, labelled by $\{1, \ldots, N\}$ down the columns from left to right.

Denote by $\ell = \lambda_1$. Then $e^\ell = 0$.

Denote by $\overline{H}_d := \mathbb{C}[x_1, \ldots, x_d] \rtimes \mathbb{C}S_d$ the semi-direct product algebra formed by the natural action of S_d on the polynomial algebra $\mathbb{C}[x_1, \ldots, x_d]$. We denote by $\mathbb{C}_\ell[x_1, \ldots, x_d]$ the truncated polynomial algebra $\mathbb{C}[x_1, \ldots, x_d]/\langle x_1^\ell, \ldots, x_n^\ell\rangle$, and then the algebra $\overline{H}_d^\ell := \mathbb{C}_\ell[x_1, \ldots, x_d] \rtimes \mathbb{C}S_d$ is a quotient of $\overline{H}_d$. Both $\overline{H}_d$ and $\overline{H}_d^\ell$ are $\mathbb{N}$-graded algebras with $\deg \sigma = 0, \forall \sigma \in S_n$, and $\deg x_i = 2, \forall i$.

We define an $\overline{H}_d$-module structure on $V^{\otimes d}$ by letting

$$x_i = 1^{\otimes(i-1)} \otimes e \otimes 1^{\otimes(d-i)}, \quad 1 \le i \le d.$$

The action of $\overline{H}_d$ factors through $\overline{H}_d^\ell$. Letting $\mathfrak{g}_e$ act as the subalgebra of $\mathfrak{g}$, we clearly have a $(U(\mathfrak{g}_e), \overline{H}_d^\ell)$-bimodule structure on $V^{\otimes d}$:

$$U(\mathfrak{g}_e) \overset{\overline{\Phi}_d}{\curvearrowright} V^{\otimes d} \overset{\overline{\Psi}_d}{\curvearrowleft} \overline{H}_d^\ell. \tag{4.17}$$

The Vust duality (cf. [50, Theorem 2.4]) states that the images of $U(\mathfrak{g}_e)$ and $\overline{H}_d^\ell$ form mutual centralizers in $\mathrm{End}(V^{\otimes d})$, i.e.

$$\overline{\Phi}_d(U(\mathfrak{g}_e)) = \mathrm{End}_{\overline{H}_d^\ell}(V^{\otimes d}), \quad \mathrm{End}_{U(\mathfrak{g}_e)}(V^{\otimes d})^{\mathrm{op}} = \overline{\Psi}_d(\overline{H}_d^\ell).$$

In general, these images are not semisimple algebras.

4.7.3 Higher level Schur duality. Recall from Section 4.2.3 the definition of $\mathcal{W}_\chi$ as $\mathfrak{m}$-invariants in $U(\mathfrak{p})$, where $\mathfrak{p}$ is given in (4.10). Let $\eta : U(\mathfrak{p}) \to U(\mathfrak{p})$ be the algebra automorphism defined by

$$\eta(e_{i,j}) = e_{i,j} + \delta_{i,j}(\lambda_1 - \lambda'_{\mathrm{col}(j)} - \lambda'_{\mathrm{col}(j)+1} - \cdots - \lambda'_\ell), \quad e_{i,j} \in \mathfrak{p},$$

where λ' denotes the conjugate partition of λ. For simplicity of some later formulas, we will adopt the definition of $\mathcal{W}_\chi$ by a twist of automorphism η as follows:

$$\mathcal{W}_\chi = \{y \in U(\mathfrak{p}) \mid [a, \eta(y)] \in I_\chi, \forall a \in \mathfrak{m}\}.$$

Clearly, this new definition is isomorphic to the one in Section 4.2.3.

The higher level Schur duality is a filtered deformation of the Vust duality (4.17) of the following form:

$$\mathcal{W}_\chi \overset{\Phi_d}{\curvearrowright} V^{\otimes d} \overset{\Psi_d}{\curvearrowleft} H_d^{[\lambda]}$$

where $U(\mathfrak{g}_e)$ is replaced by the W-algebra $\mathcal{W}_\chi$ and $\overline{H}_d^\ell$ by a cyclotomic Hecke algebra $H_d^{[\lambda]}$. Recall that the degenerate affine Hecke algebra H_d (introduced by Drinfeld [89]) is generated by its polynomial subalgebra $\mathbb{C}[x_1, \ldots, x_d]$ and a subalgebra $\mathbb{C}S_d$, and that it satisfies the additional relation for $1 \le i, j \le d$:

$$s_i x_j = x_j s_i \ (j \neq i, i+1), \quad s_i x_{i+1} = x_i s_i + 1.$$

The cyclotomic Hecke algebra $H_d^{[\lambda]}$ is defined to be the quotient of H_d by the two-sided ideal generated by $\prod_{i=1}^l (x_1 - \lambda'_i + \lambda_1)$.

The remainder of this section is to explain the two (commuting) actions on $V^{\otimes d}$ of $\mathcal{W}_\chi$ and H_d; the action of H_d factors through $H_d^{[\lambda]}$.

4.7.4 A theorem of Arakawa and Suzuki. Given a $\mathfrak{g}$-module M, we consider the endomorphism of $\mathfrak{g}$-module $x : M \otimes V \to M \otimes V$, where x acts by $\Omega = \sum_{1 \le i,j \le N} E_{ij} \otimes E_{ji}$. We also let $s = 1 \otimes \Omega : M \otimes V \otimes V \to M \otimes V \otimes V$, which can be easily seen to coincide with the permutation of the two copies of V. Abstractly, we may regard x as an endomorphism of the functor $- \otimes V$ and s as an endomorphism of the functor $(- \otimes V)^2$. This provides an example of $\mathfrak{sl}_2$-categorification of Chuang and Rouquier [78].

Now let $x_i = 1^{d-i} x 1^{i-1} (1 \le i \le d)$ and $s_j = 1^{d-j-1} s 1^{j-1} (1 \le j \le d-1)$ be the endomorphisms of the dth power $(- \otimes V)^d$. Equivalently, by the natural isomorphism $(- \otimes V)^d \cong - \otimes V^{\otimes d}$, we have the corresponding endomorphisms

$$\hat{x}_i = \sum_{k=0}^{i-1} \Omega^{(k,i)} = \Omega^{(0,i)} + \sum_{1 \le k < i} s_{ki}, \qquad \hat{s}_j = \Omega^{(j,j+1)}$$

on $M \otimes V^{\otimes d}$ for a $\mathfrak{g}$ module M, where s_{ki} denotes the permutation of the kth and ith copies of V. Here we have labeled the tensor factors from 0 to d, and the notation $\Omega^{(a,b)}$ means $\sum_{i,j} 1 \otimes \cdots E_{ij} \otimes \cdots \otimes E_{ji} \cdots \otimes 1$ with the two nontrivial factors at ath and bth places.

According to a theorem of Arakawa and Suzuki [14], these $\hat{x}_i$'s and $\hat{s}_j$'s satisfy the relations of the degenerate affine Hecke algebra H_d. Hence we have obtained a $(\mathfrak{g}, H_d)$-bimodule structure on $M \otimes V^{\otimes d}$ for any $\mathfrak{g}$-module M.

Exercise 4.7.1 Prove that $\hat{x}_i$ for $1 \le i \le d$ and $\hat{s}_j$ for $1 \le j \le d-1$ satisfy the defining relations of the degenerate affine Hecke algebra H_d.

4.7.5 Higher level Schur duality. Recall the Skryabin equivalence of categories from Section 4.4.2: Wh : $\mathfrak{g}\text{-Wmod}^{\chi} \to \mathcal{W}_{\chi}\text{-mod}$ and $Q_{\chi} \otimes_{\mathcal{W}_{\chi}} - : \mathcal{W}_{\chi}\text{-mod} \to \mathfrak{g}\text{-Wmod}^{\chi}$. Via such an equivalence, we can transfer functors on the category $\mathfrak{g}\text{-Wmod}^{\chi}$ to functors on $\mathcal{W}_{\chi}\text{-mod}$.

Given a finite-dimensional $\mathfrak{g}$-module X, the functor $- \otimes X : \mathfrak{g}\text{-Wmod}^{\chi} \to \mathfrak{g}\text{-Wmod}^{\chi}$ gives rise to, via the Skryabin equivalence, an exact functor

$$- \circledast X : \mathcal{W}_{\chi}\text{-mod} \longrightarrow \mathcal{W}_{\chi}\text{-mod}, \quad M \mapsto \mathrm{Wh}((Q_{\chi} \otimes_{\mathcal{W}_{\chi}} M) \otimes X).$$

These functors satisfy the naturality $(M \circledast X) \circledast Y \xrightarrow{\sim} M \circledast (X \otimes Y)$ for another finite dimensional $\mathfrak{g}$-module Y. For our purpose, we will mainly consider the cases when X and Y are tensor powers of V.

By using the functors $- \circledast V^{\otimes d}$ on $\mathcal{W}_{\chi}\text{-mod}$ in place of the functors $- \otimes V^{\otimes d}$ on $\mathfrak{g}\text{-Wmod}^{\chi}$, we obtain via the Skryabin equivalence, for any $\mathcal{W}_{\chi}$-module M, a $(\mathcal{W}_{\chi}, H_d)$-bimodule structure on $M \circledast V^{\otimes d}$. We will be mostly interested in the case when $M = \mathbb{C}$, the restriction to $\mathcal{W}_{\chi}$ of the trivial $U(\mathfrak{p})$-module, but the arguments below are equally valid for a $U(\mathfrak{p})$-module M.

Suppose that M is actually a $\mathfrak{p}$-module, and view M and $M \otimes X$ as $\mathcal{W}_{\chi}$-modules by restricting from $U(\mathfrak{p})$ to $\mathcal{W}_{\chi}$. Consider the map

$$\hat{\mu}_{M,X} : (Q_{\chi} \otimes_{\mathcal{W}_{\chi}} M) \otimes X \to M \otimes X$$

which sends $(\eta(u)1_{\chi} \otimes m) \otimes x \mapsto um \otimes x$ for $u \in U(\mathfrak{p})$, $m \in M$ and $x \in X$. The restriction of this map to the subspace $\mathrm{Wh}((Q_{\chi} \otimes_{\mathcal{W}_{\chi}} M) \otimes X)$ obviously defines a homomorphism of $\mathcal{W}_{\chi}$-modules

$$\mu_{M,X} : M \circledast X \longrightarrow M \otimes X.$$

It turns out that this homomorphism is actually an isomorphism [50] (also cf. [114]). Roughly speaking, the isomorphism can be obtained by using the following $\mathfrak{m}$-module isomorphism of Skryabin:

$$Q_{\chi} \otimes_{\mathcal{W}_{\chi}} M \cong M \otimes_{\mathbb{C}} U(\mathfrak{m})^{\star}$$

where $U(\mathfrak{m})^{\star} = \varprojlim_{k} U(\mathfrak{m})/I^{\mathfrak{m}}_{\chi,k}$, and $I^{\mathfrak{m}}_{\chi,k}$ is the left ideal of $U(\mathfrak{m})$ generated by $(a - \chi(a))^k, \forall a \in \mathfrak{m}$.

In case when $M = \mathbb{C}$, we have obtained an isomorphism of $(\mathcal{W}_{\chi}, H_d)$-bimodules

$$\mathbb{C} \circledast V^{\otimes d} \xrightarrow{\cong} V^{\otimes d}. \tag{4.18}$$

Recall the standard basis $v_i, 1 \leq i \leq N$ for $V = \mathbb{C}^N$. Then $V^{\otimes d}$ has a basis $v_{\mathbf{i}} = v_{i_1} \otimes \ldots \otimes v_{i_d}$, where $\mathbf{i} = (i_1, \ldots, i_d)$ with $1 \leq i_1, \ldots, i_d \leq N$. Denote by $\mathtt{L}(i)$ the number in the box immediately to the left of the ith box in the pyramid λ if it exists.

Lemma 4.7.2 *The element $x_1 \in H_d$ sends each $v_{\mathbf{i}} \in V^{\otimes d}$ to*

$$v_{\mathtt{L}\mathbf{i}} + (\lambda'_{col(i_1)} - \lambda_1) v_{\mathbf{i}} - \sum_{\substack{1<k\leq d \\ col(i_k)<col(i_1)}} v_{\mathbf{i}\cdot(1,k)},$$

where $v_{\mathtt{L}\mathbf{i}} = v_{\mathtt{L}(i_1)} \otimes v_{i_2} \otimes \ldots \otimes v_{i_d}$ if $\mathtt{L}(i_1)$ is defined; otherwise, set $v_{\mathtt{L}\mathbf{i}} = 0$.

Sketch of a proof By examining step by step how the isomorphism (4.18) is obtained, the action of x_1 is traced back to the corresponding action in Section 4.7.4. □

Theorem 4.7.3 ([50]) *The action of H_d on $V^{\otimes d}$ factors through $H_d^{[\lambda]}$ and this gives rise to commuting actions:*

$$\mathcal{W}_\chi \overset{\Phi_d}{\curvearrowright} V^{\otimes d} \overset{\Psi_d}{\curvearrowleft} H_d^{[\lambda]}.$$

The maps Φ_d and Ψ_d satisfy the double centralizer property, i.e.

$$\Phi_d(\mathcal{W}_\chi) = \mathrm{End}_{H_d^{[\lambda]}}(V^{\otimes d}),$$
$$\mathrm{End}_{\mathcal{W}_\chi}(V^{\otimes d})^{\mathrm{op}} = \Psi_d(H_d^{[\lambda]}).$$

Sketch of a proof The first statement follows from a direct computation of the minimal polynomial of x_1 by using Lemma 4.7.2.

The commutativity of the two actions follow from the theorem of Arakawa and Suzuki via the Skryabin equivalence.

The double centralizer property follows from the associated graded version (i.e. the Vust duality in Section 4.7.2) and a standard filtered algebra argument. □

Remark 4.7.4 The higher level Schur duality has applications to the BGG category $\mathcal{O}$ for $\mathfrak{g}$; see [50] for more details. The higher level Schur duality also has a combinatorial version in terms of generalized Khovanov algebras which enabled Brundan and Stroppel [51] to give a combinatorial description of maximal parabolic category $\mathcal{O}$ in type A.

4.8 W-(super)algebras in positive characteristic

In this section, we will present the W-algebras and W-superalgebras in positive characteristic, following [193, 216, 217].

4.8.1 Restricted Lie superalgebras. Let F be an algebraically closed field of characteristic $p > 2$. A Lie superalgebra $\mathfrak{g} = \mathfrak{g}_{\bar{0}} \oplus \mathfrak{g}_{\bar{1}}$ over F is a *restricted Lie superalgebra*, if there is a pth power map $\mathfrak{g}_{\bar{0}} \to \mathfrak{g}_{\bar{0}}, x \mapsto x^{[p]}$, such that the even subalgebra $\mathfrak{g}_{\bar{0}}$ is a restricted Lie algebra and the odd part $\mathfrak{g}_{\bar{1}}$ is a restricted module by the adjoint action of $\mathfrak{g}_{\bar{0}}$. The notion of restricted Lie algebras was introduced by Jacobson, cf. Jantzen [133] for an excellent review of modular representations of Lie algebras.

Example 4.8.1 The p-power map on the general linear Lie algebra $\mathfrak{gl}_n$ is given by $x \mapsto \overbrace{xx\cdots x}^{p}$, the pth product for the underlying associative algebra.

The general linear Lie superalgebra $\mathfrak{gl}_{m|n}$, which consists of $(m+n)\times(m+n)$ matrices and whose even subalgebra is isomorphic to $\mathfrak{gl}_m \oplus \mathfrak{gl}_n$, is a restricted Lie superalgebra.

All the Lie (super)algebras considered here will be restricted. If one is only interested in Lie algebras, one can always set $\mathfrak{g}_{\bar{1}} = 0$ below.

We list all the basic classical Lie superalgebras over F with restrictions on p as follows (the general linear Lie superalgebra, though not simple, is also included).

Lie superalgebra	Characteristic of K
$\mathfrak{gl}_{m\|n}$	$p > 2$
$\mathfrak{sl}_{m\|n}$	$p > 2, p \nmid (m-n)$
$B(m,n), C(n), D(m,n)$	$p > 2$
$D(2,1;\alpha)$	$p > 3$
$F(4)$	$p > 15$
$G(3)$	$p > 15$

The queer Lie superalgebra $\mathfrak{q}_n$ is a Lie subalgebra of the general linear Lie superalgebra $\mathfrak{gl}_{n|n}$, which consists of matrices of the form:

$$\begin{pmatrix} A & B \\ B & A \end{pmatrix},$$

where A and B are arbitrary $n \times n$ matrices. Note that the even subalgebra $(\mathfrak{q}_n)_{\bar{0}}$ is isomorphic to $\mathfrak{gl}_n$ and the odd part $(\mathfrak{q}_n)_{\bar{1}}$ is another isomorphic copy of $\mathfrak{gl}_n$ under the adjoint action of $(\mathfrak{q}_n)_{\bar{0}}$.

4.8.2 Reduced enveloping superalgebras. Let $\mathfrak{g}$ be a restricted Lie superalgebra. For each $x \in \mathfrak{g}_{\bar{0}}$, the element $x^p - x^{[p]} \in U(\mathfrak{g})$ is central by definition of the pth power map. We refer to $\mathcal{Z}_p(\mathfrak{g}) = K\langle x^p - x^{[p]} \,|\, x \in \mathfrak{g}_{\bar{0}}\rangle$ as the p-center of $U(\mathfrak{g})$. Let $x_1, \ldots, x_s$ (resp. $y_1, \ldots, y_t$) be a basis of $\mathfrak{g}_{\bar{0}}$ (resp. $\mathfrak{g}_{\bar{1}}$). As a consequence of the PBW theorem for $U(\mathfrak{g})$, $\mathcal{Z}_p(\mathfrak{g})$ is a polynomial algebra isomorphic to $K[x_i^p - x_i^{[p]} \,|\, i = 1, \ldots, s]$, and the enveloping superalgebra $U(\mathfrak{g})$ is free over $\mathcal{Z}_p(\mathfrak{g})$ with basis

$$\{x_1^{a_1} \cdots x_s^{a_s} y_1^{b_1} \cdots y_t^{b_t} \,|\, 0 \le a_i < p;\ b_j = 0, 1 \text{ for all } i, j\}.$$

Exercise 4.8.2 Prove that the p-center $\mathcal{Z}_p(\mathfrak{g})$ is a polynomial algebra isomorphic to $K[x_i^p - x_i^{[p]} \,|\, i = 1, \ldots, s]$, and that $U(\mathfrak{g})$ is free over $\mathcal{Z}_p(\mathfrak{g})$ with basis

$$\{x_1^{a_1} \cdots x_s^{a_s} y_1^{b_1} \cdots y_t^{b_t} \,|\, 0 \le a_i < p;\ b_j = 0, 1 \text{ for all } i, j\}.$$

Let V be a simple $U(\mathfrak{g})$-module. By Schur's lemma, the central element $x^p - x^{[p]}$ for $x \in \mathfrak{g}_{\bar{0}}$ acts by a scalar $\zeta(x)$, which can be written as $\chi_V(x)^p$ for some $\chi_V \in \mathfrak{g}_{\bar{0}}^*$. We call χ_V the *p-character* of the module V. We often regard $\chi \in \mathfrak{g}^*$ by letting $\chi(\mathfrak{g}_{\bar{1}}) = 0$.

Fix $\chi \in \mathfrak{g}_{\bar{0}}^*$. Let J_χ be the ideal of $U(\mathfrak{g})$ generated by the even central elements $x^p - x^{[p]} - \chi(x)^p$. The quotient algebra $U_\chi(\mathfrak{g}) := U(\mathfrak{g})/J_\chi$ is called *the reduced enveloping superalgebra* with p-character χ. It follows from Exercise 4.8.2 that the superalgebra $U_\chi(\mathfrak{g})$ has a linear basis

$$\{x_1^{a_1} \cdots x_s^{a_s} y_1^{b_1} \cdots y_t^{b_t} \,|\, 0 \le a_i < p;\ b_j = 0, 1 \text{ for all } i, j\}.$$

In particular, $\dim U_\chi(\mathfrak{g}) = p^{\dim \mathfrak{g}_{\bar{0}}} 2^{\dim \mathfrak{g}_{\bar{1}}}$.

A superalgebra analogue of Schur's Lemma states that the endomorphism ring of an irreducible module of an associative superalgebra is either one-dimensional or two-dimensional (in the latter case it is isomorphic to a Clifford algebra), cf. e.g. Kleshchev [147, Chap. 12]. An irreducible module is *of type* $\mathtt{M}$ if its endomorphism ring is one-dimensional and it is *of type* $\mathtt{Q}$ otherwise.

4.8.3 The good $\mathbb{Z}$-gradings. Let $\mathfrak{g}$ be one of the basic classical Lie superalgebras. The Lie superalgebra $\mathfrak{g}$ admits a nondegenerate invariant even bilinear form $(\cdot|\cdot)$, whose restriction on $\mathfrak{g}_{\bar{0}}$ gives an isomorphism $\mathfrak{g}_{\bar{0}} \xrightarrow{\sim} \mathfrak{g}_{\bar{0}}^*$. Let $\chi \in \mathfrak{g}_{\bar{0}}^*$ be a nilpotent element, that is, it is the image of some nilpotent element $e \in \mathfrak{g}_{\bar{0}}$ under the above isomorphism. Then $\mathfrak{g}_\chi := \{x \in \mathfrak{g} \mid \chi([x,y]) = 0, \forall y \in \mathfrak{g}\}$ is equal to the usual centralizer $\mathfrak{g}_e$. We shall denote $\underline{\dim}\, \mathfrak{g} = \dim \mathfrak{g}_{\bar{0}} | \dim \mathfrak{g}_{\bar{1}}$.

On the other hand, $\mathfrak{q}_n$ admits a nondegenerate invariant odd bilinear form $(\cdot|\cdot)$, which induces an isomorphism $(\mathfrak{q}_n)_{\bar{1}} \xrightarrow{\sim} (\mathfrak{q}_n)_{\bar{0}}^*$. Hence a nilpotent p-character χ corresponds to an odd nilpotent element $e \in (\mathfrak{q}_n)_{\bar{1}}$.

Proposition 4.8.3 *Let $\mathfrak{g}$ be a basic classical Lie superalgebra or a queer Lie superalgebra. Then there exists a good $\mathbb{Z}$-grading $\mathfrak{g} = \oplus_{j\in\mathbb{Z}} \mathfrak{g}_j$ as defined in 4.1.4. This $\mathbb{Z}$-grading is compatible with the $\mathbb{Z}_2$-grading, that is, $\mathfrak{g}_j = \mathfrak{g}_{j,\bar{0}} \oplus \mathfrak{g}_{j,\bar{1}}$ where $\mathfrak{g}_{j,a} = \mathfrak{g}_j \cap \mathfrak{g}_a$ for $a \in \mathbb{Z}_2, j \in \mathbb{Z}$. In particular, we have a refinement of* (4.6)*:*

$$\underline{dim}\, \mathfrak{g}_e = \underline{dim}\, \mathfrak{g}_0 + \underline{dim}\, \mathfrak{g}_1. \tag{4.19}$$

Sketch of a proof For Lie superalgebras of type A, B, C, D and type Q, we use the classification of nilpotent orbits for classical Lie algebras which leads to an elementary way of defining $\mathbb{Z}$-gradings, cf. Jantzen [134, Chapter 1]. For the exceptional Lie superalgebras, the assumptions on p in the Table of 4.8.1 ensure the existence of an $\mathfrak{sl}_2$-triple $\{e,h,f\}$ and the semisimplicity of the adjoint action on $\mathfrak{g}$ by the $\mathfrak{sl}_2$-triple (see Carter [53]). The action of Fh on $\mathfrak{g}$ lifts to a torus action which provides the desired grading on $\mathfrak{g}$. □

4.8.4 The subalgebra $\mathfrak{m}$. Let $\mathfrak{g}$ be one of the basic classical Lie superalgebras or $\mathfrak{q}_n$ with a good $\mathbb{Z}$-grading $\mathfrak{g} = \oplus_{j\in\mathbb{Z}} \mathfrak{g}_j$. Define a bilinear form $\langle\cdot,\cdot\rangle$ on $\mathfrak{g}_{-1}$ by

$$\langle x, y\rangle := \chi([x,y]), \quad x, y \in \mathfrak{g}_{-1}. \tag{4.20}$$

The following is a super analogue of Lemma 4.2.1.

Exercise 4.8.4 Show that (4.20) defines an even, skew-supersymmetric, nondegenerate form $\langle\cdot,\cdot\rangle$ on $\mathfrak{g}_{-1}$, that is, a non-degenerate bilinear form such that $\langle \mathfrak{g}_{-1,\bar{0}}, \mathfrak{g}_{-1,\bar{1}}\rangle = 0$, and that the restriction of the form to $\mathfrak{g}_{-1,\bar{0}}$ (resp. $\mathfrak{g}_{-1,\bar{1}}$) is skew-symmetric (resp. symmetric).

It follows that $\dim \mathfrak{g}_{-1,\bar{0}}$ is even. Take $\mathfrak{l}_{\bar{0}} \subseteq \mathfrak{g}_{-1,\bar{0}}$ to be a maximal isotropic subspace with respect to $\langle\cdot,\cdot\rangle$. It satisfies $\dim \mathfrak{l}_{\bar{0}} = \frac{1}{2}\dim \mathfrak{g}_{-1,\bar{0}}$.

Denote $r = \dim \mathfrak{g}_{-1,\bar{1}}$. There is a basis $v_1, \ldots, v_r$ of $\mathfrak{g}_{-1,\bar{1}}$ under which the symmetric form $\langle\cdot,\cdot\rangle$ has matrix form $E_{1r} + E_{2,r-1} + \ldots + E_{r1}$. If r is even, take $\mathfrak{l}_{\bar{1}} \subseteq \mathfrak{g}_{-1,\bar{1}}$ to be the subspace spanned by $v_1, \ldots, v_{\frac{r}{2}}$. If r is odd, take $\mathfrak{l}_{\bar{1}} \subseteq \mathfrak{g}_{-1,\bar{1}}$ to be the subspace spanned by $v_1, \ldots, v_{\frac{r-1}{2}}$. Set $\mathfrak{l} = \mathfrak{l}_{\bar{0}} \oplus \mathfrak{l}_{\bar{1}}$ and introduce the subalgebras

$$\mathfrak{m} = \mathfrak{l} \bigoplus \bigoplus_{j\geq 2} \mathfrak{g}_{-j},$$

$$\mathfrak{m}' = \begin{cases} \mathfrak{m} \oplus Fv_{\frac{r+1}{2}}, & \text{for } r \text{ odd} \\ \mathfrak{m}, & \text{for } r \text{ even.} \end{cases}$$

The subalgebra $\mathfrak{m}$ is p-nilpotent, and the linear function χ vanishes on the p-closure of $[\mathfrak{m}, \mathfrak{m}]$. It follows that $U_\chi(\mathfrak{m})$ has the trivial module, denoted by F_χ, as its only irreducible module and $U_\chi(\mathfrak{m})/N_\mathfrak{m} \cong F_\chi$ where $N_\mathfrak{m}$ is the Jacobson radical of $U_\chi(\mathfrak{m})$.

Proposition 4.8.5 *Let $\mathfrak{g}$ be a basic classical Lie superalgebra or $\mathfrak{q}_n$. Then every $U_\chi(\mathfrak{g})$-module is $U_\chi(\mathfrak{m})$-free.*

Proposition 4.8.5 in the case when $\mathfrak{g}$ is the Lie algebra of a reductive algebraic group was first established [193] using the machinery of support varieties (cf. [103, 133]) in an essential way. The proof of Proposition 4.8.5 in [216] follows an idea of Skryabin [209], completely bypassing the machinery of support varieties (which has yet to be developed for modular Lie superalgebras).

4.8.5 The super Kac-Weisfeiler conjecture. The following theorem in the present form is due to [216, 217], which generalize the original Kac-Weisfeiler conjecture [220] (theorem of Premet [193]) when $\mathfrak{g}$ is a Lie algebra of a reductive groups. Moreover, based on a deformation approach of Premet and Skryabin, a new proof [226] has optimally improved the assumption on p (that is, one may assume $p > 3$ for the last two exceptional superalgebras in the Table of 4.8.1).

Theorem 4.8.6 *Let $\mathfrak{g}$ be a basic classcial Lie superalgebra or $\mathfrak{q}_n$, and let $\chi \in \mathfrak{g}_{\bar{0}}^*$ be nilpotent. Let $d_i = \dim \mathfrak{g}_i - \dim(\mathfrak{g}_\chi)_i$, $i \in \mathbb{Z}_2$. Then the dimension of every $U_\chi(\mathfrak{g})$-module M is divisible by $p^{\frac{d_0}{2}} 2^{\lfloor \frac{d_1}{2} \rfloor}$, where $\lfloor \frac{d_1}{2} \rfloor$ denotes the least integer which is $\geq \frac{d_1}{2}$.*

Proof By (4.19) and since $\underline{\dim}\, \mathfrak{g}_j = \underline{\dim}\, \mathfrak{g}_{-j}$, we have

$$\underline{\dim}\, \mathfrak{g} - \underline{\dim}\, \mathfrak{g}_\chi = 2 \sum_{j \geq 2} \underline{\dim}\, \mathfrak{g}_{-j} + \underline{\dim}\, \mathfrak{g}_{-1}. \tag{4.21}$$

In particular, $r := \dim \mathfrak{g}_{-1,\bar{1}}$ and d_1 have the same parity. It follows now from the definition of $\mathfrak{m}$ that either (1) $\frac{d_0}{2} | \frac{d_1}{2} = \underline{\dim}\, \mathfrak{m}$ when d_1 is even, or (2) $\frac{d_0}{2} | \frac{d_1 - 1}{2} = \underline{\dim}\, \mathfrak{m}$ when d_1 is odd.

In case (1), the theorem follows immediately from Proposition 4.8.5. Note that $d_1 = d_0$ and hence the case (2) never occurs for $\mathfrak{g} = \mathfrak{q}_n$.

In case (2), the induced $U_\chi(\mathfrak{m}')$-module $V = U_\chi(\mathfrak{m}') \otimes_{U_\chi(\mathfrak{m})} F$ is 2-dimensional, irreducible, and admits an odd automorphism of order 2 induced from $v_{\frac{r+1}{2}}$. By Frobenius reciprocity, it is the only irreducible $U_\chi(\mathfrak{m}')$-module. Thus, $U_\chi(\mathfrak{m}')/N_{\mathfrak{m}'}$ is isomorphic to the simple associative superalgebra $\mathfrak{q}_1^{\text{as}}$ (which coincides with $\mathfrak{q}_1$ as a vector superspace), and the unique simple $U_\chi(\mathfrak{m}')$-module V is of type Q.

Hence, for each $U_\chi(\mathfrak{g})$-module M, the subspace $M^{\mathfrak{m}}$ of $\mathfrak{m}$-invariants in M , which coincides with $M^{\mathfrak{n}}$, is a module over the superalgebra $U_\chi(\mathfrak{m}')/N_{\mathfrak{m}} \cong \mathfrak{q}_1^{\text{as}}$. Since the (unique) simple module of $\mathfrak{q}_1^{\text{as}}$ is two-dimensional, $\dim M^{\mathfrak{m}}$ is divisible by 2. Now the isomorphism $M \cong U_\chi(\mathfrak{m})^* \otimes M^{\mathfrak{m}}$ implies the desired divisibility. □

4.8.6 The modular W-superalgebras. Denote by $\mathcal{Q}_\chi$ the induced $U_\chi(\mathfrak{g})$-module $U_\chi(\mathfrak{g}) \otimes_{U_\chi(\mathfrak{m})} F_\chi$.

Definition 4.8.7 The *(modular) W-superalgebra* associated to the pair $(\mathfrak{g}, \chi)$ is the following associative F-superalgebra

$$W_\chi = \mathrm{End}_{U_\chi(\mathfrak{g})}(\mathcal{Q}_\chi)^{\mathrm{op}}.$$

Theorem 4.8.6 can be somewhat strengthened in the following form [216, 217], which is a superalgebra generalization of a theorem of Premet [194, Theorem 2.3].

Theorem 4.8.8 *Set $\delta = \dim U_\chi(\mathfrak{m})$. Then $\mathcal{Q}_{\mathfrak{m}}$ is a projective $U_\chi(\mathfrak{g})$-module and*

$$U_\chi(\mathfrak{g}) \cong M_\delta(W_\chi).$$

Proof Let $V_1, \ldots, V_s$ (resp. $W_1, \ldots, W_t$) be all inequivalent simple $U_\chi(\mathfrak{g})$-modules of type $\mathtt{M}$ (resp. of type $\mathtt{Q}$). Let P_i (resp. Q_j) denote the projective cover of V_i (resp. W_j). By Proposition 4.8.5, V_i and W_j are free over $U_\chi(\mathfrak{m})$. It follows by Frobenius reciprocity that

$$\dim \operatorname{Hom}_{\mathfrak{g}}(\mathcal{Q}_\chi, V_i) = \dim \operatorname{Hom}_{\mathfrak{m}}(F_\chi, V_i) =: a_i.$$

By Frobenius reciprocity,

$$\dim \operatorname{Hom}_{\mathfrak{g}}(\mathcal{Q}_\chi, W_j) = \dim \operatorname{Hom}_{\mathfrak{m}}(F_\chi, W_j) = \dim \operatorname{Hom}_{\mathfrak{m}}(U_\chi(\mathfrak{m}), W_j)$$

which has to be an even number, say $2b_j$, since as a type $\mathtt{Q}$ module W_j admits an odd involution commuting with $\mathfrak{m}$. It follows that the ranks of the free $U_\chi(\mathfrak{m})$-modules V_i and W_j are a_i and $2b_j$ respectively. Put

$$P = (\bigoplus_{i=1}^{s} P_i^{a_i}) \bigoplus (\bigoplus_{j=1}^{t} Q_j^{b_j}).$$

Then P is projective and has the same head as $\mathcal{Q}_{\mathfrak{m}}$. So there is a surjective homomorphism $\psi : P \to \mathcal{Q}_\chi$.

Since $\dim V_i = \delta a_i$ and $\dim W_j = 2\delta b_j$, by Wedderburn theorem for superalgebras (cf. Kleshchev [147, Theorem 12.2.9]) the left regular $U_\chi(\mathfrak{g})$-module is isomorphic to P^δ. The equality of dimensions

$$\dim P = \dim U_\chi(\mathfrak{g})/\delta = \dim \mathcal{Q}_{\mathfrak{m}}$$

implies that ψ is an isomorphism. Finally,

$$\begin{aligned} U_\chi(\mathfrak{g}) &\cong \operatorname{End}_{U_\chi(\mathfrak{g})}(U_\chi(\mathfrak{g}))^{\mathrm{op}} \cong \operatorname{End}_{U_\chi(\mathfrak{g})}(P^\delta)^{\mathrm{op}} \\ &\cong (M_\delta(\operatorname{End}_{U_\chi(\mathfrak{g})}(P)))^{\mathrm{op}} \cong M_\delta(W_\chi). \end{aligned}$$

This completes the proof of the theorem. □

Remark 4.8.9 In light of Gan-Ginzburg's definition of W-algebras over $\mathbb{C}$, it makes better sense to define a (modular) W-superalgebra over F as

$$W'_\chi := (U_\chi(\mathfrak{g})/\mathfrak{I}_\chi)^{\operatorname{ad}\mathfrak{n}} \equiv \{\bar{y} \in U_\chi(\mathfrak{g})/\mathfrak{I}_\chi \mid [a, y] \in \mathfrak{I}_\chi, \forall a \in \mathfrak{n}\},$$

where $\mathfrak{I}_\chi$ denotes the left ideal of $U_\chi(\mathfrak{g})$ generated by $N_{\mathfrak{m}}$, with multiplication $\bar{y}_1\bar{y}_2 = \overline{y_1y_2}$, for $\bar{y}_1, \bar{y}_2 \in \mathcal{W}'_\chi$. In the case when d_1 (or equivalently, $\dim \mathfrak{g}_{-1,\bar{1}}$) is odd, W'_χ is in general a subalgebra of $W_\chi \cong (U_\chi(\mathfrak{g})/\mathfrak{I}_\chi)^{\operatorname{ad}\mathfrak{m}}$.

Similar subtle differences appear in various constructions of W-superalgebras over $\mathbb{C}$.

4.9 Further work and open problems

4.9.1 Representations of $\mathcal{W}_\chi$ of type A. The representation theory of finite W-algebras has been most adequately developed for $\mathfrak{g} = \mathfrak{gl}_N$. Brundan and Kleshchev [48] showed that the finite W-algebras were quotient algebras of what they introduced and called shifted Yangians, associated to a given pyramid π. In the special case when π is of rectangular shape, such a connection between finite W-algebras and the truncated Yangians (which are quotients of the Yangians introduced by Drinfeld) was found by Ragoucy and Sorba [198], and the truncated Yangians were studied by Cherednik [76]. Also see Brown [44] for a generalization of [198] to other classical Lie algebras.

Such a connection with shifted Yangians was subsequently explored systematically in [49], where the finite-dimensional $\mathcal{W}_\chi$-modules are classified. Moreover, a highest weight representation theory of $\mathcal{W}_\chi$ is developed, and a Kazhdan-Lusztig solution to the finite-dimensional irreducible character problem of $\mathcal{W}_\chi$ is obtained. A formulation of Kazhdan-Lusztig conjectures for general finite W-algebras was given by de Vos and van Driel [85].

Remark 4.9.1 The affine Yangian associated to the affine algebra of $\mathfrak{g}$ have been studied by Guay [118]. We remark that the affine Yangian for $\mathfrak{gl}_1$ should be identified with the $\mathcal{W}_{1+\infty}$ algebra. The main result of [97] can be reformulated as follows: *The vertex algebra for the truncated affine Yangian of $\mathfrak{gl}_1$ at level N is isomorphic to the affine W-algebra associated to $\mathfrak{gl}_N$ and a regular nilpotent element at level N.*

4.9.2 Finite-dimensional $\mathcal{W}_\chi$-modules and primitive ideals. The importance of finite-dimensional $\mathcal{W}_\chi$-modules was realized by Premet [195] in connection to primitive ideals for $U(\mathfrak{g})$, where he showed that the corresponding $\mathfrak{g}$-modules under the Skryabin equivalence have the closure $\overline{\mathcal{O}}_e$ of a nilpotent orbit as their associated varieties and $\frac{1}{2}\dim\overline{\mathcal{O}}_e$ as their Gelfand-Kirillov dimensions. Further different approaches toward connections between finite-dimensional $\mathcal{W}_\chi$-modules and primitive ideals are developed in [111, 157, 158, 196, 197] (see [195, Conjecture 3.2] for motivation).

Among all finite-dimensional $\mathcal{W}_\chi$-modules, the one-dimensional modules stand out (cf. [195, Conjecture 3.1]), and they are studied intensively in [157, 160, 197, 115].

The upshot of all these developments is that finite-dimensional modules are shown to exist for every $\mathcal{W}_\chi$, and one-dimensional modules are shown to exist (modulo some open cases in E_8). The existence of one-dimensional W-modules implies that the dimensional bound provided by Kac-Weisfeiler conjecture is optimal for characteristic $p \gg 0$ [197].

4.9.3 Category $\mathcal{O}$. The category $\mathcal{O}$ for finite W-algebras was well studied in [49] in type A. One needs to overcome substantial technical difficulties to formulate the category $\mathcal{O}$ for finite W-algebras in other types, see [47, 159, 114, 195, 218] for research in this direction.

4.9.4 Open problems. As in any reasonable representation theory, one of the basic problems in W-algebras is the following.

Problem 4.9.2 Find a (Kazhdan-Lusztig type) solution to the irreducible character problem for category $\mathcal{O}$ of $\mathcal{W}_\chi$-modules.

As we have seen that finite W-superalgebras are natural subjects of study in connections to the natural yet nontrivial superalebra generalization of the Kac-Weisfeiler conjecture [216, 217]. We remark that the queer Lie superalgebra $\mathfrak{q}_n$ is often referred to as the truly superanalogue of $\mathfrak{gl}_n$.

Problem 4.9.3 Develop systematically the structures and representation theory of finite W-superalgebras over $\mathbb{C}$, for both basic classical Lie superalgebras and the queer Lie superalgebras.

Various aspects of finite W-superalgebras over $\mathbb{C}$ are currently being developed in the Virginia dissertation of Yun-Ning Peng.

Problem 4.9.4 Formulate the notion of quantum finite W-algebras associated to quantum groups $U_q(\mathfrak{g})$ and then develop its representation theory (for both q being generic and q being a root of unity).

In the case of regular nilpotent elements, there has been a very interesting work of Sevostyanov [207].

It is shown [192] that the algebra of invariants of the centralizer $\mathfrak{g}_e$ in the symmetric algebra of $\mathfrak{g}_e$ is often a polynomial algebra in rank($\mathfrak{g}$) generators, even though $\mathfrak{g}_e$ is in general non-reductive; also see [45] for a constructive approach in type A. One should compare with the fact that the center of $\mathcal{W}_\chi$ is always isomorphic to $\mathcal{Z}(\mathfrak{g})$, which is independent of χ (see [50] for type A and [195, Footnote 2] in general).

We pose the problem of finding a skew analogue of the above results.

Problem 4.9.5 When is the algebra of $\mathfrak{g}_e$-invariants in the exterior algebra of $\mathfrak{g}_e$ an exterior algebra in rank($\mathfrak{g}$) generators?

Note that the answer is affirmative when $e = 0$ or e is regular nilpotent by classical results. A counterpart of this problem for $\mathcal{W}_\chi$ would naturally lead to the following.

Problem 4.9.6 Develop a cohomology theory of finite W-algebras.

The connection between shifted Yangians and finite W-algebras of type A is provided by explicit formulas [48]. Being explicit, these formulas greatly facilitate the detailed study in type A, but on the other hand, they seem to be too rigid to allow direct generalizations to other types or to the affine setting. This leads to the following.

Problem 4.9.7 Redevelop the connections between Yangians and finite W-algebras in type A from the BRST approach. Formulate a generalization of Remark 4.9.1, which should be viewed as an affine analogue of the above connections in type A.

Weiqiang Wang
Department of Mathematics
University of Virginia
Charlottesville, VA 22904 USA
ww9c@virginia.edu

Fields Institute Communications
Volume **59**, 2011

CHAPTER 5

Extended Affine Lie Algebras – An Introduction to Their Structure Theory

Erhard Neher

Abstract. We give an introduction to the structure theory of extended affine Lie algebras, which provide a common framework for finite-dimensional simple, affine and toroidal Lie algebras.

Introduction

Extended affine Lie algebras form a category of Lie algebras containing finite-dimensional simple, affine, toroidal and some other interesting classes of Lie algebras.

Like finite-dimensional simple Lie algebras, extended affine Lie algebras are defined by a set of axioms prescribing their internal structure, rather than a potentially elusive presentation. The structure of extended affine Lie algebras is now well understood, and is quite similar to the construction of affine Lie algebras: All of them are obtained from a generalized loop algebra, a so-called invariant Lie torus, by taking a central extension and adding some derivations:

central extension of L (another Lie torus) —(add derivations)⇝ extended affine Lie algebra

↑ (from) invariant Lie torus L

Invariant Lie tori have been classified. Although there are some rather sophisticated examples, many of them have a concrete matrix realization or can be described in terms of familiar objects like finite-dimensional simple Lie algebras and Laurent polynomial rings. This makes extended affine Lie algebras easily accessible. Since they are an emerging new area, there are many open questions, opportunities for research and applications, for example in physics. A short history of extended affine Lie algebras is given in Section 5.2.1, in particular it describes the role physicists have played.

The goal of these notes is to provide a survey of the structure theory of extended affine Lie algebras, accessible to graduate students. Very little is known in the representation theory of extended affine Lie algebras. The reader can find some hints in Chapter 6 of this book.

The emphasis of these notes is on examples. They are not an exposition of the structure theory containing all proofs. Such an exposition will hopefully appear elsewhere. Thus, while we have endeavored to present a complete picture of the theory by giving precise definitions and statements of theorems, most of the proofs have been left out. But references to proofs are provided, as far as possible.

Outline. Section 5.1 reviews the construction of affine Kac-Moody algebras and discusses some natural generalizations, like toroidal algebras. It also contains an exposition of central extensions of Lie algebras, which are crucial for the theory. The following Section 5.2 starts with the definition of an extended affine Lie algebra and then presents some easily proven properties. We also give examples of extended affine Lie algebras: finite-dimensional split simple, affine Kac-Moody and untwisted multi-loop algebras. Part of the axioms for an extended affine Lie algebra is the existence of a root space decomposition. Section 5.3 describes the structure of the roots occurring in an extended affine Lie algebra, naturally called extended affine root systems. They turn out to be special types of so-called affine reflection systems, which is a more easily accessible class of "root systems". The bulk of Section 5.3 is therefore devoted to presenting the definition, examples and the structure of affine reflection systems, and it is only at the very end that we specialize to extended affine root systems. In Section 5.4 we reverse the picture above: We start with an extended affine Lie algebra and, using the structure of affine reflection systems, we associate to it an ideal, the so-called core, and its central quotient, the centreless core. It turns out that the core is a Lie torus and that the centreless core is an invariant Lie torus. This section also presents properties of Lie tori, examples and parts of their classification. Finally, in Section 5.5 we survey the general construction of extended affine Lie algebras, as summarized in the picture above.

Prerequisites. We assume that the reader is familiar with the basic structure theory of complex finite-dimensional semisimple Lie algebras, as for example developed in [126]. Some familiarity with affine Kac-Moody algebras, e.g., Chapters 7 and 8 of [137], is helpful but not essential, since Section 5.1.1 will give a short review of the necessary background. Similarly, knowing split simple Lie algebras will facilitate reading the notes, but is not required. A short summary of the facts used here is presented in Section 5.2.3.

Notation and setting. With some rare exceptions (in Sections 5.1.1, 5.5.1, and 5.5.2), all vector spaces and algebras are defined over a field F of characteristic 0. *We will not assume that F is algebraically closed*, since this is not needed and would not do proper justice to the theory to be explained here. Thus, F could be, but need not be the field $\mathbb{C}$ of complex numbers or the field $\mathbb{R}$ of real numbers or the field of rational numbers $\mathbb{Q}$ or ... Unless specified otherwise, linear maps will always be F-linear. All unadorned tensor products will be over F.

With the exception of some remarks, all algebras will be associative or Lie algebras. For an F-algebra A we denote by $\mathrm{Der}_F(L)$ the Lie algebra of all derivations of L (recall that an F-linear map $d : L \to L$ is a *derivation* if $d([l_1, l_2]) = [d(l_1), l_2] + [l_1, d(l_2)]$ holds for all $l_1, l_2 \in L$). We let $Z(L) = \{z \in L : [z, L] = 0\}$ denote the *centre* of a Lie algebra L. We will say that a Lie algebra L is *centreless*, if $Z(L) = 0$, and call L *perfect* if $L = [L, L]$. If K is a subspace of a Lie algebra E, the *centralizer* of K in E is $C_E(K) = \{c \in E : [c, K] = 0\}$. The symbol $\mathfrak{g}$ will always denote a split simple finite-dimensional Lie algebra.

The algebras and vector spaces considered here will often be graded by some abelian group, usually denoted Λ and always written additively. A Λ-*grading* of a vector space V by the abelian group Λ is a decomposition $V = \bigoplus_{\lambda\in\Lambda} V^\lambda$ into subspaces V^λ. Suppose V is such a Λ-graded vector space. Then the *Λ-support* of V is defined as $\operatorname{supp}_\Lambda V = \{\lambda \in \Lambda : V^\lambda \neq 0\}$. A *graded subspace of* V is a subspace U of V satisfying $U = \bigoplus_\lambda (U \cap V^\lambda)$. We will say that V has *finite bounded dimension* if there exists a constant M such that for all $\lambda \in \Lambda$ we have $\dim V^\lambda \leq M$. Note that this is a stronger condition than requiring that V has *finite homogeneous dimension*, which by definition just means that every V^λ, $\lambda \in \Lambda$, is finite-dimensional.

Given two Λ-graded vector spaces $V = \bigoplus_\lambda V^\lambda$ and $W = \bigoplus_\lambda W^\lambda$, we say an F-linear map $f : V \to W$ has *degree* λ if $f(V^\mu) \subseteq W^{\lambda+\mu}$ holds for all $\mu \in \Lambda$. We denote by $\operatorname{Hom}_F(V,W)^\lambda$ the linear maps of degree λ and put

$$\operatorname{grHom}_F(V,W) = \bigoplus_{\lambda\in\Lambda} \operatorname{Hom}_F(V,W)^\lambda \quad \text{and} \quad \operatorname{grEnd}_F(V) = \operatorname{grHom}_F(V,V).$$

We note that $\operatorname{grEnd}_F(V)$ is a Λ-graded associative algebra with respect to composition of maps. We give F the trivial grading $F = F^0$ and define the *graded dual space of* V as

$$V^{\mathrm{gr}*} = \operatorname{grHom}_F(V,F) = \bigoplus_{\lambda\in\Lambda} (V^{\mathrm{gr}*})^\lambda.$$

Observe that $(V^{\mathrm{gr}*})^\lambda$ consists of those linear forms $\varphi : V \to F$ which satisfy $\varphi(V^\mu) = 0$ whenever $\lambda + \mu \neq 0$ and can therefore be identified with the usual dual space $(V^{-\lambda})^*$.

Given a symmetric bilinear form on a vector space V, an endomorphism d of V is called *skew-symmetric* if $(d(v) \mid v) = 0$ for all $v \in V$. Since we assume that our base field has characteristic 0, this is equivalent to the condition $(d(v_1) \mid v_2) + (v_1 \mid d(v_2)) = 0$ for all $v_1, v_2 \in V$. A bilinear form is *nondegenerate* if $(v \mid u) = 0$ for all $u \in V$ implies $v = 0$.

If A is an algebra, a *Λ-grading of the algebra* A is a Λ-grading of the underlying vector space A, say $A = \bigoplus_{\lambda\in\Lambda} A^\lambda$, for which in addition $A^\lambda A^\mu \subseteq A^{\lambda+\mu}$ holds for all $\lambda, \mu \in \Lambda$. Two Λ-graded algebras, say $A = \bigoplus_{\lambda\in\Lambda} A_\lambda$ and $A' = \bigoplus_{\lambda\in\Lambda} A'_\lambda$ are called *graded-isomorphic* if there exists an algebra isomorphism $f : A \to A'$ respecting the homogenous subspaces, i.e., $f(A_\lambda) = A'_\lambda$ for all $\lambda \in \Lambda$. We will often deal with algebras with two gradings, e.g., in Section 5.4. In this case, it is convenient to use superscripts and subscripts to distinguish them.

The author would like to thank all the participants of the summer school for their interest and questions, and Bruce Allison and Juana Sánchez Ortega for their remarks on an earlier version of these notes.

Acknowledgments. This work is partially supported by the Natural Sciences and Engineering Research Council (NSERC) of Canada through the author's Discovery Grant.

5.1 Affine Lie algebras and some generalizations

We will always assume that F is a field of characteristic 0. Occasionally we will need some roots of unity in F, so certainly an algebraically closed field like $\mathbb{C}$ will do.

We denote by $\mathfrak{g}$ a split simple finite-dimensional Lie algebra over F. For example, if F is algebraically closed then this just means that $\mathfrak{g}$ is a simple and finite-dimensional. Their structure theory is explained in most standard textbooks, for example in [126]. For more general fields, an example of a split simple $\mathfrak{g}$ is the

Lie algebra $\mathfrak{sl}_n(F)$ of $n \times n$-matrices over F which have trace 0. These types of Lie algebra are for example investigated in [40, Ch. VII], [87, Ch. 1] or [128, Ch. IV].

5.1.1 Realization (construction of affine Kac-Moody Lie algebras). In this section we will describe the structure of affine Kac-Moody algebras in a way which is appropriate for the envisioned generalization later. The reader can find more details in [137], or consult Sections 6.2.1–6.2.3 of this book.

Let $\zeta \in F$ be a primitive mth root of 1. In other words, the multiplicative subgroup of F generated by ζ is isomorphic to $\mathbb{Z}/m\mathbb{Z}$. For example, in $F = \mathbb{C}$ we can take $\zeta = \exp(2\pi i/m)$.

Let σ be an automorphism of $\mathfrak{g}$ of finite order $m \in \mathbb{N}$. Thus, the subgroup $\langle\sigma\rangle$ of the automorphism group of $\mathfrak{g}$ is isomorphic to $\mathbb{Z}/m\mathbb{Z}$. For example, if $\mathfrak{g} = \mathfrak{sl}_n(F)$ an example of such an automorphism is $\sigma(x) = axa^{-1}$, where a is an $n \times n$-matrix of order m, and an example of such a matrix is $a = \zeta E_n$ where E_n is the $n \times n$ identity matrix.

Observe that σ is diagonalizable. Indeed, its minimal polynomial divides the polynomial $t^m - 1$ and therefore has no multiple roots in F. For a general field F this would of course only say that σ is a semisimple endomorphism. But since we assumed that F contains all roots of unity which we need, σ is diagonalizable over F. To describe its eigenspaces we require some notation. In anticipation of the later developments we put

$$\Lambda = \mathbb{Z} \quad \text{and} \quad \bar{\Lambda} = \mathbb{Z}/m\mathbb{Z},$$

and denote the canonical map $\Lambda \to \bar{\Lambda}$ by $\lambda \mapsto \bar{\lambda}$. That σ is diagonalizable, means

$$\mathfrak{g} = \textstyle\bigoplus_{\bar{\lambda}\in\bar{\Lambda}} \mathfrak{g}_{\bar{\lambda}} \quad \text{for } \mathfrak{g}_{\bar{\lambda}} = \{x \in \mathfrak{g} : \sigma(x) = \zeta^{\lambda} x\} \tag{5.1}$$

Of course, some of the $\mathfrak{g}_{\bar{\lambda}}$ could be zero. The eigenspaces of σ are precisely the non-zero subspaces $\mathfrak{g}_{\bar{\lambda}}$. It is also appropriate to note that $\mathfrak{g}_{\bar{\lambda}}$ is well-defined: if $\bar{\lambda} = \bar{\mu}$ then $\zeta^{\lambda} = \zeta^{\mu}$. Finally we point out that the decomposition (5.1) is a $\bar{\Lambda}$-grading, which means that it satisfies

$$[\mathfrak{g}_{\bar{\lambda}}, \mathfrak{g}_{\bar{\mu}}] \subseteq \mathfrak{g}_{\bar{\lambda}+\bar{\mu}} \quad \text{for all } \bar{\lambda}, \bar{\mu} \in \bar{\Lambda}. \tag{5.2}$$

Let $F[t^{\pm 1}]$ be the ring of Laurent polynomials. This is a unital associative commutative F-algebra with F-basis $\{t^{\lambda} : \lambda \in \mathbb{Z}\}$ and multiplication rule $t^{\lambda}t^{\mu} = t^{\lambda+\mu}$.

The *loop algebra* associated to the data $(\mathfrak{g}, \sigma)$ is the Lie algebra

$$\mathcal{L} = L(\mathfrak{g}, \sigma) = \textstyle\bigoplus_{\lambda\in\Lambda} \mathfrak{g}_{\bar{\lambda}} \otimes Ft^{\lambda} \tag{5.3}$$

with product

$$[u_{\bar{\lambda}} \otimes t^{\lambda},\, v_{\bar{\mu}} \otimes t^{\mu}] = [u_{\bar{\lambda}}, v_{\bar{\mu}}] \otimes t^{\lambda+\mu}. \tag{5.4}$$

We will sometimes use more precise terminology: If $\sigma = \mathrm{Id}$, i.e., $m = 1$, we will call $L(\mathfrak{g}, \mathrm{Id}) = \mathfrak{g} \otimes F[t^{\pm 1}]$ the *untwisted loop algebra*, and we will call $L(\mathfrak{g}, \sigma)$ a *twisted loop algebra* if it is clear that $\sigma \neq \mathrm{Id}$ and we want to emphasize this.

We point out that we consider $L(\mathfrak{g}, \sigma)$ as a Lie algebra over F. It is therefore infinite-dimensional. It is also important to note that $\mathcal{L}$ is a Λ-graded algebra, whose homogenous spaces are $\mathcal{L}^{\lambda} = \mathfrak{g}_{\bar{\lambda}} \otimes Ft^{\lambda}$ for $\lambda \in \Lambda$. For the reader with some background in algebraic geometry, a more geometric definition of $L(\mathfrak{g}, \sigma)$ is the following: $L(\mathfrak{g}, \sigma)$ is (isomorphic to) the Lie algebra of equivariant maps $F^{\times} \to \mathfrak{g}$, where σ acts on $F^{\times}$ by $\sigma(x) = \zeta x$.

Let κ be the Killing form of $\mathfrak{g}$, i.e., $\kappa(u,v) = \operatorname{tr}(\operatorname{ad} u \circ \operatorname{ad} v)$, and define

$$\psi : \mathcal{L} \times \mathcal{L} \to F, \quad \psi(u \otimes t^\lambda, v \otimes t^\mu) = \lambda\, \delta_{\lambda,-\mu}\, \kappa(u,v) \tag{5.5}$$

where $\delta_{\lambda,-\mu}$ is the Kronecker delta: It has the value 1 if $\lambda = -\mu$ and is zero otherwise.

Exercise 5.1.1 Check that the map ψ of (5.5) is a 2-*cocycle* of $\mathcal{L}$, i.e. an F-bilinear map satisfying

$$\psi(l,l) = 0 = \psi([l_1,l_2],l_3) + \psi([l_2,l_3],l_1) + \psi([l_3,l_1],l_2) \tag{5.6}$$

for $l, l_i \in \mathcal{L}$.

As a consequence of Exercise 5.1.1, we can enlarge our Lie algebra $\mathcal{L}$ by adjoining a 1-dimensional space, denoted Fc here:

$$\tilde{\mathcal{L}} = \tilde{\mathcal{L}}(\mathfrak{g},\sigma) = L(\mathfrak{g},\sigma) \oplus Fc \tag{5.7}$$

is a Lie algebra over F with respect to the product

$$[l_1 \oplus s_1 c,\, l_2 \oplus s_2 c]_{\tilde{\mathcal{L}}} = [l_1, l_2]_L \oplus \psi(l_1,l_2)c$$

for $l_i \in \mathcal{L}$ and $s_i \in F$. We have added subscripts on the products to emphasize where the product is calculated, in $\tilde{\mathcal{L}}$ or in $\mathcal{L}$. It is obvious from the product formula, that it is important to know in which Lie algebra the product is being calculated. But in the future we will leave out the subscripts, if it is clear in which algebra the product is calculated.

The equations (5.6) are exactly what is needed to make $\tilde{\mathcal{L}}$ a Lie algebra. The map

$$\mathfrak{u} : \tilde{\mathcal{L}} \to \mathcal{L}, \quad \mathfrak{u}(l \oplus sc) = l$$

is a surjective Lie algebra homomorphism with kernel $\ker(\mathfrak{u}) = Fc = Z(\tilde{\mathcal{L}})$, the centre of $\tilde{\mathcal{L}}$. In other words, $\mathfrak{u}$ is a central extension (see 5.1.3 for a short review of central extensions). In fact, $\mathfrak{u}$ is the "biggest" central extension, the so-called universal central extension, see [110] and [221] for a proof.

The Lie algebra $\tilde{\mathcal{L}}$ has a canonical derivation d, the so-called *degree derivation*

$$d\big((u \otimes t^\lambda) \oplus sc\big) = \lambda u \otimes st^\lambda, \quad (\lambda \in \mathbb{Z}, u \in \mathfrak{g}_{\bar{\lambda}}, s \in F). \tag{5.8}$$

Hence we can form the semidirect product $\hat{\mathcal{L}} = L(\mathfrak{g},\sigma)\hat{} = \tilde{\mathcal{L}} \rtimes Fd$ with product

$$[\tilde{l}_1 \oplus s_1 d,\, \tilde{l}_2 \oplus s_2 d]_{\hat{\mathcal{L}}} = [\tilde{l}_1, \tilde{l}_2]_{\tilde{\mathcal{L}}} + s_1 d(\tilde{l}_2) - s_2 d(\tilde{l}_1)$$

for $\tilde{l}_i \in \tilde{\mathcal{L}}$ and $s_i \in F$. In untangled form,

$$\hat{\mathcal{L}} = \big(\textstyle\bigoplus_{\lambda \in \mathbb{Z}} (\mathfrak{g}_{\bar{\lambda}} \otimes Ft^\lambda)\big) \oplus Fc \oplus Fd \tag{5.9}$$

is the Lie algebra with product

$$\begin{aligned}&[u_{\bar{\lambda}} \otimes t^\lambda \oplus s_1 c \oplus s_1' d,\, v_{\bar{\mu}} \otimes t^\mu \oplus s_2 c \oplus s_2' d] \\ &\quad = \big([u_{\bar{\lambda}}, v_{\bar{\mu}}] \otimes t^{\lambda+\mu} + \mu s_1' v_{\bar{\mu}} \otimes t^\mu - \lambda s_2' u_{\bar{\lambda}} \otimes t^\lambda\big) \oplus \lambda\, \delta_{\lambda,-\mu}\, \kappa(u_{\bar{\lambda}}, v_{\bar{\mu}})\, c. \end{aligned} \tag{5.10}$$

Exercise 5.1.2 Show $[\hat{\mathcal{L}}, \hat{\mathcal{L}}] = \tilde{\mathcal{L}}$ and $Z(\tilde{\mathcal{L}}) = Fc = Z(\hat{\mathcal{L}})$.

The importance of the Lie algebras $L(\mathfrak{g},\sigma)\hat{}$ stems from the following.

Theorem 5.1.3 (**Realization Theorem** [137, Th. 7.4, Th. 8.3, Th. 8.5]) *Suppose F is algebraically closed.*

(a) *The Lie algebra $L(\mathfrak{g}, \sigma)\hat{}$ is an affine Kac-Moody Lie algebra, and every affine Kac-Moody Lie algebra is isomorphic (as F-algebra) to some $L(\mathfrak{g}, \sigma)\hat{}$.*

(b) *$L(\mathfrak{g}, \sigma)\hat{} \cong L(\mathfrak{g}, \sigma')\hat{}$ where σ' is a diagram automorphism with respect to some Cartan subalgebra of $\mathfrak{g}$.*

We note that diagram automorphisms have order 1, 2 or 3, with the latter case only occurring for $\mathfrak{g}$ of type D_4.

5.1.2 Multiloop and toroidal Lie algebras. We will discuss some (straightforward) generalizations of $\mathcal{L} = L(\mathfrak{g}, \sigma)$, the central extension $\tilde{\mathcal{L}}$ and the big Lie algebra $\hat{\mathcal{L}}$.

The first idea is to replace the Laurent polynomial ring $F[t^{\pm 1}]$ by a ring with similar properties. Instead of one variable we will use the Laurent polynomial ring $F[t_1^{\pm 1}, \dots, t_n^{\pm 1}]$ in n variables. This ring has indeed very similar properties to the ring $F[t^{\pm 1}]$. We put $\Lambda = \mathbb{Z}^n$ and define

$$t^\lambda = t_1^{\lambda_1} \cdots t_n^{\lambda_n} \quad \text{for } \lambda = (\lambda_1, \dots, \lambda_n) \in \Lambda$$

Then $\{t^\lambda : \lambda \in \Lambda\}$ is an F-basis of $F[t_1^{\pm 1}, \dots, t_n^{\pm 1}]$ and the multiplication rule in $F[t_1^{\pm 1}, \dots, t_n^{\pm 1}]$ is $t^\lambda t^\mu = t^{\lambda+\mu}$, which is the "same" as in the 1-variable case. Also, $F[t_1^{\pm 1}, \dots, t_n^{\pm 1}]$ is still a unital commutative associative F-algebra. We can therefore define the *untwisted multiloop algebra*, the "several variables" generalization of the untwisted loop algebra of Section 5.1.1 as

$$L(\mathfrak{g}) = \mathfrak{g} \otimes F[t_1^{\pm 1}, \dots, t_n^{\pm 1}], \tag{5.11}$$

which becomes a Lie algebra with respect to the product

$$[u \otimes t^\lambda, v \otimes t^\mu] = [u, v] \otimes t^{\lambda+\mu}$$

for $u, v \in \mathfrak{g}$ and $\lambda, \mu \in \mathbb{Z}^n$. We will meet this Lie algebra again in Example 5.4.33.

To continue the analogy we let $\boldsymbol{\sigma} = (\sigma_1, \dots, \sigma_n)$ be a family of n commuting finite order automorphisms of $\mathfrak{g}$, say σ_i has order $m_i \in \mathbb{N}_+$. Let $\zeta_i \in F$ be a primitive m_i-th root of 1 (recall that we assumed that F has an ample supply of them). We put

$$\bar{\Lambda} = (\mathbb{Z}/m_1\mathbb{Z}) \oplus \cdots \oplus (\mathbb{Z}/m_n\mathbb{Z})$$

and let $\lambda \mapsto \bar{\lambda}$ be the obvious map. The automorphisms σ_i are simultaneously diagonalizable:

$$\mathfrak{g} = \bigoplus\nolimits_{\bar{\lambda} \in \bar{\Lambda}} \mathfrak{g}_{\bar{\lambda}}, \quad \mathfrak{g}_{\bar{\lambda}} = \{u \in \mathfrak{g} : \sigma_i(u) = \zeta_i^{\lambda_i} u, 1 \le i \le n\}. \tag{5.12}$$

As in the one-variable case, the decomposition (5.12) is a $\bar{\Lambda}$-grading: $[\mathfrak{g}_{\bar{\lambda}}, \mathfrak{g}_{\bar{\mu}}] \subseteq \mathfrak{g}_{\bar{\lambda}+\bar{\mu}}$ for $\bar{\lambda}, \bar{\mu} \in \bar{\Lambda}$. It follows from this that

$$L(\mathfrak{g}, \boldsymbol{\sigma}) = \bigoplus\nolimits_{\lambda \in \Lambda} \mathfrak{g}_{\bar{\lambda}} \otimes F t^\lambda \tag{5.13}$$

is a subalgebra of $\mathfrak{g} \otimes F[t_1^{\pm 1}, \dots, t_n^{\pm 1}]$, called the *multiloop algebra associated to $\mathfrak{g}$ and $\boldsymbol{\sigma}$*. If all $\sigma_i = \mathrm{Id}_{\mathfrak{g}}$ it coincides with the untwisted multiloop algebra defined above. Multiloop algebras are investigated in the papers [4, 5, 7, 8, 9, 10].

Following our procedure in Section 5.1.1 we should now make a central extension to get a bigger Lie algebra $\tilde{\mathcal{L}}$ and then add some derivations:

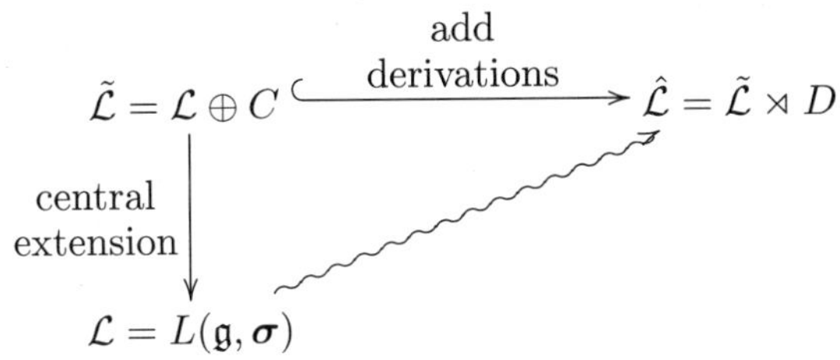

(5.14)

To define the Lie algebra product on $\tilde{\mathcal{L}}$ we would use a 2-cocycle $\psi : \mathcal{L} \times \mathcal{L} \to C$ where C is some vector space and then put

$$[l_1 \otimes c_1, l_1 \otimes c_2]_{\tilde{\mathcal{L}}} = [l_1, l_2]_{\mathcal{L}} \oplus \psi(l_1, l_2) \tag{5.15}$$

for $l_i \in \mathcal{L}$ and $c_i \in C$. The Lie algebra $\hat{\mathcal{L}}$ should be a semidirect product with D acting on $\tilde{\mathcal{L}}$ by derivations.

But here is where the problems start, or things become interesting depending on one's taste. In the one-variable case the 2-cocycle ψ of (5.5) was the only possible choice up to scalars, i.e., the universal central extension $\tilde{\mathcal{L}}$ of $\mathcal{L}$ had a 1-dimensional centre $C = Fc$. This is no longer true in the case of several variables. It is not so surprising that there exists a 2-cocycle with values in F^n: We can simply use the same formula as in (5.5).

Exercise 5.1.4 Let $\mathcal{L} = L(\mathfrak{g}, \boldsymbol{\sigma})$ be a multiloop algebra and embed $\Lambda \subseteq F^n$ canonically. Then $\psi : \mathcal{L} \times \mathcal{L} \to F^n$, given by

$$\psi(u \otimes t^\lambda, v \otimes t^\mu) = \delta_{\lambda+\mu,0}\, \kappa(u, v)\, \lambda\,, \tag{5.16}$$

is a 2-cocycle of $\mathcal{L}$.

However, this is still not the "biggest" possible choice. Rather, the centre of the universal central extension is infinite-dimensional if $n \geq 2$, and the so-called *universal* 2*-cocycle*, i.e., the 2-cocycle used in (5.15) to describe the universal central extension $\hat{\mathcal{L}}$ of $\mathcal{L}$, is described in the following result.

Theorem 5.1.5 ([188]) *Let $\mathcal{L} = L(\mathfrak{g}, \boldsymbol{\sigma})$ be a multiloop algebra. We embed $\Lambda \subseteq F^n$ canonically, put $\Gamma = m_1\mathbb{Z} \oplus \cdots \oplus m_n\mathbb{Z}$ and let $C = \bigoplus_{\gamma \in \Gamma} C_\gamma$ where $C_\gamma = F^n/F\gamma$. Then the universal* 2*-cocycle is the map $\psi_{\mathrm{u}} : \mathcal{L} \times \mathcal{L} \to C$ whose γ-component is*

$$\psi_{\mathrm{u}}(u \otimes t^\lambda, v \otimes t^\mu)_\gamma = \kappa(u, v)\delta_{\lambda+\mu,-\gamma}\, \bar{\lambda} \in C_\gamma. \tag{5.17}$$

Observe that (5.16) is just the 0-component of (5.17). The theorem is well-known in the untwisted case (all $\sigma_i = \mathrm{Id}_{\mathfrak{g}}$, so $\Gamma = \Lambda$), in which it can be deduced from the description of the universal central extension of the Lie algebra $\mathfrak{g} \otimes A$ where A is any unital commutative associative F-algebra, see [145] and [175]. (In these references the centre C of the universal central extension is described as Ω_A/dA where Ω_A is the module of Kähler differentials, which is also the same as the first cyclic homology group $\mathrm{HC}_1(A)$.)

In the untwisted case, the universal central extension $\hat{\mathcal{L}}$ was termed the *n-toroidal Lie algebra* based on $\mathfrak{g}$. The reader should however be warned that this terminology is not standard. It is sometimes used for the Lie algebra $\tilde{\mathcal{L}}$ with the

2-cocycle of Exercise 5.1.4, and sometimes also for the Lie algebras of the form $\hat{\mathcal{L}} = \mathcal{L} \oplus C \oplus D$ for an appropriate subalgebra D of derivations, e.g. in [86].

Thus, there are many possibilities for C in the diagram (5.14), and it is not clear which one is the best possible choice. In fact, we will later in 5.5.4 allow almost any central extension.

Assuming that we have settled for some space C, which D should we take? For simplicity we will discuss this only in the untwisted case. If $n = 1$ we added the degree derivation d defined in (5.8). This is far from being an arbitrary derivation. The full derivation algebra of the Lie algebra $\mathfrak{g} \otimes A$ for A is described in [25, Th. 1]:

$$\begin{aligned} \mathrm{Der}_F(\mathfrak{g} \otimes A) &= \big(\mathrm{Der}_F(\mathfrak{g}) \otimes A\big) \oplus \big(F\,\mathrm{Id} \otimes \mathrm{Der}_F(A)\big) \\ &= \quad \mathrm{IDer}(\mathfrak{g} \otimes A) \oplus \big(F\,\mathrm{Id} \otimes \mathrm{Der}_F(A)\big). \end{aligned} \tag{5.18}$$

where $\mathrm{Der}_F(\mathfrak{g}) \otimes A$ and $F\,\mathrm{Id} \otimes \mathrm{Der}_F(A)$ act on $\mathfrak{g} \otimes A$ in the obvious way.

Since $\mathfrak{g} \otimes A$ is perfect, up to a canonical isomorphism, this is then also the derivation algebra of the universal central extension of $\mathfrak{g} \otimes A$ (see for example [25, Th. 2.2]). From

$$\mathrm{Der}\, F[t^{\pm 1}] = F[t^{\pm 1}]d$$

we see that we added a rather special derivation, one which can be used to define the Λ-grading of $\mathcal{L}$. This will be made more precise in 5.5.1 and 5.5.3 where we will describe Fd as the space $\mathrm{SCDer}_F(\mathcal{L})$ of skew centroidal derivations of $\mathcal{L}$, see also Exercise 5.1.7.

We can do something similar in the multi-variable case. Define the i-th *degree derivation* ∂_i of $L(\mathfrak{g}) \oplus C$ by

$$\partial_i(u \otimes t^\lambda \oplus c) = \lambda_i\, u \otimes t^\lambda \quad \text{for } \lambda = (\lambda_1, \dots, \lambda_n) \in \Lambda = \mathbb{Z}^n \tag{5.19}$$

and put

$$\mathcal{D} = \mathrm{Span}_F\{\partial_i : 1 \le i \le n\},$$

the space of *degree derivations*. Possible (interesting) choices for D are:

(i) $D = \mathcal{D}$,

(ii) $F[t_1^{\pm 1}, \dots, t_n^{\pm 1}]\mathcal{D}$ (in physics parlance: "all vector fields"), and

(iii) $\bigoplus_{\lambda \in \Lambda} F t^\lambda \{\sum_{i=1}^n s_i \partial_i : \sum_i s_i = 0\}$ (the "divergence 0 vector fields").

It will turn out (Theorem 5.5.15) that for the Lie algebras which we are going to study in the next sections, the choices (1) and (3) are the correct ones. In addition, there will be a surprise: semidirect products in (5.14) will not be enough!

Exercise 5.1.6 Recall that a bilinear form $(\cdot|\cdot)$ on a Lie algebra L is called *invariant* (= *associative*) if $([l_1, l_2] \mid l_3) = (l_1 \mid [l_2, l_3])$ holds for all $l_i \in L$. Show:

(a) The set $\mathrm{IF}(L)$ of invariant bilinear forms on L is a vector space with respect to the obvious scalar multiplication and addition defined by $(\beta_1 + \beta_2)(l_1, l_2) = \beta_1(l_1, l_2) + \beta_2(l_1, l_2)$ for $\beta_i \in \mathrm{IF}(L)$ and $l_i \in L$.

(b) If L is perfect, any invariant bilinear form is symmetric.

(c) Let S be a unital associative F-algebra. A bilinear form b on S is called *invariant* if $b(s_1 s_2, s_3) = b(s_1, s_2 s_3) = b(s_2, s_3 s_1)$ for $s_i \in S$.

(i) The set $\mathrm{IF}(S)$ of invariant bilinear forms on S is a vector space with respect to the obvious operations.

(ii) Any linear form $\lambda \in S^*$ with $\lambda([S, S]) = 0$ gives rise to an invariant bilinear form b_λ on S, defined by $b_\lambda(s_1, s_2) = \lambda(s_1 s_2)$.

(iii) The map $(S/[S,S])^* \to \mathrm{IF}(S)$, given by $\lambda \mapsto b_\lambda$, is a vector space isomorphism.

(d) Let L be a perfect Lie algebra with a 1-dimensional space $\mathrm{IF}(L)$, say $\mathrm{IF}(L) = F\kappa$. Also, let S be a unital associative commutative F-algebra. We consider $L \otimes S$ as Lie algebra with respect to the product $[l_1 \otimes s_1, l_2 \otimes s_2] = [l_1, l_2] \otimes s_2 s_2$, cf. (5.5). For $\lambda \in \mathrm{IF}(S)$ define a bilinear form $\kappa \otimes \lambda$ on $L \otimes S$ by

$$(\kappa \otimes \lambda)\,(l_1 \otimes s_2,\, l_2 \otimes s_2) = \kappa(l_1, l_2)\,\lambda(s_1, s_2).$$

Then $\kappa \otimes \lambda \in \mathrm{IF}(L \otimes S)$ and the map $\mathrm{IF}(S) \to \mathrm{IF}(L \otimes S)$, given by $\lambda \mapsto \kappa \otimes \lambda$, is an isomorphism of vector spaces.

Exercise 5.1.7 Define the ith degree derivation ∂_i of the Laurent polynomial ring $S = F[t_1^{\pm 1}, \dots, t_n^{\pm 1}]$ by $\partial_i(t^\lambda) = \lambda_i t^\lambda$, so that the derivation ∂_i of (5.19) becomes $\partial_i(u \otimes t^\lambda) = u \otimes \partial_i(t^\lambda) = (\mathrm{Id} \otimes \partial_i)(u \otimes t^\lambda)$ (this double meaning of ∂_i should not create any confusion). Show:

(a) The derivation algebra $\mathrm{Der}_F(S)$ of S is given by

$$\mathrm{Der}_F(S) = S\mathcal{D} = \bigoplus\nolimits_{\lambda \in \mathbb{Z}^n} F t^\lambda \mathcal{D}$$

where, as above, $\mathcal{D} = \mathrm{Span}_F\{\partial_i : 1 \le i \le n\}$. The derivation algebra is a $\mathbb{Z}^n$-graded Lie algebra with Lie algebra product determined by

$$[t^\lambda \partial_i, t^\mu \partial_j] = t^{\lambda+\mu}(\mu_i \partial_j - \lambda_j \partial_i).$$

Thus, for $n = 1$ we obtain the usual *Witt algebra*, see for example [174, 1.4].

(b) $(t^\lambda \mid t^\mu) = \delta_{\lambda+\mu,0}$ defines a nondegenerate symmetric bilinear form $(\cdot|\cdot)$ on S which is invariant as defined in Exercise 5.1.6.

(c) Let $\mathrm{SDer}_F(S)$ be the subalgebra of derivations of S, which are skew-symmetric with respect to the form $(\cdot|\cdot)$ of (b). Then

$$\mathrm{SDer}_F(S) = \bigoplus\nolimits_{\lambda \in \mathbb{Z}^n} F t^\lambda \big\{ \textstyle\sum_{i=1}^n s_i \partial_i : \sum_i s_i \lambda_i = 0 \big\}.$$

In particular, for $n = 1$ we get $\mathrm{SDer}_F(F[t^{\pm 1}]) = Fd$ for $d = \partial_1$.

5.1.3 Appendix on central extensions of Lie algebras. Central extensions will turn out to be an important tool in the construction of extended affine Lie algebras. Although this provides one with a bigger and hence potentially more complicated Lie algebra, central extensions turn up naturally in the general theory and the biggest of them (the universal central extension) is in fact quite "nice". For example, universal central extensions often have a simpler presentation and a much richer representation theory than the original Lie algebra. In this appendix we review the necessary background.

Definition 5.1.8 (Extensions) An *extension of a Lie algebra* L is a surjective homomorphism $f : K \to L$ of Lie algebras. A *homomorphism* from an extension $f : K \to L$ to another extension $f' : K' \to L$ is a Lie algebra homomorphism $g : K \to K'$ satisfying $f = f' \circ g$. In other words, the diagram below is commutative.

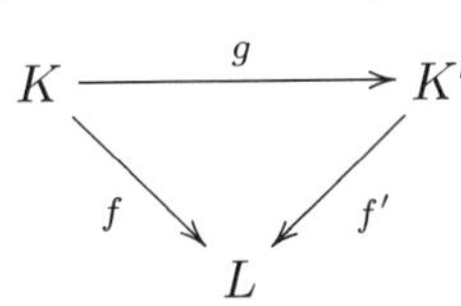

(5.20)

We will use *abelian extensions*, i.e., extensions $f : K \to L$ with $\ker f$ an abelian ideal in the construction of an extended affine Lie algebra in Section 5.5.4.

Definition 5.1.9 (Central extensions) A *central extension* of L is an extension $f: K \to L$ whose kernel $\ker f$ is contained in the centre $Z(K)$ of K. A central extension $f: K \to L$ is called a *covering* if K is perfect, i.e., $K = [K, K]$. It is traditional (but not always advisable) to not specify the morphism f and simply say that *K is a central extension of L* or a *covering*.

A central extension $\mathfrak{u}: \mathfrak{L} \to L$ is called a *universal central extension* if there exists a unique homomorphism from $\mathfrak{u}: \mathfrak{L} \to L$ to any other central extension $f: K \to L$ of L. It is obvious from the universal property that two universal central extensions of L are isomorphic as central extensions and hence in particular their underlying Lie algebras are isomorphic. We denote the universal central extension of L by $\mathfrak{u}: \mathfrak{uce}(L) \to L$ or simply $\mathfrak{uce}(L)$.

Theorem 5.1.10 ([213, Prop. 1.3], [110, §1]) *A Lie algebra L has a universal central extension if and only if L is perfect. In this case, the universal central extension $\mathfrak{u}: \mathfrak{uce}(L) \to L$ is perfect too, i.e., $\mathfrak{u}$ is a covering.*

The process of taking universal central extensions stops at $\mathfrak{uce}(L)$, due to the following equivalent conditions for a Lie algebra L:

(i) $\mathrm{Id}: L \to L$ is a universal central extension, i.e., $\mathfrak{uce}(L) = L$,

(ii) every central extension $f: K \to L$ is uniquely a direct product $K = \tilde{L} \times \ker f$ such that $f|_{\tilde{L}}$ is an isomorphism between $\tilde{L}$ and L.

If (i) and (ii) hold, one calls L *centrally closed*.

Examples 5.1.11 (a) It is an immediate corollary of the Levi-Malcev Theorem that every finite-dimensional semisimple Lie algebra is centrally closed ([40, VII, §6.8, Cor. 3] or [219, Cor. 7.9.5]).

(b) An example of a universal central extension is the Virasoro algebra: It is the universal central extension of the Witt algebra $\mathrm{Der}_F(F[t^{\pm 1}])$, see for example [174, I.9, Prop. 4]. Hence the Virasoro algebra is centrally closed, while $\mathrm{Der}_F(F[t^{\pm 1}])$ is not. On the other hand, the higher rank Witt algebra $\mathrm{Der}_F(F[t_1^{\pm 1}, \ldots, t_n^{\pm 1}])$, $n > 1$, is centrally closed ([199, V, Th. 5.1]).

Definition 5.1.12 (Central extensions via 2-cocycles.) We have already seen in 5.1.1 that one can construct central extensions of a Lie algebra L by using *2-cocycles*, which, we recall, are bilinear maps $\psi: L \times L \to C$ into a vector space C satisfying for all $l, l_1, l_2, l_3 \in L$

$$\psi(l, l) = 0 \quad \text{and} \quad \psi([l_1, l_2], l_3) + \psi([l_2, l_3], l_1) + \psi([l_3, l_1], l_2) = 0. \tag{5.21}$$

The first equation is of course equivalent to $\psi(l_1, l_2) = -\psi(l_2, l_1)$. Given a 2-cocycle $\psi: L \times L \to C$, the algebra

$$K = L \oplus C \quad \text{with} \quad [l_1 \oplus c_1, \, l_2 \oplus c_2]_K = [l_1, l_2]_L \oplus \psi(l_1, l_2) \tag{5.22}$$

($l_i \in L$, $c_i \in C$) is a Lie algebra and $\mathrm{pr}_L: K \to L$, $\mathrm{pr}_L(l \oplus c) = l$, is a central extension of L, which we will denote by $\mathrm{E}(L, C, \psi)$ or $\mathrm{E}(L, \psi)$ for short.

Conversely, given a central extension $f: K \to L$, let $s: L \to K$ be a *section of f* in the category of vector spaces, i.e., a linear map $s: L \to K$ such that $f \circ s = \mathrm{Id}_L$. Such a section always exists: We can choose a subspace L' of K, which is complementary to $C = \ker f$, and take $s = (f|_{L'})^{-1}$ which makes sense since $(f|L'): L' \to L$ is an invertible linear map (but in general not a Lie algebra homomorphism since L' need not be a subalgebra). Given a section s, the map

$$\psi_s: L \times L \to C, \quad \psi_s(l_1, l_2) = [s(l_1), s(l_2)]_K - s([l_1, l_2]_L) \tag{5.23}$$

turns out to be a 2-cocycle. Moreover, the map

$$K \to L \oplus C, \quad x \mapsto f(x) \oplus \big(x - (s \circ f)(x)\big) = f(x) \oplus x_C,$$

where x_C is the C-component of $x \in K$, is an isomorphism from the central extension $f : K \to L$ to the central extension $\mathrm{E}(L, \ker f, \psi_s)$. To summarize, modulo some verifications left as an exercise, we have proven the following well-known result.

Proposition 5.1.13 *For any 2-cocycle ψ the construction* (5.22) *is a central extension $\mathrm{E}(L, \psi)$ of L and, conversely, every central extension L is isomorphic as central extension to some $\mathrm{E}(L, \psi)$.*

Exercise 5.1.14 Let $\psi : L \times L \to C$ be a 2-cocycle and let C' be a subspace of C satisfying $\psi(L, L) := \mathrm{Span}_F\{\psi(l_1, l_2) : l_i \in L\} \subseteq C'$. Then $\mathrm{E}(L, C', \psi)$ is also a central extension, and if $\mathrm{E}(L, C, \psi)$ is a covering then $C = \psi(L, L)$.

Examples 5.1.15 (a) Any Lie algebra L has many uninteresting central extensions. One can simply take the direct product of L with an abelian Lie algebra, i.e., $L \times C$ with product $[(l_1, c_1), (l_2, c_2)] = ([l_1, l_2], 0)$ for $l_i \in L$, $c_i \in C$, and consider the canonical projection $\mathrm{pr}_L : L \times C \to L$, which is a central extension (but not a covering, unless L is perfect and $C = \{0\}$). Observe that the canonical inclusion $\mathrm{inc} : L \to L \times C$ is a section of pr_L, not only in the category of vector spaces, but even in the category of Lie algebras. Its associated 2-cocycle is $\psi_{\mathrm{inc}} = 0$.

(b) Let $h : L \to C$ be a linear map into some vector space C. Then $\beta_h : L \times L \to C$, $\beta_h(l_1, l_2) = h([l_1, l_2])$ is a 2-cocycle, a so-called 2-*coboundary*.

The two examples are related in the following exercise.

Exercise 5.1.16 For a central extension $f : K \to L$ of L with $C = \ker f$ the following are equivalent:

(i) The extension $f : K \to L$ is split in the category of Lie algebras, i.e, there exists a section $L \to K$ of f, which is a Lie algebra homomorphism.
(ii) For any section s of f the associated 2-cocycle ψ_s is a 2-coboundary.
(iii) There exists a section s of f, for which the associated 2-cocycle ψ_s is a 2-coboundary.
(iv) As central extension, f is isomorphic to the central extension $\mathrm{pr}_L : L \oplus C \to L$.

If these conditions are fulfilled, one calls f a *split extension*.

Exercise 5.1.17 Let $\psi : L \times L \to C$ be a 2-cocycle and let $\pi : C \to C'$ be a linear map. Show:

(a) $\psi' = \pi \circ \psi$ is a 2-cocycle of L and the map

$$\mathrm{E}(\pi) : \mathrm{E}(L, C, \psi) \to \mathrm{E}(L, C', \psi'), \quad l \oplus c \mapsto l \oplus \pi(c)$$

is a homomorphism of central extensions of L:

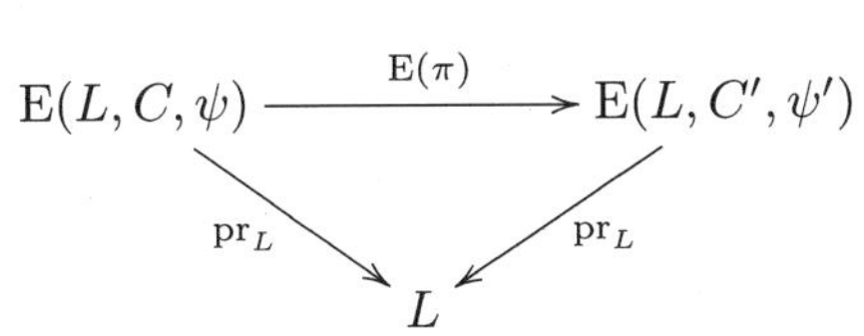

(b) If π is surjective, the map $\mathrm{E}(\pi)$ is a central extension of $L' = \mathrm{E}(L, C', \pi \circ \psi)$, which as central extension of L' has the form $\mathrm{E}(L', C'', \psi'')$ for

$$\psi''(l_1 \oplus c_1', l_2 \oplus c_2') = \big((\mathrm{Id} - \gamma \circ \pi) \circ \psi\big)(l_1, l_2),$$

where $\gamma : C' \to C$ is a section of π with $\gamma(C') = C''$.

(c) Conversely, suppose $f' : L' \to L$ is a central extension and $f : \mathrm{E}(L, C, \psi) \twoheadrightarrow L'$ is a surjective homomorphism of central extensions. Then $\pi = f|C$ maps C onto $C' = \ker f'$ and there exists a unique isomorphism of extensions $\Phi : L' \to \mathrm{E}(L, C', \psi')$, $\psi' = \pi \circ \psi$ such that all triangles in the diagram below commute:

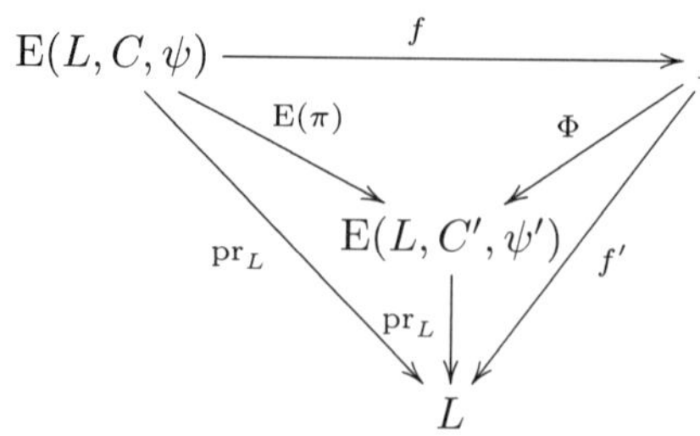

Exercise 5.1.18 Let $C = C_1 \oplus C_2$ be a vector space direct sum and denote by $\pi_i : C \to C_i$ the canonical projections. Let $\psi : L \times L \to C$ be a 2-cocycle with the property that $\pi_2 \circ \psi$ is a 2-coboundary. Then for $\psi_1 = \pi_1 \circ \psi$,

$$\mathrm{E}(L, C, \psi) \cong \mathrm{E}(L, C_1, \psi_1) \times C_2$$

as central extensions of L (even as central extensions of the Lie algebra $\mathrm{E}(L, C_1, \psi_1)$).

Example 5.1.19 Let $(\cdot|\cdot) : L \times L \to F$ be a symmetric bilinear form, which is invariant, see Exercise 5.1.6. We denote by $\mathrm{SDer}_F(L)$ the subalgebra of $\mathrm{Der}_F(L)$ which consists of all skew-symmetric derivations, where a derivation $d \in \mathrm{Der}_F(L)$ is called *skew-symmetric* if $(d(l_1) \mid l_2) + (l_1 \mid d(l_2)) = 0$ for all $l_1, l_2 \in L$. Observe that the inner derivations are an ideal of $\mathrm{SDer}_F(L)$:

$$\mathrm{IDer}(L) = \{\mathrm{ad}\, l : l \in L\} \lhd \mathrm{SDer}_F(L). \tag{5.24}$$

Let D be a subspace of $\mathrm{SDer}_F(L)$ and let D^* be its dual space. Then the map $\psi_D : L \times L \to D^*$, given by

$$\psi_D(l_1, l_2)(d) = \big(d(l_1) \mid l_2\big) \tag{5.25}$$

for $l_i \in L$ and $d \in D$, is a 2-cocycle of L.

Exercise 5.1.20 Show: (a) (5.25) defines indeed a 2-cocycle.

(b) ψ_D for $D \subseteq \mathrm{IDer}(L)$ is a 2-coboundary.

(c) If $\tilde{D}$ is a subspace of D, then the central extension $\mathrm{E}(L, D^*, \psi_D)$ of L factors through the central extension $\mathrm{E}(L, \tilde{D}^*, \psi_{\tilde{D}})$ of L,

$$\mathrm{E}(L, D^*, \psi_D) \twoheadrightarrow \mathrm{E}(L, \tilde{D}^*, \psi_{\tilde{D}}) \twoheadrightarrow L.$$

Definition 5.1.21 (Graded central extensions) Let Λ be an abelian group, and let $L = \bigoplus_{\lambda \in \Lambda} L^\lambda$ be a Λ-graded Lie algebra. We say that $f : K \to L$ is a *Λ-graded central extension of* L if K is a Λ-graded Lie algebra and f is a central extension which is at the same time a homomorphism of Λ-graded algebras: $f(K^\lambda) \subseteq L^\lambda$ for all $\lambda \in \Lambda$. A Λ-graded central extension $f : K \to L$ is called a *Λ-covering*, if f is a covering, i.e., K is perfect. We note that an arbitrary central extension of a graded Lie algebra need not be a graded central extension.

A *homomorphism* of a Λ-graded central extension $f : K \to L$ to another Λ-graded central extension $f' : K' \to L$ is a homomorphism $g : K \to K'$ of Λ-graded Lie algebras satisfying $f = f' \circ g$, cf. (5.20).

To define graded central extensions of a Λ-graded Lie algebra L via a 2-cocycle, we need (obviously) a Λ-*graded* 2-*cocycle*, i.e., a 2-cocycle $\psi : L \times L \to C$ into a Λ-graded vector space $C = \bigoplus_{\lambda \in \Lambda} C^\lambda$ which is graded of degree 0,

$$\psi(L^\lambda, L^\mu) \subseteq C^{\lambda+\mu} \quad \text{for all } \lambda, \mu.$$

For a graded 2-cocycle ψ the Lie algebra $K = L \oplus C$ of (5.22) is naturally Λ-graded by

$$K^\lambda = L^\lambda \oplus C^\lambda$$

and the central extension $\mathrm{pr}_L : K \to L$ is a Λ-graded central extension. Conversely, if $f : K \to L$ is a Λ-graded central extension, we can choose a section $s : L \to K$ of the underlying vector spaces of degree 0, meaning $s(L^\lambda) \subseteq K^\lambda$. The 2-cocycle associated to s in (5.23) is then a graded 2-cocycle. Thus, Proposition 5.1.13 holds in an analogous way for graded central extensions.

The following proposition also shows that one does not have to introduce a new object of a "graded universal central extension".

Proposition 5.1.22 ([184, 1.16]) *Let* $L = \bigoplus_{\lambda \in \Lambda} L^\lambda$ *be a* Λ*-graded perfect Lie algebra. Then its universal central extension* $\mathfrak{u} : \mathfrak{uce}(L) \to L$ *is* Λ*-graded, hence a* Λ*-covering. Moreover,* $\ker \mathfrak{u}$ *is a graded subspace of* $\mathfrak{uce}(L)$.

Example 5.1.23 We also have the graded versions of the Example 5.1.15 (details left to the reader) and the Example 5.1.19, whose details follow.

Let $L = \bigoplus_{\lambda \in \Lambda} L^\lambda$ be a Λ-graded Lie algebra and let $(\cdot|\cdot)$ be an invariant bilinear form on L, which is Λ-*graded* in the following sense:

$$(L^\lambda \mid L^\mu) = 0 \quad \text{if } \lambda + \mu \neq 0. \tag{5.26}$$

We define the Λ-graded subalgebra of $\mathrm{grEnd}_F(L)$

$$\mathrm{grSDer}_F(L) = \mathrm{grEnd}_F(L) \cap \mathrm{SDer}_F(L) = \bigoplus_{\lambda \in \Lambda} \big(\mathrm{SDer}_F(L)\big)^\lambda \tag{5.27}$$

where $(\mathrm{SDer}_F(L))^\lambda$ consists of all skew-symmetric derivations of degree λ. If $D \subseteq \mathrm{grSDer}_F(L)$ is a graded subspace of $\mathrm{grSDer}_F(L)$, the 2-cocycle ψ_D of (5.25) is Λ-graded and maps $L \times L$ into $D^{\mathrm{gr}\,*}$, thus giving rise to a graded central extension $\mathrm{E}(L, D^{\mathrm{gr}\,*}, \psi_D)$ of L.

Exercise 5.1.24 Show that the 2-cocycles ψ of (5.5), (5.16) and (5.17) can be obtained in the form (5.25), i.e., find an invariant bilinear form on $\mathcal{L} = L(\mathfrak{g}, \sigma)$ resp. $\mathcal{L} = L(\mathfrak{g}, \boldsymbol{\sigma})$ and a subspace $D \subseteq \mathrm{SDer}_F(\mathcal{L})$ such that ψ and ψ_D yield isomorphic central extensions of $\mathcal{L}$.

It is not so surprising that the 2-cocycles we used in Sections 5.1.1 and 5.1.2 can all be obtained in the form ψ_D for $D \subseteq \mathrm{grSDer}_F(L)$. This is a special case of the following general result.

Theorem 5.1.25 ([188]) *Let* $L = \bigoplus_{\lambda \in \Lambda} L^\lambda$ *be a* Λ*-graded Lie algebra, which*

(i) *is perfect and finitely generated as Lie algebra,*
(ii) *has finite homogeneous dimension:* $\dim L^\lambda < \infty$ *for all* $\lambda \in \Lambda$*, and*
(iii) *has an invariant nondegenerate* Λ*-graded symmetric bilinear form.*

(a) *Then* $\mathrm{Der}_F(L) = \mathrm{grDer}_F(L)$ *is* Λ*-graded and has finite homogeneous dimension, whence the same is true for* $\mathrm{SDer}_F(L)$.

(b) *The universal central extension* $\mathfrak{uce}(L)$ *has finite homogenous dimension with respect to the* Λ*-grading of* (5.1.22). *Moreover,*

$$\mathfrak{uce}(L) \cong \mathrm{E}(L, D^{\mathrm{gr}\,*}, \psi_D)$$

as central extensions of L, *where* D *is any graded subspace of* $\mathrm{SDer}_F(L)$ *which complements* $\mathrm{IDer}(L)$ *in* $\mathrm{SDer}_F(L)$, *and* ψ_D *is the 2-cocycle of* (5.25).

Remarks 5.1.26 (a) Theorem 5.1.5 is an application of Theorem 5.1.25, as is Theorem 5.4.13(c).

(b) The Exercise 5.1.18 gives some indication why it is sufficient to take a subspace of $\mathrm{SDer}_F(L)$ complementing $\mathrm{IDer}(L)$ and not an arbitrary subspace of $\mathrm{SDer}_F(L)$.

Exercise 5.1.27 In the setting of Theorem 5.1.25, every Λ-graded central covering of L is isomorphic as central extension to a central extension $\mathrm{E}(L, B^{\mathrm{gr}\,*}, \psi_B)$ for some graded subspace B of D.

5.2 Extended affine Lie algebras: Definition and first examples

Rather than constructing Lie algebras in a concrete way as we have done in Section 5.1, in this section we will define extended affine Lie algebras by a set of axioms and give examples. We will see that these examples encompass all the examples of Section 5.1 (with the exception of the choice 2. for D in 5.1.2).

As before we will consider Lie algebras over an arbitrary field F of characteristic 0, but we will no longer assume that F has enough roots of unity (multiloop algebras will not be play a role here), except in Section 5.2.4 where $F = \mathbb{C}$).

5.2.1 Definition of an extended affine Lie algebra. An *extended affine Lie algebra*, or EALA for short, is a pair (E, H) consisting of a Lie algebra E over F and subalgebra H satisfying the following axioms (EA1) – (EA6).

(EA1): *E has an invariant nondegenerate symmetric bilinear form* $(\cdot|\cdot)$.

(EA2): *H is nontrivial finite-dimensional toral and self-centralizing subalgebra of E.*

Before we can state the other four axioms, we need to draw some consequences of the axioms (EA1) and (EA2). But first we give explanations of some of the notions used. In these notes a *toral subalgebra*, sometimes also called an *ad-diagonalizable subalgebra,* is a subalgebra H which induces a decomposition of E via the adjoint representation of H:

$$E = \bigoplus\nolimits_{\alpha \in H^*} E_\alpha, \tag{5.28}$$

$$E_\alpha = \{e \in E : [h, e] = \alpha(h)e \text{ for all } h \in H\}. \tag{5.29}$$

We will refer to (5.28) as the *root space decomposition* of (E, H) and to the nonzero subspaces E_α as the *root spaces*. Such a subalgebra H is necessarily abelian, whence $H \subseteq E_0 = \{e \in E : [h, e] = 0 \text{ for all } h \in H\}$. That H is also required to be *self-centralizing* means

$$H = E_0.$$

Now to the consequences of (EA1) and (EA2). Because of invariance of the bilinear form $(\cdot|\cdot)$, we have

$$(E_\alpha \mid E_\beta) = 0 \quad \text{if } \alpha + \beta \neq 0, \tag{5.30}$$

in particular the restriction of the bilinear form to $E_0 = H$ is nondegenerate. Because of this and finite-dimensionality of H, every linear form $\alpha \in H^*$ is represented by a unique $t_\alpha \in H$, defined by the condition that $(t_\alpha \mid h) = \alpha(h)$ holds for all $h \in H$. This allows us to transport the restricted form $(\cdot|\cdot) \mid H \times H$ to a symmetric bilinear form on H^*, also denoted $(\cdot|\cdot)$ and defined by

$$(\alpha \mid \beta) = (t_\alpha \mid t_\beta), \quad \alpha, \beta \in H^*. \tag{5.31}$$

This transport of bilinear forms is a standard procedure in the theory of semisimple Lie algebras, see for example [126, §8]. We can now define

$$\begin{aligned} R &= \{\alpha \in H^* : E_\alpha \neq 0\} \quad (\textit{set of roots of } (E,H)), \\ R^0 &= \{\alpha \in R : (\alpha \mid \alpha) = 0\} \quad (\textit{null roots}), \\ R^{\mathrm{an}} &= \{\alpha \in R : (\alpha \mid \alpha) \neq 0\} \quad (\textit{anisotropic roots}). \end{aligned} \tag{5.32}$$

We prefer to call R the set of roots of (E, H) and not the "root system" since we want to restrict the latter term for root systems in the usual sense, see Example 5.3.3. We point out that by definition 0 is a root,

$$0 \in R^0 \subseteq R.$$

This is the customary convention for EALAs and has some notational advantages.

We define the *core of* (E, H) as the subalgebra E_c of E generated by all anisotropic root spaces:

$$E_c = \langle \textstyle\bigcup_{\alpha \in R^{\mathrm{an}}} E_\alpha \rangle_{\mathrm{subalg}}$$

We can now state the remaining four axioms.

(EA3): *For every* $\alpha \in R^{\mathrm{an}}$ *and* $x_\alpha \in E_\alpha$, *the operator* $\operatorname{ad} x_\alpha$ *is locally nilpotent on* E.

(EA4): R^{an} *is connected* in the sense that for any decomposition $R^{\mathrm{an}} = R_1 \cup R_2$ with $(R_1 \mid R_2) = 0$ we have $R_1 = \emptyset$ or $R_2 = \emptyset$.

(EA5): *The centralizer of the core* E_c *of* E *is contained in* E_c: $\{e \in E : [e, E_c] = 0\} \subseteq E_c$.

(EA6): *The subgroup* $\Lambda = \operatorname{Span}_{\mathbb{Z}}(R^0) \subseteq H^*$ *generated by* R^0 *in* $(H^*, +)$ *is a free abelian group of finite rank.* In other words, $\Lambda \cong \mathbb{Z}^n$ for some $n \in \mathbb{N}$ (including $n = 0$!).

The term *locally nilpotent* in (EA3) means that for every $e \in E$ there exists an $n \in \mathbb{N}$, possibly depending on e, such that $(\operatorname{ad} x_\alpha)^n(e) = 0$. The property (EA5) is called *tameness*. The condition $[e, E_c] = 0$ is of course equivalent to $[e, E_\alpha] = 0$ for all $\alpha \in R^{\mathrm{an}}$. The rationale for this axiom is the following. The subalgebra E_c is in fact an ideal of E (Theorem 5.4.14). Hence we have a representation ρ of E on E_c, given by $\rho(e)(x_c) = [e, x_c]$ for $e \in E$ and $x_c \in E_c$. The kernel of the representation ρ is the centralizer of E_c in E. Therefore tameness means that $\ker \rho \subseteq E_c$. The idea here is that the core E_c should control E. We will make this more precise in Section 5.5.4. The rank of the free abelian group Λ in axiom (EA6) is called the *nullity* of (E, H). It is invariant under isomorphisms. We will describe EALAs of nullity 0 and 1 below.

Although the structure of an EALA requires the existence of an invariant nondegenerate symmetric bilinear form $(\cdot|\cdot)$ in the axiom (EA1), which is then used to define the anisotropic roots, it turns out that this bilinear form is really not so important. Because of this, we have defined an EALA as a pair (E, H) and not as a triple $(E, H, (\cdot|\cdot))$ as it is for example done in [11]. Consequently, an *isomorphism* from an EALA (E, H) to another EALA (E', H') is a Lie algebra isomorphism $f : E \to E'$ such that $f(H) = H'$. It is immediate that any isomorphism induces a bijection f' between the set of roots R and R' of (E, H) and (E', H') respectively. It then follows that f' maps R^{an} onto R'^{an}, whence also R^0 onto R'^0. One can then show that f' preserves the forms on $X = \mathrm{Span}_F(R)$ and $X' = \mathrm{Span}_F(R')$ up to scalars.

For $F = \mathbb{C}$ one can define a special class of EALAs. We call a pair (E, H) a *discrete EALA* if it satisfies the axioms (EA1) – (EA5) and in addition

(DE): *R is a discrete subset of H^* with respect to the natural topology of the finite-dimensional complex vector space H^*.*

It is justified to call a discrete EALA an EALA, since one can show that a discrete EALA also satisfies (EA6). Indeed, this follows from Proposition 5.3.21 and Theorem 5.3.22. However, not every EALA over $\mathbb{C}$ is a discrete EALA (see [187, 6.17]).

Some historical comments. Although there were some precursors (papers by Saito and Slodowy for nullity 2), it was in the paper [121] by the physicists Høegh-Krohn and Torrésani that the class of discrete extended affine Lie algebras was introduced, however not under this name. Rather, they were called "irreducible quasi-simple Lie algebras" and later ([29, 30]) "elliptic quasi-simple Lie algebras". The stated goal of the paper [121] was applications in quantum gauge theory. The theory developed there did however not stand up to the scrutiny of mathematicians. The errors of [121] were corrected in the AMS memoir [1] by Allison, Azam, Berman, Gao and Pianzola. There also the name "extended affine Lie algebras" appears for the first time. But not in the sense as defined above. Rather, the authors develop the basic theory of what here are called discrete EALAs. Nevertheless, [1] has become the standard reference even for the more general extended affine Lie algebras, since many of the results presented there for discrete extended affine Lie algebras easily extend to the more general setting. The definition of an extended affine Lie algebra given above is due to the author ([186]) and was motivated by the fact that all the examples presented in [1] did make sense over an arbitrary base field F and not just over $\mathbb{C}$ only. Before [186] the tameness axiom (EA5) was not part of the definition of an EALA. However, as examples show ([29, §3] or [187, 6.10]), it seems impossible to classify EALAs without (EA5). After [186], several generalizations of EALAs have been proposed. They are surveyed in [187].

5.2.2 Some elementary properties of extended affine Lie algebras. The following sections will (hopefully) show that extended affine Lie algebras share many properties with familiar Lie algebras, like finite-dimensional split simple Lie algebras or affine Kac-Moody Lie algebras. Some of these properties are immediate consequences of the axioms. The following (strongly recommended!) exercise gives an incomplete list of such properties.

Exercise 5.2.1 Let (E, H) be an EALA. We use the notation of above. Show the following:

(a) For $\alpha, \beta \in R$ we have

$$[E_\alpha, E_\beta] \subseteq E_{\alpha+\beta}. \tag{5.33}$$

Thus the root space decomposition (5.28) is a grading by the abelian group $\mathrm{Span}_{\mathbb{Z}}(R) \subseteq H^*$.

(b) H is a *Cartan subalgebra*, defined as a nilpotent subalgebra which is self-normalizing: $H = \{e \in E : [e, H] \subseteq H\}$.

(c) For $\alpha, \beta \in R$ we have $(E_\alpha \mid E_\beta) = 0$ unless $\alpha + \beta = 0$. The restriction of the bilinear form $(\cdot|\cdot)$ to $E_\alpha \times E_{-\alpha}$ is nondegenerate, i.e., if $x_\alpha \in E_\alpha$ satisfies $(x_\alpha \mid E_{-\alpha}) = 0$ then $x_\alpha = 0$. In particular, $R = -R$.

(d) For $\alpha \in R$ and $x_\alpha \in E_\alpha$ and $y_{-\alpha} \in E_{-\alpha}$,

$$[x_\alpha, y_{-\alpha}] = (x_\alpha \mid y_{-\alpha})\, t_\alpha. \tag{5.34}$$

In particular, $[E_\alpha, E_{-\alpha}] = Ft_\alpha$, and if $\alpha \in R^{\mathrm{an}}$ then

$$[[E_\alpha, E_{-\alpha}], E_\alpha] = E_\alpha. \tag{5.35}$$

(e) The core E_c satisfies

$$E_c = \Big(\oplus_{\alpha \in R^{\mathrm{an}}} E_\alpha \Big) \oplus \Big(\textstyle\bigoplus_{\alpha \in R^0} (E_c \cap E_\alpha) \Big). \tag{5.36}$$

Using (5.34), one can show that EALAs are built out of "little" $\mathfrak{sl}_2$'s and Heisenberg's (albeit in a complicated way).

Proposition 5.2.2 *Let (E, H) be an extended affine Lie algebra, with anisotropic root R^{an} and null roots R^0.*

(a) *Let $\alpha \in R^{\mathrm{an}}$. Then $\dim E_\alpha = 1$, and for any $e_\alpha \in E_\alpha$ there exists $f_\alpha \in E_{-\alpha}$ such that $(e_\alpha, h_\alpha = [e_\alpha, f_\alpha], f_\alpha) \in E_\alpha \times H \times E_{-\alpha}$ is an $\mathfrak{sl}_2$-triple:*

$$E_\alpha \oplus [E_\alpha, E_{-\alpha}] \oplus E_{-\alpha} = Fe_\alpha \oplus Fh_\alpha \oplus Ff_\alpha \cong \mathfrak{sl}_2(F).$$

(b) *Let $\alpha \in R^0$. Then for any $0 \neq x_\alpha \in E_\alpha$ there exists $y_\alpha \in E_{-\alpha}$ such that $[x_\alpha, y_\alpha] = t_\alpha$ and*

$$Fx_\alpha \oplus Ft_\alpha \oplus Fy_\alpha \cong \mathfrak{h}_3,$$

the 3-dimensional Heisenberg algebra.

It is not true that $\dim E_\alpha = 1$ if $\alpha \in R^0$ (this is already not true in the examples of Sections 5.2.3 and 5.2.4). But we will show in Theorem 5.4.18 that all root spaces E_α are finite-dimensional in a rather strong way.

The following exercise shows that one can "extend" the 3-dimensional Heisenberg subalgebras in (b) above.

Exercise 5.2.3 In the setting and notation of Proposition 5.2.2(b) show that there exists $d_\alpha \in H$ such that

$$[d_\alpha, x_\alpha] = x_\alpha \quad \text{and} \quad [d_\alpha, y_\alpha] = -y_\alpha.$$

Hence

$$Fx_\alpha \oplus Ft_\alpha \oplus Fd_\alpha \oplus Fy_\alpha$$

is a 4-dimensional subalgebra. It is 2-step solvable, not nilpotent and isomorphic to the subalgebra

$$\left\{ \begin{pmatrix} 0 & a & b \\ 0 & c & d \\ 0 & 0 & 0 \end{pmatrix} : a, b, c, d \in F \right\}$$

of $\mathfrak{gl}_3(F)$.

And now an exercise which implies that in an EALA one can produce many so-called elementary automorphisms.

Exercise 5.2.4 Let M be an F-vector space. One calls an endomorphism $f \in \mathrm{End}_F(M)$ *locally nilpotent* if for every $m \in M$ there exists $n \in \mathbb{N}$, possibly depending on m, such that $f^n(m) = 0$.

(a) Show that the following conditions are equivalent for $f \in \mathrm{End}_F(M)$:

(i) f is locally nilpotent,
(ii) for every finitely spanned subspace N of M there exists a finite-dimensional subspace P of M such that $N \subseteq P$ and $f(P) \subseteq P$,
(iii) f is nilpotent on every finite-dimensional and f-invariant subspace of M.

(b) Let $f \in \mathrm{End}_F(M)$ be locally nilpotent and define the *exponential* $\exp f$ *of* f by

$$(\exp f)(m) = \textstyle\sum_{n\in\mathbb{N}} \frac{1}{n!} f^n(m),$$

for $m \in M$ (note that the sum on the right is always finite). Show that $\exp f$ is an invertible endomorphism of M with inverse given by $(\exp f)^{-1} = \exp(-f)$.

(c) Let L be a Lie algebra and let d be a locally nilpotent derivation of L. Show that then $\exp d$ is an automorphism of L, called an *elementary automorphism.*

We will next present some examples of EALAs.

5.2.3 Extended affine Lie algebras of nullity 0. Let $\mathfrak{g}$ be a finite-dimensional split simple Lie algebra with splitting Cartan subalgebra $\mathfrak{h}$, for example $\mathfrak{sl}_l(F)$ or a finite-dimensional simple Lie algebra over an algebraically closed field. We will show that then $(\mathfrak{g}, \mathfrak{h})$ is an EALA of nullity 0. The facts needed to prove this can be found in [40, VIII, §2] or in [126, §8] for F algebraically closed. Let us now verify the axioms (EA1)–(EA6).

(EA1) Up to a scalar, there exists only one invariant nondegenerate symmetric bilinear form on $\mathfrak{g}$, the Killing form κ. Hence we can (and will) take $(\cdot|\cdot) = \kappa$.

(EA2) By definition of a splitting Cartan subalgebra, the Lie algebra $\mathfrak{g}$ has a root space decomposition

$$\mathfrak{g} = \mathfrak{g}_0 \oplus \big(\textstyle\bigoplus_{\alpha\in\Phi} \mathfrak{g}_\alpha\big), \quad \mathfrak{g}_0 = \mathfrak{h},$$

where Φ is the root system of $(\mathfrak{g}, \mathfrak{h})$ (which is a reduced root system in the usual sense, see Example 5.3.3) and where the root spaces $\mathfrak{g}_\alpha$ are defined as in (5.29). Hence the set of roots R of $(\mathfrak{g}, \mathfrak{h})$ is

$$R = \{0\} \cup \Phi. \tag{5.37}$$

It is a basic fact that $\kappa(t_\alpha, t_\alpha) \neq 0$ for $t_\alpha \in \mathfrak{h}$ representing $\alpha \in \Phi$ via $\kappa(t_\alpha, h) = \alpha(h)$ for all $h \in \mathfrak{h}$. Hence, the anisotropic and null roots are

$$R^{\mathrm{an}} = \Phi \quad \text{and} \quad R^0 = \{0\}.$$

(EA3) is now obvious: From $[\mathfrak{g}_\alpha, \mathfrak{g}_\beta] \subseteq \mathfrak{g}_{\alpha+\beta}$ for $\alpha \in \Phi$ and $\beta \in R$ and finite-dimensionality of $\mathfrak{g}$, it is clear that $\mathrm{ad}\, x_\alpha$ for $x_\alpha \in \mathfrak{g}_\alpha$ is not only locally nilpotent but even (globally) nilpotent.

(EA4) is another way of saying that Φ is an irreducible root system. This is indeed the case and follows from the simplicity of $\mathfrak{g}$.

(EA5) We first need to determine the core $\mathfrak{g}_c$ of $\mathfrak{g}$. By definition, $\mathfrak{g}_c$ is the subalgebra of $\mathfrak{g}$ generated by $\bigoplus_{\alpha\in\Phi}\mathfrak{g}_\alpha$. Since $\mathfrak{h}=\sum_{\alpha\in\Phi}[\mathfrak{g}_\alpha,\mathfrak{g}_{-\alpha}]$ we have

$$\mathfrak{g}_c=\mathfrak{g}.$$

It is now a tautology that (EA5) holds, i.e., that the centralizer of the core $\mathfrak{g}_c$ is contained in $\mathfrak{g}_c=\mathfrak{g}$. Of course, we know even more: The centralizer of the core equals the centre of $\mathfrak{g}$, and is therefore $\{0\}$.

(EA6) We have $\Lambda=\langle R^0\rangle=\langle\{0\}\rangle=\{0\}$.

We have now shown:

A finite-dimensional split simple Lie algebra is an EALA of nullity 0, (5.38)

obviously a discrete EALA. We will see in Proposition 5.3.24 that the converse of (5.38) is true too. We thus know all the nullity 0 examples of EALAs, and can therefore focus on the higher nullity examples. We will consider the case of nullity 1 in the next section.

5.2.4 Affine Kac-Moody Lie algebras again. To justify the name extended *affine Lie algebra*, we will now show that any affine Kac-Moody Lie algebra is an extended affine Lie algebra. To do so, we will need some basic facts about affine Kac-Moody Lie algebras. All of them can be found in Kac's book [137]. Since this reference uses $\mathbb{C}$ as base field, we will do the same in this section. But everything we say here holds true for arbitrary algebraically closed fields of characteristic 0 (see [137, Remark 0.7]). Thus we let

$$\mathcal{L}=\bigoplus\nolimits_{n\in\mathbb{Z}}\mathfrak{g}_{\bar n}\otimes\mathbb{C}t^n \quad\text{and}$$
$$\hat{\mathcal{L}}=\hat{\mathcal{L}}(\mathfrak{g},\sigma)=\mathcal{L}\oplus\mathbb{C}c\oplus\mathbb{C}d$$

be the complex Lie algebras described in (5.9) and (5.10). Recall that $\mathfrak{g}$ is a finite-dimensional simple Lie algebra over $\mathbb{C}$ and σ is a diagram automorphism of $\mathfrak{g}$. We let $m\in\{1,2,3\}$ be the order of σ, and denote the canonical map $\mathbb{Z}\to\mathbb{Z}/m\mathbb{Z}$ by $n\mapsto\bar n$. Recall from (5.1) and (5.2) that σ induces a $\mathbb{Z}/m\mathbb{Z}$-grading of $\mathfrak{g}$, namely

$$\mathfrak{g}=\mathfrak{g}_{\bar 0}\oplus\cdots\oplus\mathfrak{g}_{\overline{m-1}}$$

where $\mathfrak{g}_{\bar n}=\{x\in\mathfrak{g}:\sigma(x)=\zeta^n x\}$ for a primitive mth root of unity ζ. For example, for $m=2$ we get a $\mathbb{Z}/2\mathbb{Z}$-grading $\mathfrak{g}=\mathfrak{g}_{\bar 0}\oplus\mathfrak{g}_{\bar 1}$ with $\mathfrak{g}_{\bar 0}=\{x\in\mathfrak{g}:\sigma(x)=x\}$ and $\mathfrak{g}_{\bar 1}=\{x\in\mathfrak{g}:\sigma(x)=-x\}$. We identify $\mathfrak{g}_{\bar 0}\equiv\mathfrak{g}_{\bar 0}\otimes\mathbb{C}t^0$.

We now verify the axioms (EA1) – (EA5) and (DE) which, we recall, implies (EA6).

(EA1) We let κ be the Killing form of $\mathfrak{g}$ and define a bilinear form $(\cdot|\cdot)$ on $\hat{\mathcal{L}}$, using the notation of (5.10),

$$\begin{aligned}&\big(u_{\bar\lambda}\otimes t^\lambda\oplus s_1c\oplus s_1'd \mid v_{\bar\mu}\otimes t^\mu\oplus s_2c\oplus s_2'd\big)\\ &\quad=\kappa(u_{\bar\lambda},v_{\bar\mu})\delta_{\lambda,-\mu}+s_1s_2'+s_2s_1'.\end{aligned}\tag{5.39}$$

The form is visibly symmetric. The reader is invited in Exercise 5.2.6 to show that it is an invariant nondegenerate symmetric bilinear form on $\hat{\mathcal{L}}$, as required in (EA1) (In fact it is up to a scalar the only such form which is graded with respect to the root space decomposition of $\hat{\mathcal{L}}$). In anticipation of the later developments, we point out that $(\cdot|\cdot)$ has the following features:

- $\hat{\mathcal{L}}$ is an orthogonal sum of $\mathcal{L}$ and $\mathbb{C}c\oplus\mathbb{C}d$: $\hat{\mathcal{L}}=\mathcal{L}\perp(\mathbb{C}c\oplus\mathbb{C}d)$,
- $\mathbb{C}c\oplus\mathbb{C}d$ is a hyperbolic plane, i.e., $(c\mid c)=0=(d\mid d)$ while $(c\mid d)=1$.

- The Laurent polynomial ring $\mathbb{C}[t^{\pm 1}]$ has a nondegenerate symmetric bilinear form ϵ given by $\epsilon(t^\lambda, t^\mu) = \delta_{\lambda,-\mu}$. It is invariant in the sense of Exercise 5.1.6(c) and is graded as defined in (5.26). For $\sigma = \mathrm{Id}_{\mathfrak{g}}$, the bilinear form on the loop algebra $\mathfrak{g} \otimes \mathbb{C}[t^{\pm 1}]$ is simply the tensor product form $\kappa \otimes \epsilon$, and for a general σ the form is obtained by restriction.

(EA2) To construct a subalgebra H as required in axiom (EA2) we start with a Cartan subalgebra $\mathfrak{h}$ of $\mathfrak{g}$. Since σ is a diagram automorphism, it leaves $\mathfrak{h}$ invariant. We let

$$\mathfrak{h}_{\bar{0}} = \mathfrak{h} \cap \mathfrak{g}_{\bar{0}} = \{h \in \mathfrak{h} : \sigma(h) = h\}$$

and put

$$H = \mathfrak{h}_{\bar{0}} \oplus \mathbb{C}c \oplus \mathbb{C}d.$$

One knows that $\mathfrak{g}_{\bar{0}}$ is a simple Lie algebra with Cartan subalgebra $\mathfrak{h}_{\bar{0}}$ ([137, Prop. 7.9]). The grading property implies that $[\mathfrak{g}_{\bar{0}}, \mathfrak{g}_{\bar{n}}] \subseteq \mathfrak{g}_{\bar{n}}$ for $n \in \mathbb{Z}$. Hence $\mathfrak{g}_{\bar{0}}$ acts on $\mathfrak{g}_{\bar{n}}$ by the adjoint action. Let $\Delta_{\bar{n}}$ be the set of weights of the $\mathfrak{g}_{\bar{0}}$-module $\mathfrak{g}_{\bar{n}}$ with respect to $\mathfrak{h}_{\bar{0}}$:

$$\begin{aligned} \mathfrak{g}_{\bar{n}} &= \textstyle\bigoplus_{\gamma \in \Delta_{\bar{n}}} \mathfrak{g}_{\bar{n},\gamma} \\ \mathfrak{g}_{\bar{n},\gamma} &= \{x \in \mathfrak{g}_{\bar{n}} : [h_{\bar{0}}, x] = \gamma(h_{\bar{0}})x \text{ for all } h_{\bar{0}} \in \mathfrak{h}_{\bar{0}}\}. \end{aligned}$$

In particular, $\Delta_{\bar{0}} \setminus \{0\}$ is the root system of $\mathfrak{g}_{\bar{0}}$ with respect to $\mathfrak{h}_{\bar{0}}$ and $\mathfrak{h}_{\bar{0}} = \mathfrak{g}_{\bar{0},0}$.

We extend $\Delta_{\bar{n}} \subseteq \mathfrak{h}_{\bar{0}}^*$ to a linear form on H by zero, i.e., for $\gamma \in \Delta_{\bar{n}}$ we put

$$\gamma(h_{\bar{0}} \oplus sc \oplus s'd) = \gamma(h_{\bar{0}})$$

for $s, s' \in \mathbb{C}$ and define a linear form δ on H by

$$\delta(h_{\bar{0}} \oplus sc \oplus s'd) = s'.$$

Then for $\gamma \in \Delta_{\bar{n}}$, $n \in \mathbb{Z}$, we have

$$\begin{aligned} \hat{\mathcal{L}}_{\gamma \oplus n\delta} &= \{u \in \hat{\mathcal{L}} : [h, u] = (\gamma \oplus n\delta)(h)u \text{ for all } h \in H\} \\ &= \begin{cases} \mathfrak{g}_{\bar{n},\gamma} \otimes t^n, & \gamma \oplus n\delta \neq 0, \\ H, & \gamma \oplus n\delta = 0. \end{cases} \end{aligned} \tag{5.40}$$

Hence $\hat{\mathcal{L}} = \bigoplus_{\alpha \in R} \hat{\mathcal{L}}_\alpha$ has a root space decomposition with respect to H with set of roots

$$R = \{\gamma \oplus n\delta : \gamma \in \Delta_{\bar{s}},\ \bar{n} = \bar{s}, 0 \leq s < m\}. \tag{5.41}$$

This establishes (EA2).

To check the other axioms we first need to determine which of the roots in R are the null respectively anisotropic roots. Following the procedure in Section 5.2.1, we consider the restriction of the bilinear form $(\cdot|\cdot)$ to H. With obvious notation this is

$$(h_{\bar{0}} \oplus s_1 c \oplus s_1' d \mid h_{\bar{0}}' \oplus s_2 c \oplus s_2' d) = \kappa(h_{\bar{0}}, h_{\bar{0}}') + s_1 s_2' + s_2 s_1'.$$

Since $\kappa|_{\mathfrak{h}_{\bar{0}} \times \mathfrak{h}_{\bar{0}}}$ is nondegenerate, this is indeed a nondegenerate symmetric bilinear form on H, as it should be. Let $t_\gamma \in \mathfrak{h}_{\bar{0}}$ be the element representing $\gamma \in \mathfrak{h}_{\bar{0}}^*$: $\kappa(t_\gamma, h_{\bar{0}}) = \gamma(h_{\bar{0}})$ for all $h_{\bar{0}} \in \mathfrak{h}_{\bar{0}}$. For the canonical extension of γ to a linear form of H, also denoted by γ, we then get $(t_\gamma \mid h) = \gamma(h)$ for all $h \in H$. Moreover $(c \mid h_{\bar{0}} \oplus sc \oplus s'd) = s' = \delta(h_{\bar{0}} \oplus sc \oplus s'd)$ shows that δ is represented by $t_\delta = c \in H$. Therefore $\alpha = \gamma \oplus n\delta \in R$ is represented by

$$t_{\gamma \oplus n\delta} = t_\gamma \oplus nc.$$

Now observe $(t_{\gamma \oplus n\delta} \mid t_{\gamma \oplus n\delta}) = (t_\gamma \oplus nc \mid t_\gamma \oplus nc) = \kappa(t_\gamma, t_\gamma)$. It is of course well-known that $\kappa(t_\gamma, t_\gamma) \neq 0$ for $0 \neq \gamma \in \Delta_{\bar{0}}$. But one can (easily) show that this also holds for any $0 \neq \gamma \in \Delta_{\bar{n}}$. The anisotropic and null roots are therefore

$$R^{\mathrm{an}} = \{\gamma \oplus n\delta \in R : \gamma \neq 0\} \quad \text{and} \quad R^0 = \mathbb{Z}\delta, \tag{5.42}$$

which in the theory of affine Kac-Moody algebras are usually called *real* and *imaginary roots*. We are now set for the verification of the remaining axioms.

(EA3) holds in the stronger form: $\operatorname{ad} \hat{\mathcal{L}}_\lambda$, $\alpha \in R^{\mathrm{an}}$, is nilpotent. (We have already seen the same phenomenon in the Example 5.2.3 of a finite-dimensional split simple Lie algebra. Perhaps the reader wonders if this is true in general. The answer is yes.)

(EA4) The verification of (EA4) is left to the reader.

(EA5) The core of $\hat{\mathcal{L}}$ is $\hat{\mathcal{L}}_c = \big(\bigoplus_{n\in\mathbb{Z}} \mathfrak{g}_{\bar{n}} \otimes \mathbb{C}t^n\big) \oplus \mathbb{C}c$, and therefore equals the derived algebra $[\hat{\mathcal{L}}, \hat{\mathcal{L}}]$ of $\hat{\mathcal{L}}$. The centralizer of $\hat{\mathcal{L}}_c$ in $\hat{\mathcal{L}}$, in fact the centre of $\hat{\mathcal{L}}$ is $\mathbb{C}c \subseteq \hat{\mathcal{L}}_c$, see Exercise 5.1.2.

(DE) In this example the subgroup $\Lambda = \langle R^0 \rangle$ equals $R^0 = \mathbb{Z}d$ and is a discrete subset of H^*.

We have now shown one implication of the following result.

Theorem 5.2.5 ([6]) *A complex Lie algebra E is a discrete EALA of nullity 1 if and only if E is an affine Kac-Moody Lie algebra.*

Exercise 5.2.6 Check the following details of the construction above.

(a) (5.39) defines an invariant symmetric bilinear form on $\hat{\mathcal{L}}$.

(b) $\hat{\mathcal{L}}$ has a root space decomposition whose root spaces are given by (5.40) and whose set of roots is (5.41).

(c) $\kappa(t_\gamma, t_\gamma) \neq 0$ for any $0 \neq \gamma \in \Delta_{\bar{n}}$.

(d) (EA4) holds for $(\hat{\mathcal{L}}, H)$.

5.2.5 Higher nullity examples. We have seen all examples of EALAs of nullity 0 and 1. In this section we will construct examples of higher nullity. To simplify things we consider untwisted algebras (no non-trivial finite order automorphism are involved). We can therefore go back to our standard setting: $\mathfrak{g}$ is a split simple Lie algebra over a field F of characteristic 0.

As in Section 5.1.2 let $F[t_1^{\pm 1}, \dots, t_n^{\pm n}]$ be the Laurent polynomial ring in n variables and let

$$L = L(\mathfrak{g}) = \mathfrak{g} \otimes F[t_1^{\pm 1}, \dots, t_n^{\pm n}]$$

be the associated untwisted multiloop algebra. We have seen in Exercise 5.1.4 that L has a 2-cocycle $\psi : \mathcal{L} \times \mathcal{L} \to F^n =: \mathcal{C}$, given by (5.16): $\psi(u \otimes t^\lambda, v \otimes t^\mu) = \delta_{\lambda+\mu,0}\, \kappa(u, v)\, \lambda$. We can therefore define the central extension

$$K = L \oplus \mathcal{C}$$

with product (5.22). In (5.19) we have defined degree derivations ∂_i, $i = 1, \dots, n$, of K. Let

$$\mathcal{D} = \operatorname{Span}_F\{\partial_1, \dots, \partial_n\} \tag{5.43}$$

and define the Lie algebra E as the semidirect product,

$$E = \big(L(\mathfrak{g}) \oplus \mathcal{C}\big) \rtimes \mathcal{D}.$$

Let $\mathfrak{h}$ be a Cartan subalgebra of $\mathfrak{g}$ and put

$$H = \mathfrak{h} \oplus \mathcal{C} \oplus \mathcal{D}.$$

We claim that *(E, H) is an EALA of nullity n.*

(EA1) We will mimic the construction of an invariant nondegenerate symmetric bilinear form in Section 5.2.4 and require

- $(L(\mathfrak{g}) \mid \mathcal{C} \oplus \mathcal{D}) = 0$.
- $\mathcal{C} \oplus \mathcal{D}$ is a hyperbolic space with $(\mathcal{C} \mid \mathcal{C}) = 0 = (\mathcal{D} \mid \mathcal{D})$ and

  $$(\textstyle\sum_i s_i c_i \mid \sum_i s_i' \partial_i) = \sum_i s_i s_i',$$

  where $c_1, \dots, c_n$ is the canonical basis of F^n. Thus $\mathcal{C} \oplus \mathcal{D}$ is the orthogonal sum of the n hyperbolic planes $Fc_i \oplus F\partial_i$.
- On $L(\mathfrak{g})$ the form is the tensor product form of the Killing form κ of $\mathfrak{g}$ and the natural invariant bilinear form on $F[t_1^{\pm 1}, \dots, t_n^{\pm 1}]$.

Putting all these requirements together, we arrive at the global formula for an invariant bilinear form on E, which is completely analogous to (5.39):

$$\begin{aligned} &\big(u \otimes t^\lambda \oplus \textstyle\sum_i s_i c_i \oplus \sum_j s_j' \partial_j \mid v \otimes t^\mu \oplus \sum_i t_i c_i \oplus \sum_j t_j' \partial_j\big) \\ &\qquad = \kappa(u, v)\delta_{\lambda, -\mu} + \textstyle\sum_i (s_i t_i' + t_i s_i'). \end{aligned} \tag{5.44}$$

(EA2) Let $\mathfrak{h}$ be a splitting Cartan subalgebra and let Φ be the usual root system of $(\mathfrak{g}, \mathfrak{h})$, thus $0 \notin \Phi$. We put $\Delta = \{0\} \cup \Phi$ and then have the root space decomposition $\mathfrak{g} = \bigoplus_{\gamma \in \Delta} \mathfrak{g}_\gamma$ with $\mathfrak{g}_0 = \mathfrak{h}$. We embed $\Delta \hookrightarrow H^*$ by requiring $\gamma \mid \mathcal{C} \oplus \mathcal{D} = 0$ for $\gamma \in \Delta$. Also we embed $\Lambda = \mathbb{Z}^n \hookrightarrow H^*$ by $\lambda(\mathfrak{h} \oplus \mathcal{C}) = 0$ and $\lambda(\partial_i) = \lambda_i$ for $\lambda = (\lambda_1, \dots, \lambda_n) \in \Lambda$. Then E has the root space decomposition $E = \bigoplus_{\alpha \in R} E_\alpha$ with root spaces

$$E_{\gamma \oplus \lambda} = \mathfrak{g}_\gamma \otimes t^\lambda \quad (\gamma \oplus \lambda \neq 0), \qquad E_0 = H, \tag{5.45}$$

$$R^{\mathrm{an}} = \Phi \times \Lambda, \qquad R^0 = \Lambda. \tag{5.46}$$

It is now not difficult to verify the remaining axioms (EA3) – (EA5) and (DE). Thus:

Lemma 5.2.7 *The pair (E, H) constructed above is a discrete EALA of nullity n.*

There is however no analogue of Proposition 5.38 and Theorem 5.2.5: There are many more EALAs of nullity $n \geq 2$. We have just seen the "tip of the iceberg"! Other examples can be found in Ch.III of [1], some of them involving heavy-duty nonassociative algebras, like octonion algebras and Jordan algebras over Laurent polynomial rings!

Exercise 5.2.8 Supply the missing details of the proof that (E, H) above is a discrete EALA of nullity n. In particular, prove:

(a) (5.44) defines an invariant nondegenerate symmetric bilinear form on E.

(b) The root spaces of (E, H) and the anisotropic and null roots are as stated in (5.45) and (5.46).

5.3 The structure of the roots of an EALA

In this section we will describe the structure of the set of roots R of an EALA (E, H), defined in (5.32). We have already seen some examples: R can be a finite irreducible reduced root system (5.37), R can be an affine root system (5.41), i.e., the set of roots of an affine Kac-Moody Lie algebra, or R can be of the form $R = S \times \mathbb{Z}^n$ where $S \setminus \{0\}$ is a finite irreducible reduced root system (5.46). Thus any description of the general case has to encompass all these different examples.

It turns out that the roots of an EALA form an extended affine root system and that the latter is naturally described as a special case of affine reflection systems. We therefore first introduce the latter, describe their structure and then specialize later to extended affine root systems. Affine reflection systems are themselves special cases of reflection systems, whose theory is developed in [156].

5.3.1 Affine reflection systems: Definition. Throughout this section we work with a triple $(R, X, (\cdot|\cdot))$ where

- X is a finite-dimensional vector space over a field F of characteristic 0,
- $(\cdot|\cdot)$ is a symmetric bilinear form on X and
- $R \subseteq X$.

For any such triple $(R, X, (\cdot|\cdot))$ we define

$$\begin{aligned} X^0 &= \{x \in X : (x \mid X) = 0\}, \text{ the radical of } (\cdot|\cdot), \\ R^0 &= \{\alpha \in R : (\alpha|\alpha) = 0\}, \quad (\textit{null roots}) \\ R^{\mathrm{an}} &= \{\alpha \in R : (\alpha|\alpha) \neq 0\}, \quad (\textit{anisotropic roots}) \\ \langle x, \alpha^\vee \rangle &= 2\frac{(x|\alpha)}{(\alpha|\alpha)}, \quad (x \in X \text{ and } \alpha \in R^{\mathrm{an}}) \\ s_\alpha(x) &= x - \langle x, \alpha^\vee \rangle \alpha. \end{aligned} \tag{5.47}$$

By definition we therefore have $R = R^0 \cup R^{\mathrm{an}}$. The map $s_\alpha : X \to X$ is a *reflection in* α, i.e., $s_\alpha^2 = \mathrm{Id}_X$ and $\{x \in X : s_\alpha(x) = -x\} = F\alpha$. It is also orthogonal with respect to $(\cdot|\cdot)$: $(s_\alpha(x) \mid s_\alpha(y)) = (x \mid y)$ for all $x, y \in X$.

We call $(R, X, (\cdot|\cdot))$, or just R for short, an *affine reflection system* if

(AR1): $0 \in R$ *and* R *spans* X,
(AR2): $s_\alpha(R) = R$ *for all* $\alpha \in R^{\mathrm{an}}$,
(AR3): *for every* $\alpha \in R^{\mathrm{an}}$ *the set* $\langle R, \alpha^\vee \rangle$ *is finite and contained in* $\mathbb{Z}$, *and*
(AR4): $R^0 = R \cap X^0$.

An affine reflection system is said to be

- *reduced* if for every $\alpha \in R^{\mathrm{an}}$ and $c \in F$: $c\alpha \in R^{\mathrm{an}} \iff c = \pm 1$,
- *connected* if for any decomposition $R^{\mathrm{an}} = R_1 \cup R_2$ with $(R_1 \mid R_2) = 0$ we have $R_1 = \emptyset$ or $R_2 = \emptyset$.

The *nullity* of (R, X) is the rank of the torsion-free abelian group $\mathbb{Z}[R^0] = \mathrm{Span}_{\mathbb{Z}}(R^0)$ generated by R^0 in $(X, +)$. Thus, by definition,

$$\text{nullity of } (R, X) = \dim_{\mathbb{Q}}(\mathbb{Z}[R^0] \otimes_{\mathbb{Z}} \mathbb{Q}) = \dim_F(\mathbb{Z}[R^0] \otimes_{\mathbb{Z}} F).$$

Since the vector space $\mathbb{Z}[R^0] \otimes_{\mathbb{Z}} F$ maps onto $\mathrm{Span}_F(R^0)$, the nullity of (R, Z) is bounded below by $\dim_F \mathrm{Span}_F(R^0)$. It is in general not equal to it. But this is of course so for nullity 0: (R, X) has nullity 0 if and only if $R^0 = \{0\} \iff \dim_F \mathrm{Span}_F(R^0) = 0$.

Remarks 5.3.1 - For a large part of the theory it is not necessary that X be finite-dimensional, see [156]. But assuming this right from the start, simplifies the presentation.

- We need the bilinear form $(\cdot|\cdot)$ to define R^0 and the reflections. But although we will sometimes write $(R, X, (\cdot|\cdot))$, we will not consider $(\cdot|\cdot)$ as part of the structure of an affine reflection system. For example, in the definition of an isomorphism below we will not require that the bilinear forms are preserved. See [156], where this point of view is emphasized.

- The requirement $0 \in R$ is in line with the previous section, in which 0 was considered a root of an EALA. This conflicts with the traditional approach to root systems in which 0 is not a root, see for example [39], [126] or [137]. The question whether 0 is a root or is not a root, has lead to heated debates. In the author's opinion, there are some advantages of considering 0 as a root, which however can only be fully seen when one develops the theory for affine reflection systems. But perhaps the reader can be convinced by the natural (?) example $(R, X) = (\{0\}, \{0\})$ of an affine reflection system.

- The condition $\langle \beta, \alpha^\vee \rangle \in \mathbb{Z}$ in axiom (AR3) makes sense since every field of characteristic 0 contains (an isomorphic copy of) the field of rational numbers, which allows us to identify $\mathbb{Z} \equiv \mathbb{Z}1_F$.

- By definition $\langle X^0, \alpha^\vee \rangle = 0$ for all $\alpha \in R^{\mathrm{an}}$. Hence $s_\alpha(x^0) = x^0$ for $x^0 \in X^0$. Also, the inclusion $R \cap X^0 \subseteq R^0$ in (AR4) is always true. Therefore the axioms (AR2)–(AR4) can be replaced by the following conditions

(AR2)′ $s_\alpha(R^{\mathrm{an}}) = R^{\mathrm{an}}$ for all $\alpha \in R^{\mathrm{an}}$,
(AR3)′ for every $\alpha \in R^{\mathrm{an}}$ the set $\langle R^{\mathrm{an}}, \alpha^\vee \rangle \subseteq \mathbb{Z}$ is finite,
(AR4)′ $R^0 \subseteq X^0$.

This new set of axioms makes it (even more) clear that the conditions on R^0 are rather weak: We (may) need R^0 to span X from (AR1), we need $R^0 \subseteq X^0$ for (AR4)′ and we need $0 \in R$, which is no condition since one can always add 0 to R^0. We will see this phenomena re-appearing in the examples, e.g., in Example 5.3.2, and in the definition of an extension datum in Definition 5.3.9.

- The definition of a connected affine reflection system is the same as the axiom (EA4) in the definition of an EALA.

- The definition of an affine reflection system given in [156] is not the same as the one given here. The equivalence of two definitions follows from [156, Prop. 5.4].

An *isomorphism* from an affine reflection system $(R, X, (\cdot|\cdot))$ to another affine reflection system $(R', X'(\cdot|\cdot)')$ is a vector space isomorphism $f : X \to X'$ satisfying

$$f(R^{\mathrm{an}}) = R'^{\mathrm{an}} \quad \text{and} \quad f(R^0) = R'^0$$

If such a map exists, $(R, X, (\cdot|\cdot))$ and $(R', X, (\cdot|\cdot)')$ are called *isomorphic*. One can show, as a corollary of the Structure Theorem 5.3.10, that an isomorphism f also satisfies $f \circ s_\alpha = s_{f(\alpha)} \circ f$ for all $\alpha \in R^{\mathrm{an}}$, equivalently,

$$\langle x, \alpha^\vee \rangle = \langle f(x), f(\alpha)^\vee \rangle$$

for all $x \in X$ and $\alpha \in R^{\mathrm{an}}$. This is always fulfilled if f is an isometry for $(\cdot|\cdot)$ and $(\cdot|\cdot)'$ respectively. But in general an isomorphism is not necessarily an isometry. For example, one can always multiply the bilinear form $(\cdot|\cdot)$ by a non-zero scalar without changing $\langle x, \alpha^\vee \rangle$.

Since a reflection s_α is an isometry, it follows from (AR2) and (AR4) that s_α leaves R^{an} and R^0 invariant and is thus an automorphism of (R, X). The subgroup $W(R)$ of the automorphism group of (R, X) generated by all reflections s_α, $\alpha \in R^{\mathrm{an}}$, is (obviously) called the *Weyl group* of (R, X). (It will not play a big role in this section.)

5.3.2 Examples of affine reflection systems. We will now give some immediate examples of affine reflection systems.

Example 5.3.2 (*The real part of an affine reflections system*) Let $(R, X, (\cdot|\cdot))$ be an affine reflection system. Then

$$\mathrm{Re}(R) = \{0\} \cup R^{\mathrm{an}}, \quad \mathrm{Re}(X) = \mathrm{Span}_F(R^{\mathrm{an}}),$$
$$(\cdot|\cdot)_{\mathrm{Re}} = (\cdot|\cdot)_{\mathrm{Re}(X)\times \mathrm{Re}(X)}$$

defines an affine reflection system, called the *real part of* (R, X), with

$$\mathrm{Re}(R)^{\mathrm{an}} = R^{\mathrm{an}}, \quad \mathrm{Re}(R)^0 = \{0\},$$

in particular $\mathrm{Re}(R)$ has nullity 0.

Observe that $(\cdot|\cdot)_{\mathrm{Re}}$ need not be nondegenerate, see Example 5.3.4 for an example.

The fact that one can "throw away" the non-zero null roots and still have an affine reflection system indicates that one has little control over the null roots in a general affine reflection system. This will be made even more evident in the concept of an extension datum 5.3.9, used in the general Structure Theorem 5.3.10 for affine reflection systems. It is therefore natural to define subclasses of affine reflection system by imposing conditions on the null roots. For example, we will do so when we define extended affine root systems in Section 5.3.4.

A finite root system is an affine reflection system of nullity 0, as we will show now.

Example 5.3.3 (*Finite root systems*) Let Φ be a (finite) root system à la Bourbaki [39, VI, §1.1]. Recall that this means that Φ is a subset of an F-vector space Y satisfying the axioms (RS1)–(RS3) below.

(RS1): Φ is finite, $0 \notin \Phi$ and Φ spans Y.

(RS2): For every $\alpha \in \Phi$ there exists a linear form $\alpha^\vee \in Y^*$ such that $\alpha^\vee(\alpha) = 2$ and $s_\alpha(\Phi) = \Phi$, where s_α is the reflection of Y defined by $s_\alpha(y) = y - \alpha^\vee(y)\alpha$,

(RS3): for every $\alpha \in \Phi$ the set $\alpha^\vee(\Phi)$ is contained in $\mathbb{Z}$.

Observe that the reflection s_α defined in (RS2) satisfies $s_\alpha(\alpha) = -\alpha$ and $s_\alpha(y) = y$ for $\alpha^\vee(y) = 0$. It therefore seems to depend on α and the linear form $\alpha^\vee$. However, since Φ is finite, there exists at most one reflection s with $s(\Phi) = \Phi$ and $s(\alpha) = -\alpha$ ([39, VI, §1.1, Lemme 1]). It is therefore not necessary to indicate $\alpha^\vee$ in the notation of s_α.

Note that we do not assume that Φ is reduced. This more general concept of a root system is necessary for the Structure Theorem 5.3.10 of affine reflection systems. The reader who is only familiar with the theory of reduced finite root systems, as for example developed in [126, Ch. III], can perhaps be comforted by the fact that the difference is not very big. Indeed, every finite root system is a direct sum of connected (= irreducible) root systems and there is only one irreducible non-reduced root system of rank l, namely

$$\mathrm{BC}_l = \mathrm{B}_l \cup \mathrm{C}_l = \{\pm\varepsilon_i : 1 \le i \le l\} \cup \{\pm\varepsilon_i \pm \varepsilon_j : 1 \le i, j \le l\}$$

where here and in the following $\varepsilon_1, \dots, \varepsilon_l$ is the standard basis of F^l. (Note $0 \in \mathrm{BC}_l$ in anticipation of the convention introduced below.)

In the context of finite-dimensional Lie algebras, non-reduced root systems arise naturally as the roots of a finite-dimensional semisimple Lie algebra L with respect to a maximal ad-diagonalizable subalgebra $H \subseteq L$ which is not self-centralizing, hence not a Cartan algebra. In particular, non-reduced root systems do not occur over an algebraically closed field. However, they do occur in the context of infinite-dimensional Lie algebras, even over algebraically closed fields, see Example 5.3.6.

Given a finite root system (Φ, Y), define

$$S = \{0\} \cup \Phi \quad \text{and} \quad (x \mid y) = \textstyle\sum_{\alpha \in \Phi} \alpha^\vee(x)\, \alpha^\vee(y) \tag{5.48}$$

for $x, y \in Y$. Then $(\cdot|\cdot)$ is a nondegenerate symmetric bilinear form on Y with respect to which all reflections s_α are isometric ([39, VI, §1.1, Prop. 3]). Moreover, $(\alpha \mid \alpha)$ is a positive integer for every $\alpha \in \Phi$ (viewing $\mathbb{Q} \subseteq F$ canonically) and

$$\langle y, \alpha^\vee \rangle = \alpha^\vee(y) = 2\,\tfrac{(y|\alpha)}{(\alpha|\alpha)}$$

for all $y \in Y$. Hence s_α as defined in (RS2) is also given by the formula (5.47). We have $S^0 = \{0\} = X^0 = X^0 \cap S$. Since $\langle \Phi, \alpha^\vee \rangle \subseteq \mathbb{Z}$ we have shown that

$(S, Y, (\cdot|\cdot))$ *as defined in* (5.48) *is a finite affine reflection system of nullity* 0.

We will characterize finite root systems within the category of affine reflection systems in Corollary 5.3.11.

In the following we will always assume that a finite root system contains 0. We will usually use the symbol S for a finite root system, and put

$$S^\times = S \setminus \{0\} = \Phi.$$

We will also need the following subsets of roots of a finite root system S:

S_{div} is the set of divisible roots, where $\alpha \in S$ is called *divisible* if $\alpha/2 \in S$. In particular $0 \in S_{\mathrm{div}}$. We put $S^\times_{\mathrm{div}} = S_{\mathrm{div}} \cap S^\times = S_{\mathrm{div}} \setminus \{0\}$.

$S_{\mathrm{ind}} = S \setminus S^\times_{\mathrm{div}}$, the subsystem of *indivisible roots*.

We also need the fact that there exists a unique symmetric bilinear form $(\cdot|\cdot)_u$ on Y which is invariant under the Weyl group $W(S)$ and which satisfies $2 \in \{(\alpha|\alpha)_u : 0 \neq \alpha \in C\} \subseteq \{2, 4, 6, 8\}$ for every connected component C of S. This follows easily from [39, Prop. 7]. Observe that

$$S^\times_{\mathrm{div}} = \{\alpha \in S : (\alpha|\alpha)_u = 8\}.$$

We use $(\cdot|\cdot)_u$ to define short and long roots:

$S_{\mathrm{sh}} = \{\alpha \in S : (\alpha|\alpha)_u = 2\}$ is the set of *short roots*.

$S_{\mathrm{lg}} = \{\alpha \in S : (\alpha|\alpha)_u \in \{4, 6\}\}$ is the set of *long roots* in S.

Thus $S_{\mathrm{lg}} = S \setminus (S_{\mathrm{sh}} \cup S_{\mathrm{div}})$. For example, for $S = \mathrm{BC}_l$ we have

$$\begin{aligned} \mathrm{BC}_{l,\mathrm{sh}} &= \{\pm\varepsilon_i : 1 \le i \le l\}, \\ \mathrm{BC}^\times_{l,\mathrm{div}} &= \{\pm 2\varepsilon_i : 1 \le i \le l\}, \\ \mathrm{BC}_{l,\mathrm{lg}} &= \{\pm\varepsilon_i \pm \varepsilon_j : 1 \le i \neq j \le l\}, \end{aligned}$$

in particular $\mathrm{BC}_{1,\mathrm{lg}} = \emptyset$, and if S is *simply laced*, i.e., $S^\times = S_{\mathrm{sh}}$, then $S_{\mathrm{div}} = \{0\}$ and $S_{\mathrm{lg}} = \emptyset$.

Example 5.3.4 (*Untwisted affine reflection systems*) Let $(S, Y, (\cdot|\cdot)_Y)$ be a finite root system. Hence $0 \in S$ and $\Phi = S \setminus \{0\}$, as stipulated in Example 5.3.3. Also, let Z be an n-dimensional F-vector space, say with a basis $\varepsilon_1, \dots, \varepsilon_n$. We define

$$\begin{aligned} X &= Y \oplus Z, \\ \Lambda &= \mathbb{Z}\varepsilon_1 \oplus \cdots \oplus \mathbb{Z}\varepsilon_n \subseteq Z, \\ R &= \textstyle\bigcup_{\xi \in S}\{\xi \oplus \lambda : \lambda \in \Lambda\} \subseteq Y \oplus Z, \\ (x_1 \mid x_2)_X &= (y_1 \mid y_2)_Y \quad \text{for } x_i = y_i \oplus z_i \text{ with } y_i \in Y \text{ and } z_i \in Z. \end{aligned}$$

By construction we then have

$$X^0 = Z, \quad R^0 = \Lambda, \quad R^{\text{an}} = \textstyle\bigcup_{\xi \in \Phi} \xi \oplus \Lambda$$

where of course $\xi \oplus \Lambda = \{\xi \oplus \lambda : \lambda \in \Lambda\}$. For $\alpha = \xi \oplus \lambda \in R^{\text{an}}$ with $\xi \in S$ and $\lambda \in \Lambda$ the reflection s_α satisfies

$$s_\alpha(y \oplus z) = s_\xi(y) \oplus (z - \langle y, \xi^\vee \rangle \lambda). \tag{5.49}$$

We will leave it to the reader to verify that

$$(R, X) \textit{ is an affine reflection system of nullity } n. \tag{5.50}$$

Observe that (R, X) is the set of roots of the EALA constructed in Section 5.2.5, see in particular (5.46).

Observe that $\operatorname{Span}_F(R^{\text{an}}) = X = \operatorname{Re}(X)$ in case $S \neq \{0\}$. This shows that the form $(\cdot|\cdot)_{\text{Re}}$ of the real part $\operatorname{Re}(R)$ of R need not be nondegenerate.

Exercise 5.3.5 Show the claim in (5.50), and also that (R, X) is reduced resp. connected if and only if (S, Y) is so.

Example 5.3.6 (*Affine root systems*) By definition, an *affine root system* is the set of roots of an affine Kac-Moody Lie algebra, which we studied in Section 5.1.1 and then again in Section 5.2.4, where we showed that an affine Kac-Moody algebra is an EALA of nullity 1. Our goal here is not surprising. We want to show that

$$\textit{an affine root system is an affine reflection system of nullity } 1. \tag{5.51}$$

Let us first collect the data necessary to prove this. We use the notation established in Section 5.2.4. Thus, $\hat{\mathcal{L}} = \hat{\mathcal{L}}(\mathfrak{g}, \sigma)$ is an affine Kac-Moody Lie algebra over $\mathbb{C}$, σ is a diagram automorphism of the simple finite-dimensional Lie algebra $\mathfrak{g}$ of order $m \in \{1, 2, 3\}$, and $\Delta_{\bar{s}}$ denotes the set of weights of the $(\mathfrak{g}_{\bar{0}}, \mathfrak{h}_{\bar{0}})$-module $\mathfrak{g}_{\bar{s}} \subseteq \mathfrak{g}$, $s = 0, \dots, m-1$. One knows that $\Delta_{\bar{0}}$ is a reduced irreducible root system in $\mathfrak{h}_{\bar{0}}^* =: Y$. The roots of $\hat{\mathcal{L}}$ with respect to $H = \mathfrak{h}_{\bar{0}} \oplus \mathbb{C}c \oplus \mathbb{C}d$ are

$$R = \{\gamma \oplus n\delta : \gamma \in \Delta_{\bar{s}},\ \bar{n} = \bar{s},\ 0 \le s < m\},$$

see (5.41), hence

$$X = \operatorname{Span}_{\mathbb{C}}(R) = Y \oplus \mathbb{C}\delta.$$

The bilinear form $(\cdot|\cdot)_X$ used to determine the (an)isotropic roots in R has the form

$$(x_1 \mid x_2)_X = (y_1 \mid y_2)_Y$$

where $x_i = y_i \oplus a_i\delta$ with $y_i \in Y$ and $a_i \in \mathbb{C}$, and where $(\cdot|\cdot)_Y$ is the nondegenerate symmetric bilinear form on Y, obtained by transporting the Killing form $\kappa\,|_{\mathfrak{h}_{\bar{0}} \times \mathfrak{h}_{\bar{0}}}$ from $\mathfrak{h}_{\bar{0}}$ to Y. It follows that

$$X^0 = \mathbb{C}\delta \quad \text{and} \quad R^{\text{an}} = \{\gamma \oplus n\delta \in R : \gamma \neq 0\}.$$

We can now verify the axioms (AR1)–(AR4).

(AR1) holds by definition. (AR2) is a consequence of [137, Prop. 3.7(b)]. Concerning (AR3), it follows from the structure of $(\cdot|\cdot)_X$ that

$$\langle x, \alpha^\vee\rangle = \langle y, \gamma^\vee\rangle \quad \text{for } x = y \oplus a\delta \in X \text{ and } \alpha = \gamma \oplus n\delta \in R^{\mathrm{an}}. \tag{5.52}$$

This implies that $\langle R, \alpha^\vee\rangle$ is a finite set since

$$S = \Delta_{\bar 0} \cup \cdots \cup \Delta_{\overline{m-1}}$$

is a finite set (S is actually a finite root system; for $m > 1$ see the table below). Moreover $\langle R, \alpha^\vee\rangle \subseteq \mathbb{Z}$ because $\hat{\mathcal{L}}$ is an integrable $\hat{\mathcal{L}}$-module ([137, Lemma 3.5]). Thus (AR3) holds, and (AR4) follows from (5.52) and $(\gamma \mid \gamma) = 0 \Leftrightarrow \gamma = 0$ for $\gamma \in S$. This proves (5.51).

To motivate the definition of extension data in Definition 5.3.9 and the Structure Theorem 5.3.10 for affine reflection systems, we will now look at R and S more closely. In the untwisted case, i.e., $m = 1$, we have of course

$$\Delta_{\bar 0} = S, \quad R = S \times \mathbb{Z}\delta \quad (m = 1).$$

Thus R is an untwisted affine root system of nullity 1, a special case of the Example 5.3.4. For $m = 2, 3$ the structure of $\Delta_{\bar s}$ and S is summarized in the table below. Proofs can be extracted from [137, 7.9, 7.8, 8.3].

$(\mathfrak{g}, m)$	$\Delta_{\bar 0}$	$\Delta_{\bar 1}$	S
$(\mathrm{A}_{2l}, 2), l \geq 1$	A_1 or $\mathrm{B}_l (l \geq 2)$	$\Delta_{\bar 0} \cup \{\pm 2\varepsilon_i : 1 \leq i \leq l\}$	BC_l
$(\mathrm{A}_{2l-1}, 2), l \geq 2$	C_l	$\{0\} \cup \mathrm{C}_{l,\mathrm{sh}}$	C_l
$(\mathrm{D}_{l+1}, 2), l \geq 3$	B_l	$\{0\} \cup \mathrm{B}_{l,\mathrm{sh}}$	B_l
$(\mathrm{E}_6, 2)$	F_4	$\{0\} \cup \mathrm{F}_{4,\mathrm{sh}}$	F_4
$(\mathrm{D}_4, 3)$	G_2	$\{0\} \cup \mathrm{G}_{2,\mathrm{sh}}$	G_2

(5.53)

We can now rewrite R. For subsets $T \subseteq S$ and $\Xi \subseteq \mathbb{Z}\delta$ we put

$$T \oplus \Xi = \{\tau \oplus n\delta : \tau \in T, \, n\delta \in \mathbb{Z}\delta\}$$

and abbreviate $\mathrm{B}_1 = \mathrm{A}_1 = \{0, \pm\alpha\}$. For $m = 2$ we get

$$\begin{aligned} R &= (\Delta_{\bar 0} \oplus 2\mathbb{Z}\delta) \cup \big(\Delta_{\bar 1} \oplus (1 + 2\mathbb{Z})\delta\big) \\ &= \begin{cases} (\{0\} \oplus \mathbb{Z}\delta) \cup \big((\mathrm{B}_l \setminus \{0\} \oplus \mathbb{Z}\delta\big) \cup \big(\mathrm{BC}^\times_{\mathrm{div}} \oplus (1 + 2\mathbb{Z})\delta\big), & \mathfrak{g} = \mathrm{A}_{2l} \\ (\{0\} \oplus \mathbb{Z}\delta) \cup \quad (S_{\mathrm{sh}} \oplus \mathbb{Z}\delta) \quad \cup \quad (S_{\mathrm{lg}} \oplus 2\mathbb{Z}\delta), & \mathfrak{g} \neq \mathrm{A}_{2l}\,. \end{cases} \end{aligned}$$

For $(\mathfrak{g}, m) = (\mathrm{D}_4, 3)$ one knows $\Delta_{\bar 1} = \Delta_{\bar 2} = \{0\} \cup \mathrm{G}_{2,\mathrm{sh}}$, whence

$$\begin{aligned} R &= (\Delta_{\bar 0} \oplus 3\mathbb{Z}\delta) \cup \big(\Delta_{\bar 1} \oplus (1 + 3\mathbb{Z})\delta\big) \cup \big(\Delta_{\bar 2} \oplus (2 + 3\mathbb{Z})\delta\big) \\ &= (\{0\} \oplus \mathbb{Z}\delta) \cup (S_{\mathrm{sh}} \oplus \mathbb{Z}\delta) \cup (S_{\mathrm{lg}} \oplus 3\mathbb{Z}\delta). \end{aligned}$$

In all three cases R has a uniform description in terms of the root system S and subsets $\Lambda_{\mathrm{sh}}, \Lambda_{\mathrm{lg}}, \Lambda_{\mathrm{div}} \subseteq \mathbb{Z}\delta$ as

$$R = R^0 \cup (S_{\mathrm{sh}} \oplus \Lambda_{\mathrm{sh}}) \cup (S_{\mathrm{lg}} \oplus \Lambda_{\mathrm{lg}}) \cup (S^\times_{\mathrm{div}} \oplus \Lambda_{\mathrm{div}}) \tag{5.54}$$

where

$$\Lambda_{\rm sh} = \mathbb{Z}\delta = R^0, \quad \Lambda_{\rm div} = (1+2\mathbb{Z})\delta, \quad \Lambda_{\rm lg} = \begin{cases} \mathbb{Z}\delta, & \mathfrak{g} = \mathrm{A}_{2l},\ m=2, \\ 2\mathbb{Z}\delta, & \mathfrak{g} \neq \mathrm{A}_{2l},\ m=2, \\ 3\mathbb{Z}\delta, & m=3. \end{cases} \tag{5.55}$$

If we define Λ_ξ for $\xi \in S$ by $\Lambda_\xi \in \{\Lambda_0 = R^0, \Lambda_{\rm sh}, \Lambda_{\rm lg}, \Lambda_{\rm div}\}$ according to ξ belong to the corresponding subset of S, then (5.54) becomes

$$R = \textstyle\bigcup_{\xi \in S} \xi \oplus \Lambda_\xi. \tag{5.56}$$

Note that we also recover [137, Th. 5.6(b)]:

$$R \cap X^0 = R \cap \mathbb{Z}\delta.$$

Example 5.3.7 (*Type* A_1 *generalized*) We consider a final example of an affine reflection system to motivate the definition of an extension datum in Definition 5.3.9 below.

Let Z be a finite-dimensional F-vector space and define the vector space X and a symmetric bilinear form on X by

$$X = F\alpha \oplus Z, \quad (a_1\alpha \oplus z_1 \mid a_2\alpha \oplus z_2) = a_1 a_2,$$

where $0 \neq \alpha$ and $a_i \in F$. We define $R \subseteq X$ in terms of three non-empty subsets $\Lambda_0, \Lambda_\alpha, \Lambda_{-\alpha} \subseteq Z$ as follows:

$$R = \Lambda_0 \cup (\alpha \oplus \Lambda_\alpha) \cup (-\alpha \oplus \Lambda_{-\alpha}). \tag{5.57}$$

It is then immediate that

$$X^0 = Z, \quad R^0 = \Lambda_0, \quad R^{\rm an} = (\alpha \oplus \Lambda_\alpha) \cup (-\alpha \oplus \Lambda_{-\alpha}).$$

We will now discuss under which conditions $(R, X, (\cdot|\cdot))$ is an affine reflection system. Let us start with (AR2). For $s_i \in \{\pm 1\}$, $\mu \in \Lambda_{s_1\alpha}$ and $\lambda \in \Lambda_{s_2\alpha}$ we have

$$\begin{aligned} \langle s_1\alpha \oplus \mu, (s_2\alpha \oplus \lambda)^\vee \rangle &= 2\, s_1 s_2, \\ s_{s_2\alpha\oplus\lambda}(s_1\alpha \oplus \mu) &= -s_1\alpha \oplus (\mu - 2 s_1 s_2 \lambda) \end{aligned}$$

Hence, all reflections $s_{s_2\alpha\oplus\lambda}$ leave R invariant if and only if $\mu - 2s_1s_2\lambda \in \Lambda_{-s_1\alpha}$ for μ, λ as above, i.e., in obvious short form $\Lambda_{s_1\alpha} - 2s_1s_2\Lambda_{s_2\alpha} \subseteq \Lambda_{-s_1\alpha}$. In particular,

$$\begin{aligned} &\Lambda_\alpha - 2\Lambda_\alpha \subseteq \Lambda_{-\alpha}, \quad \Lambda_{-\alpha} - 2\Lambda_{-\alpha} \subseteq \Lambda_\alpha \quad \text{for } s_1 = s_2, \\ &\Lambda_{-\alpha} + 2\Lambda_\alpha \subseteq \Lambda_\alpha \quad \text{for } s_1 = -1 = -s_2. \end{aligned}$$

For $\lambda \in \Lambda_\alpha$ we therefore get $\lambda - 2\lambda = -\lambda \in \Lambda_{-\alpha}$, whence $-\Lambda_\alpha \subseteq \Lambda_{-\alpha}$ and, analogously, $\Lambda_{-\alpha} \subseteq -\Lambda_\alpha$. We therefore obtain

$$\Lambda_{-\alpha} = -\Lambda_\alpha, \tag{5.58}$$

or with the notation of above $\Lambda_{s_1\alpha} = s_1\Lambda_\alpha$. It is now easy to see that (AR2) is equivalent to the two conditions (5.58) and

$$2\Lambda_\alpha - \Lambda_\alpha \subseteq \Lambda_\alpha. \tag{5.59}$$

It then follows that R is an affine reflection system if and only if

(i) (5.58) and (5.59) hold,
(ii) $0 \in \Lambda_0$, and
(iii) $Z = \mathrm{Span}_F(\Lambda_0 \cup \Lambda_\alpha \cup \Lambda_{-\alpha})$.

Observe the similarity with the previous examples: R has the form

$$R = \bigcup_{\xi \in S} \xi \oplus \Lambda_\xi$$

where $S = \{0, \pm\alpha\}$ is a finite root system and $(\Lambda_\xi : \xi \in S)$ is a family of subsets in X^0. However, in the previous examples the Λ_ξ were subgroups of $(Z, +)$ while here we only have the condition (5.59). Does this imply that Λ_α is a subgroup? The answer is no! For example, in $Z = F$ the subset $\Lambda_\alpha = 1 + 2\mathbb{Z} \subseteq F$ satisfies (5.59).

A subset A of an abelian group $(Z, +)$ is called a *reflection subspace* if $2a_1 - a_2 \in A$ for all $a_i \in A$ (see [155] or [187, 3.3] for a justification for this terminology). Hence, (5.59) just says that Λ_α is a reflection subspace. The structure of two special types of reflection subspaces is described in Exercise 5.3.8 below.

While in general Λ_α is far from being a subgroup, one can always "re-coordinatize" R to at least get $0 \in \Lambda_\alpha$. Namely, for a fixed $\lambda \in \Lambda_\alpha$ we have $\alpha + \Lambda_\alpha = (\alpha + \lambda) + (\Lambda_\alpha - \lambda)$. Hence, with $\tilde{\alpha} = \alpha + \lambda$ and $\Lambda_{\tilde{\alpha}} = \Lambda_\alpha - \lambda$, we obtain

$$R = \Lambda_0 \cup (\tilde{\alpha} + \Lambda_{\tilde{\alpha}}) \cup \big(- (\tilde{\alpha} + \Lambda_{\tilde{\alpha}})\big), \tag{5.60}$$

where now $\Lambda_{\tilde{\alpha}}$ not only satisfies (5.59) but also $0 \in \Lambda_{\tilde{\alpha}}$. In other words, $\Lambda_{\tilde{\alpha}}$ is a pointed reflection subspace as defined in Exercise 5.3.8 and therefore also satisfies $\Lambda_{\tilde{\alpha}} = -\Lambda_{\tilde{\alpha}}$.

The process of re-coordinatization works well in this example. The reason is that the finite root system S in (5.60) is reduced. Re-coordinatization will not work if S is not reduced, as for example in the case $(\mathfrak{g}, m) = (\mathrm{A}_{2l}, 2)$ of Example 5.3.6. This "explains" why in the property (ED2) of an extension datum in 5.3.9 we require $0 \in \Lambda_\xi$ only for an indivisible root $\xi \in S$.

Exercise 5.3.8 Let A be a subset of an abelian group $(Z, +)$. As above we put $2A - A = \{2a_1 - a_2 : a_i \in A\}$. We denote by $\Lambda = \mathrm{Span}_\mathbb{Z}(A)$ the $\mathbb{Z}$-span of A in Z. A subset $A \subseteq Z$ is called *symmetric* if $A = -A$.

(a) The following equivalent conditions characterize symmetric reflection subspaces $A \subseteq Z$:
  (i) $2A - A \subseteq A$ and $A = -A$,
  (ii) $2\lambda + a \in A$ for every $\lambda \in \Lambda$ and $a \in A$,
  (iii) A is a union of cosets modulo 2Λ,
  (iv) $a_1 - 2a_2 \in A$ for all $a_i \in A$.

(b) The following are equivalent for $A \subseteq Z$:
  (i) $0 \in A$ and $A - 2A \subseteq A$,
  (ii) $0 \in A$ and $2A - A \subseteq A$,
  (iii) $2\mathbb{Z}[A] \subseteq A$ and $2\mathbb{Z}[A] - A \subseteq A$,
  (iv) A is a union of cosets modulo $2\mathbb{Z}[A]$, including the trivial coset $2\mathbb{Z}[A]$.

  In this case A is called a *pointed* reflection subspace.

(c) Every pointed reflection subspace is symmetric.

(d) If A is a symmetric reflection subspace then $A + A$ is a pointed reflection subspace.

5.3.3 The structure theorem of affine reflection systems. After the many examples in Section 5.3.2, the following definition should not be too surprising.

Definition 5.3.9 Let S be a finite root system as defined in Example 5.3.3. Recall $S^\times = S \setminus \{0\}$ and $S_{\text{ind}} = \{0\} \cup \{\alpha \in S : \alpha/2 \notin S\} = S \setminus S^\times_{\text{div}}$. Also, let Z be a finite-dimensional F-vector space. An *extension datum of type* (S, Z), sometimes simply called an *extension datum*, is a family $(\Lambda_\xi : \xi \in S)$ of subsets $\Lambda_\xi \subseteq Z$ satisfying the axioms (ED1)–(ED3) below.

(ED1): For $\eta, \xi \in S^\times$, $\mu \in \Lambda_\eta$ and $\lambda \in \Lambda_\xi$ we have $\mu - \langle \eta, \xi^\vee \rangle \xi \in \Lambda_{s_\xi(\eta)}$, in obvious short form

$$\Lambda_\eta - \langle \eta, \xi^\vee \rangle \Lambda_\xi \subseteq \Lambda_{s_\xi(\eta)}.$$

(ED2): $0 \in \Lambda_\xi$ for $\xi \in S_{\text{ind}}$, and $\Lambda_\xi \neq \emptyset$ for $\xi \in S \setminus S_{\text{ind}} = S^\times_{\text{div}}$.
(ED3): $Z = \operatorname{Span}_F \left(\bigcup_{\xi \in S} \Lambda_\xi \right)$.

The axiom (ED1) is trivially true for $\eta = 0$ since $\langle \eta, \xi^\vee \rangle = 0$ and $s_\xi(0) = 0$. Also, if $S^\times_{\text{div}} = \emptyset$, then there is no Λ_ξ for $\xi \in S^\times_{\text{div}}$ and so the second condition in (ED2) is trivially fulfilled. (ED3) simply serves to determine Z. If it does not hold, one can simply replace Z by $\operatorname{Span}_{\mathbb{Z}}(\bigcup_{\xi \in S} \Lambda_\xi)$.

The definition of an extension datum above is a special case of the notion of an extension datum for a pre-reflection system, introduced in [156, 4.2]. (The reader will note that the axiom (ED1) in [156] simplifies since in our setting the subset S^{re} of [156] is $S^{\text{re}} = S^{\text{an}} = S \setminus \{0\}$.) The Structure Theorem 5.3.10 below is proven in [156, Th. 4.6] for extensions of pre-reflection systems. Affine reflection systems are special types of such extensions, namely finite-dimensional extensions of finite root systems.

The rationale for the concept of an extension datum is the following Structure Theorem for affine reflection systems.

Theorem 5.3.10 (Structure Theorem for affine reflection systems) (a) *Let $(S, Y, (\cdot|\cdot)_Y)$ be a finite root system and let $\mathfrak{L} = (\Lambda_\xi : \xi \in S)$ be an extension datum of type (S, Z). Define $(R, X, (\cdot|\cdot)_X)$ by*

$$X = Y \oplus Z$$
$$R = \textstyle\bigcup_{\xi \in S} \xi \oplus \Lambda_\xi \ \subseteq Y \oplus Z = X,$$
$$(y_1 \oplus z_1 \mid y_2 \oplus z_2)_X = (y_1 \mid y_2)_Y$$

for $y_i \in Y$ and $z_i \in Z$. Then $(R, X, (\cdot|\cdot)_X)$ is an affine reflection system, denoted $\mathcal{A}(S, \mathfrak{L})$, with

$$R^0 = \Lambda_0, \quad X^0 = Z \quad \text{and} \quad R^{\text{an}} = \textstyle\bigcup_{0 \neq \xi \in S} \xi \oplus \Lambda_\xi.$$

For $\alpha = \xi \oplus \lambda \in R^{\text{an}}$ and $x = y \oplus z \in X$ the reflection s_α is given by

$$s_\alpha(x) = s_\xi(y) \oplus (z - \langle y, \xi^\vee \rangle \lambda)$$

(b) *Conversely, let $(R, X, (\cdot|\cdot)_X)$ be an affine reflection system.*

(i) *Let $f : X \to X/X^0 =: Y$ be the canonical map, put $S = f(R)$ and let $(\cdot|\cdot)_Y$ be the induced bilinear form on Y, that is $(f(x_1) \mid f(x_2))_Y = (x_1 \mid x_2)_X$. Then $(S, Y, (\cdot|\cdot)_Y)$ is a finite root system, the so-called* quotient root system *of (R, X).*

(ii) *There exists a linear map $g : Y \to X$ satisfying $f \circ g = \mathrm{Id}_Y$ and $g(S_{\text{ind}}) \subseteq R$.*

(iii) *For g as in* (ii) *and $\xi \in S$ define $\Lambda_\xi \subseteq \ker(f) =: Z$ by*

$$R \cap f^{-1}(\xi) = g(\xi) \oplus \Lambda_\xi. \tag{5.61}$$

Then $\mathfrak{L} = (\Lambda_\xi : \xi \in S)$ is an extension datum of type (S, Z).

(iv) *(R, X) is isomorphic to the affine reflection system $\mathcal{A}(S, \mathfrak{L})$ constructed in* (a).

Let us note that it is not reasonable to expect $g(S) \subseteq R$ in (b.ii) above, since R may be reduced while S is not, see for example the case $(\mathfrak{g}, \mathrm{A}_{2l})$ in Example 5.3.6. The quotient root system S is uniquely determined, but not so the extension datum, see [156, Th. 4.6(c)].

Corollary 5.3.11 *An affine reflection system $(R, X, (\cdot|\cdot))$ is nondegenerate in the sense that $(\cdot|\cdot)$ is nondegenerate if and only if R is a finite root system.*

Proof If $(R, X, (\cdot|\cdot))$ is an affine reflection system with a nondegenerate form $(\cdot|\cdot)$, then $\{0\} = X^0 = \ker f$, so f is the identity. We have seen the other direction in Example 5.3.3. □

Corollary 5.3.12 ([156, Cor. 5.5]) *Let $(R, X, (\cdot|\cdot))$ be an affine reflection system over $F = \mathbb{R}$. Then there exists a positive semidefinite symmetric bilinear form $(\cdot|\cdot)_{\geq}$ on X such that $(R, X, (\cdot|\cdot)_{\geq})$ is an affine reflection system with the same anisotropic and null roots and reflections.*

The morale of the Structure Theorem is

affine reflection system	=	finite root system + extension datum

Thus properties of an affine reflection system can be described in terms of properties of its quotient root system and the associated extension datum. Some examples of this philosophy are given in the Proposition 5.3.13 and the Exercise 5.3.14 below.

Proposition 5.3.13 ([156, Cor. 5.2]) *Let R be an affine reflection system, let S be its quotient root system and let $(\Lambda_\xi : \xi \in S)$ be the associated extension datum. We define*

$$\Lambda_{\text{diff}} = \bigcup\nolimits_{0 \neq \xi \in S} \Lambda_\xi - \Lambda_\xi .$$

Then $\mathbb{Z}\Lambda_{\text{diff}} = \Lambda_{\text{diff}}$. Moreover:

(a) *R is* tame *in the sense that $R^0 \subseteq R^{\text{an}} - R^{\text{an}}$ if and only if $R^0 \subseteq \Lambda_{\text{diff}}$.*

(b) *All* root strings

$$\mathbb{S}(\beta, \alpha) = R \cap (\beta + \mathbb{Z}\alpha), \quad (\beta \in R, \alpha \in R^{\text{an}})$$

are unbroken, *i.e., $\mathbb{Z}(\beta, \alpha) = \{n \in \mathbb{Z} : \beta + n\alpha \in R\}$ is either a finite interval in $\mathbb{Z}$ or equals $\mathbb{Z}$, if and only if $\Lambda_{\text{diff}} \subseteq R^0$.*

(c) *A tame affine reflection system with unbroken root strings is symmetric.*

Exercise 5.3.14 Let R be an affine reflection system, let S be its quotient root system and let $(\Lambda_\xi : \xi \in S)$ be the associated extension datum. We use the notation of Proposition 5.3.13. Prove:

(a) R is reduced if and only if for all $0 \neq \xi \in S$ with $2\xi \in S$ we have

$$\Lambda_{2\xi} \cap 2\Lambda_\xi = \emptyset.$$

In particular, S need not be reduced for R to be reduced!

(b) R is connected iff S is connected (= irreducible) .

(c) R is symmetric, i.e., $R = -R$, iff Λ_0 is symmetric.

(d) For all $\alpha \in R^{\text{an}}$ and $\beta \in R$ the α-string through β, i.e., $\mathbb{S}(\beta, \alpha)$ has length $|\mathbb{S}(\beta, \alpha)| \leq 5$.

(e) Let $(\alpha, \beta) \in R^{\text{an}} \times R$ and define $d, u \in \mathbb{N}$ by put $-d = \min \mathbb{Z}(\beta, \alpha)$ and $u = \max \mathbb{Z}(\beta, \alpha)$. Then $u - d = \langle \beta, \alpha^\vee \rangle$.

We will now describe how the examples of affine reflection systems of Section 5.3.2 fit into the general scheme of the Structure Theorem 5.3.10 above.

Examples 5.3.15 (a) Let $\mathfrak{L} = (\Lambda_\xi : \xi \in S)$ be an extension datum of type (S, Z). Observe that the only conditions on Λ_0 are $0 \in \Lambda_0$ from (ED2) and that Λ_0 together with the other Λ_ξ's spans Z by (ED3). This is in line with our earlier observation that one has little control over the null roots R^0 of an affine reflection system. Following the Example 5.3.2 we define a new extension datum $\mathrm{Re}(\mathfrak{L}) = (\mathrm{Re}(\Lambda_\xi) : \xi \in S)$ of type $(S, \mathrm{Re}(Z))$ by

$$\mathrm{Re}(Z) = \mathrm{Span}_F\left(\bigcup_{0\neq\xi\in S} \Lambda_\xi\right), \quad \mathrm{Re}(\Lambda_\xi) = \begin{cases} \{0\} & \text{for } \xi = 0, \\ \Lambda_\xi & \text{for } \xi \neq 0. \end{cases}$$

If $\mathfrak{L}$ is the extension datum associated to the affine reflection system (R, X), then $\mathrm{Re}(\mathfrak{L})$ is the extension datum associated to the affine reflection system $\mathrm{Re}(R)$, the real part of (R, X).

(b) All $\Lambda_\xi = \{0\}$, whence $Z = \{0\}$, defines a *trivial* extension datum for any root system S. It is "used" when we view the finite root system S as an affine reflection system, as we have done in Example 5.3.3.

(c) Let Λ be a subgroup of a finite-dimensional vector space Z such that $\mathrm{Span}_F(\Lambda) = Z$. Then for any finite root system S the family $(\Lambda_\xi \equiv \Lambda : \xi \in S)$ is an extension datum of type (S, Z). It is used to construct the untwisted affine reflection systems of Example 5.3.4.

(d) Let R be an affine root system . We have seen that R is an affine reflection system. Its quotient root system S and its associated extension datum $(\Lambda_\xi : \xi \in S)$ are described in Example 5.3.6 using the very same symbols, see the formulas (5.56) and (5.61).

(e) The family $\tilde{\mathfrak{L}} = (\Lambda_0, \Lambda_{\tilde\alpha}, \Lambda_{-\tilde\alpha})$ in Example 5.3.7 is an extension datum, but not necessarily $\mathfrak{L} = (\Lambda_0, \Lambda_\alpha, \Lambda_{-\alpha})$ since 0 need not lie in $\Lambda_{\pm\alpha}$. In fact, replacing $\mathfrak{L}$ by $\tilde{\mathfrak{L}}$ was the rationale for the re-coordinatization in 5.3.7.

To describe the classification of affine reflection systems we need some more properties of the subsets Λ_ξ of an extension datum. They are given in the following exercise (just do it!). Recall from Exercise 5.3.8 that a reflection subspace A is called symmetric if $A = -A$ and is called pointed if $0 \in A$.

Exercise 5.3.16 Let $(\Lambda_\eta : \eta \in S)$ be an extension datum of type (S, Z). Show:

(a) Every Λ_ξ for $0 \neq \xi \in S$ is a symmetric reflection subspace and is even a pointed reflection subspace if $\xi \in S^\times_{\mathrm{ind}}$.

(b) For $w \in W(S)$, the Weyl group of S, we have

$$\Lambda_\xi = \Lambda_{w(\xi)}. \quad \text{In particular, } \Lambda_\xi = \Lambda_{-\xi} = -\Lambda_\xi. \tag{5.62}$$

(c) Whenever $0 \neq \xi \in S$ and $2\xi \in S$, then

$$\Lambda_{2\xi} \subseteq \Lambda_\xi.$$

(d) $\mathbb{Z}\Lambda_\xi \subseteq \Lambda_\xi$ for $\xi \in S^\times_{\mathrm{ind}}$.

Let $\mathfrak{L} = (\Lambda_\xi : \xi \in S)$ be an extension datum and assume that S is irreducible. Then $W(S)$ acts transitively on the roots of the same length ([39, VI, §1.3, Prop. 11]), i.e., on $\{0\}$, S_{sh}, S_{lg} and S_{div} (some of these sets might be empty).

Because of (5.62), there are therefore at most four different subsets Λ_0, $\Lambda_{\rm sh}$, $\Lambda_{\rm lg}$ and $\Lambda_{\rm div}$ among the Λ_ξ, defined by

$$\Lambda_\xi = \begin{cases} \Lambda_0, & \xi = 0; \\ \Lambda_{\rm sh}, & \xi \in S_{\rm sh}; \\ \Lambda_{\rm lg}, & \xi \in S_{\rm lg}; \\ \Lambda_{\rm div}, & \xi \in S_{\rm div}^\times. \end{cases} \tag{5.63}$$

Of course, $\Lambda_{\rm lg}$ or $\Lambda_{\rm div}$ only exists if the corresponding subset of roots exits. The assertions below referring to $\Lambda_{\rm lg}$ or $\Lambda_{\rm div}$ should be interpreted correspondingly.

We have seen in Exercise 5.3.16 (did you do it?), that the subsets $\Lambda_{\rm sh}$ and $\Lambda_{\rm lg}$ are pointed reflection subspaces and that $\Lambda_{\rm div}$ is a symmetric reflection subspace. Assuming only these properties, does however not give an extension datum since only parts of the axiom (ED1) are fulfilled, namely those with $\eta = \pm\xi$. We also need to evaluate what happens for $\eta \neq \pm\xi$ with $\langle \eta, \xi^\vee\rangle \neq 0$. We will do this in the following examples.

Examples 5.3.17 Let S be an irreducible root system. We suppose that we are given a pointed reflection subspace $\Lambda_{\rm sh}$ of a finite-dimensional vector space, and if $S_{\rm lg} \neq \emptyset$ or $S_{\rm div}^\times \neq \emptyset$ then also a pointed reflection subspace $\Lambda_{\rm lg}$ and a symmetric reflection subspace $\Lambda_{\rm div}$. We define Λ_ξ, $\xi \in S^\times$, by (5.63), put $\Lambda_0 = \{0\}$ and ask, when is the family

$$\mathfrak{L}_{\rm min} = (\{0\}, \Lambda_{\rm sh}, \Lambda_{\rm lg}, \Lambda_{\rm div})$$

defined in this way an extension datum in $Z = \operatorname{Span}\big(\bigcup_{\xi\in S}\Lambda_\xi\big)$? Note that we only have to check (ED1). We will consider some examples of S.

(a) $S = \mathrm{A}_1$: In this case there are no further conditions, so $\mathfrak{L}_{\rm min}$ describes all possible extension data for A_1 with $\Lambda_0 = \{0\}$.

(b) $S = \mathrm{A}_2$: In this case there exist roots η, ξ with $\langle \eta, \xi^\vee\rangle = 1$, namely those for which $\angle(\eta, \xi) = \frac{\pi}{3}$. Evaluating (ED1) for those roots yields $\Lambda_{\rm sh} - \Lambda_{\rm sh} \subseteq \Lambda_{\rm sh}$, forcing $\Lambda_{\rm sh}$ to be a subgroup of Z. Thus, $\mathfrak{L}_{\rm min}$ is an extension datum for $S = \mathrm{A}_2$ iff $\Lambda_{\rm sh}$ is a subgroup.

(c) S simply laced, $\operatorname{rank} S \geq 2$: The argument in (b) works whenever $S_{\rm sh}$ contains roots η, ξ with $\langle \eta, \xi^\vee\rangle = 1$. Since this is the case here, we get that $\mathfrak{L}_{\rm min}$ is an extension datum iff $\Lambda_{\rm sh}$ is a subgroup of Z.

(d) $S = \mathrm{B}_2 = \{\pm\varepsilon_1, \pm\varepsilon_2, \pm\varepsilon_1 \pm \varepsilon_2\}$: Here we have pointed reflection subspaces $\Lambda_{\rm sh}$ and $\Lambda_{\rm lg}$. Since non-zero roots of the same length are either proportional or orthogonal, (ED1) is fulfilled for them. Because (ED1) is invariant under sign changes, we are left to evaluate the case of two roots η, ξ of different lengths forming an obtuse angle of $\frac{3\pi}{4}$, for example $\eta = \varepsilon_1$, $\xi = \varepsilon_2 - \epsilon_1$.

If $\langle \eta, \xi^\vee\rangle = -2$, then η is long, ξ is short and so (ED1) becomes $\Lambda_{\rm lg} + 2\Lambda_{\rm sh} \subseteq \Lambda_{\rm lg}$. If η is short, ξ is long, we have $\langle \eta, \xi^\vee\rangle = -1$ and thus get the condition $\Lambda_{\rm sh} + \Lambda_{\rm lg} \subseteq \Lambda_{\rm sh}$. To summarize: $\mathfrak{L}_{\rm min}$ *is an extension datum for* $S = \mathrm{B}_2$ *iff* $\Lambda_{\rm sh}$ *and* $\Lambda_{\rm lg}$ *are pointed reflection subspaces satisfying*

$$\Lambda_{\rm lg} + 2\Lambda_{\rm sh} \subseteq \Lambda_{\rm lg} \quad \textit{and} \quad \Lambda_{\rm sh} + \Lambda_{\rm lg} \subseteq \Lambda_{\rm sh}. \tag{5.64}$$

Note that (5.64) implies $2\Lambda_{\rm sh} \subseteq \Lambda_{\rm lg} \subseteq \Lambda_{\rm sh}$.

(e) $S = \mathrm{B}_l$, $l \geq 3$. Recall $S = \{\pm\epsilon_i : 1 \leq i \leq l\} \cup \{\pm\varepsilon_i \pm \varepsilon_j : 1 \leq i \neq j \leq l\}$. Since the short roots in S are either proportional or orthogonal, (ED1) is fulfilled for all short roots η, ξ. But there exist long roots $\eta, \xi \in S$ with $\angle(\eta, \xi) = \frac{\pi}{3}$, whence

$\langle \eta, \xi^\vee \rangle = 1$ and so (ED1) reads $\Lambda_{\mathrm{lg}} - \Lambda_{\mathrm{lg}} \subseteq \Lambda_{\mathrm{lg}}$. This forces Λ_{lg} to be a subgroup. As for $S = \mathrm{B}_2$, (ED1) for roots of different lengths leads to the condition (5.64). It is then easy to check that *$\mathfrak{L}_{\min}$ is an extension datum for $S = \mathrm{B}_l$, $l \geq 3$, iff Λ_{sh} is a pointed reflection subspace, Λ_{lg} is a subgroup and* (5.64) *holds.*

Continuing in this way, one arrives at the following.

Theorem 5.3.18 (Structure of extension data) *Let S be an irreducible finite root system and define a family $\mathfrak{L}_{\min}$ as in* (5.63) *with $\Lambda_0 = \{0\}$. Then $\mathfrak{L}_{\min}$ is an extension datum if and only if Λ_{sh} and Λ_{lg} are pointed reflection subspaces, Λ_{div} is a symmetric reflection subspace and the following conditions, depending on S, hold.*

(i) *S is simply laced,* $\operatorname{rank} S \geq 1$: *No further condition for $S = \mathrm{A}_1$, but Λ_{sh} is a subgroup if* $\operatorname{rank} S \geq 2$.

(ii) $S = \mathrm{B}_l (l \geq 2)$, $\mathrm{C}_l (l \geq 3)$, F_4 : *Λ_{sh} and Λ_{lg} satisfy*

$$\Lambda_{\mathrm{lg}} + 2\Lambda_{\mathrm{sh}} \subseteq \Lambda_{\mathrm{lg}} \quad \textit{and} \quad \Lambda_{\mathrm{sh}} + \Lambda_{\mathrm{lg}} \subseteq \Lambda_{\mathrm{sh}}.$$

Moreover,
- *Λ_{lg} is a subgroup if $S = \mathrm{B}_l$, $l \geq 3$ or $S = \mathrm{F}_4$, and*
- *Λ_{sh} is a subgroup if $S = \mathrm{C}_l$ or $S = F_4$.*

(iii) $S = \mathrm{G}_2$: *Λ_{sh} and Λ_{lg} are subgroups satisfying*

$$\Lambda_{\mathrm{lg}} + 3\Lambda_{\mathrm{sh}} \subseteq \Lambda_{\mathrm{lg}} \quad \textit{and} \quad \Lambda_{\mathrm{sh}} + \Lambda_{\mathrm{lg}} \subseteq \Lambda_{\mathrm{sh}}.$$

(iv) $S = \mathrm{BC}_1$: *Λ_{sh} and Λ_{div} satisfy*

$$\Lambda_{\mathrm{div}} + 4\Lambda_{\mathrm{sh}} \subseteq \Lambda_{\mathrm{div}} \quad \textit{and} \quad \Lambda_{\mathrm{sh}} + \Lambda_{\mathrm{div}} \subseteq \Lambda_{\mathrm{sh}}.$$

(v) $S = \mathrm{BC}_l (l \geq 2)$: *Λ_{sh}, Λ_{lg} and Λ_{div} satisfy*

$$\begin{aligned} \Lambda_{\mathrm{lg}} + 2\Lambda_{\mathrm{sh}} &\subseteq \Lambda_{\mathrm{lg}}, & \Lambda_{\mathrm{sh}} + \Lambda_{\mathrm{lg}} &\subseteq \Lambda_{\mathrm{sh}}, \\ \Lambda_{\mathrm{div}} + 2\Lambda_{\mathrm{lg}} &\subseteq \Lambda_{\mathrm{div}}, & \Lambda_{\mathrm{lg}} + \Lambda_{\mathrm{div}} &\subseteq \Lambda_{\mathrm{lg}}, \\ \Lambda_{\mathrm{div}} + 4\Lambda_{\mathrm{sh}} &\subseteq \Lambda_{\mathrm{div}}, & \Lambda_{\mathrm{sh}} + \Lambda_{\mathrm{div}} &\subseteq \Lambda_{\mathrm{sh}}. \end{aligned}$$

In addition, if $l \geq 3$ then Λ_{lg} is a subgroup.

The inclusions $\Lambda_{\mathrm{div}} + 4\Lambda_{\mathrm{sh}} \subseteq \Lambda_{\mathrm{div}}$ and $\Lambda_{\mathrm{sh}} + \Lambda_{\mathrm{div}} \subseteq \Lambda_{\mathrm{sh}}$ in case (v) above are consequences of the other inclusions. Since 0 lies in Λ_{sh} and also in Λ_{lg} if it exists, the displayed inclusions in the Structure Theorem above imply

$$\Lambda_{\mathrm{div}} \subseteq \Lambda_{\mathrm{lg}} \subseteq \Lambda_{\mathrm{sh}}. \tag{5.65}$$

The details of this theorem are given in [1, II, §2] for the special case of extended affine root systems and then in [224] in general (it follows from the Structure Theorem 5.3.10 that an affine reflection system is the same as a "root system extended by a torsion-free abelian group of finite rank" in the sense of [224]). The reference [1] also contains a classification of discrete extension data for extended affine root systems of low nullity.

Exercise 5.3.19 Without looking at [1] or [224], work out some of the cases in Theorem 5.3.18 above.

5.3.4 Extended affine root systems. Let us come back to the beginning of this section. Our goal was to describe the structure of the set of roots occurring in an extended affine Lie algebra. After all the preparations in Sections 5.3.1–5.3.3, this is now easy.

We start with the same setting as in Section 5.3.1, i.e., X is a finite-dimensional vector space over a field F of characteristic 0, R is a subset of X and $(\cdot|\cdot)$ is a symmetric bilinear form on X. As in Section 5.3.1 we define $R^0 = \{\alpha \in R : (\alpha|\alpha) = 0\}$, $R^{\mathrm{an}} = \{\alpha \in R : (\alpha|\alpha) \neq 0\}$ and

$$\langle x, \alpha^\vee \rangle = 2\frac{(x|\alpha)}{(\alpha|\alpha)}, \quad (x \in X \text{ and } \alpha \in R^{\mathrm{an}}).$$

Definition 5.3.20 A triple $(R, X, (\cdot|\cdot))$ as above is called an *extended affine root system* or EARS for short, if the following seven axioms (EARS1)–(EARS7) are fulfilled.

(EARS1): $0 \in R$ and R spans X,

(EARS2): R has *unbroken finite root strings* , i.e., for every $\alpha \in R^{\mathrm{an}}$ and $\beta \in R$ there exist $d, u \in \mathbb{N} = \{0, 1, 2, \ldots\}$ such that

$$\{\beta + n\alpha : n \in \mathbb{Z}\} \cap R = \{\beta - d\alpha, \ldots, \beta + u\alpha\} \quad \text{and} \quad d - u = \langle \beta, \alpha^\vee \rangle.$$

(d stands for "down" and u for "up".)

(EARS3): $R^0 = R \cap X^0$.

(EARS4): R is reduced as defined in Section 5.3.1: for every $\alpha \in R^{\mathrm{an}}$ we have $F\alpha \cap R^{\mathrm{an}} = \{\pm\alpha\}$.

(EARS5): R is connected in the sense of Section 5.3.1: whenever $R^{\mathrm{an}} = R_1 \cup R_2$ with $(R_1 \mid R_2) = 0$, then $R_1 = \emptyset$ or $R_2 = \emptyset$.

(EARS6): R is tame, i.e., $R^0 \subseteq R^{\mathrm{an}} + R^{\mathrm{an}}$.

(EARS7): The abelian group $\mathrm{Span}_{\mathbb{Z}}(R^0)$ is free of finite rank.

In analogy with the concept of discrete EALAs we call $(R, X, (\cdot|\cdot))$ for $F = \mathbb{C}$ or $F = \mathbb{R}$ a *discrete* extended affine root system if (EARS1)–(EARS6) hold and in addition

(DE): R is a discrete subset of X, equipped with the natural topology.

As for EALAs, a discrete extended affine root system necessarily satisfies (EARS7), see Proposition 5.3.21(c) below, so that it is justified to call it an EARS.

We will immediately connect EARS to affine reflection systems:

Proposition 5.3.21 (a) *A pair* (R, X) *satisfying* (EARS1)–(EARS3) *is an affine reflection system. In particular:*

(i) *An extended affine root system is an affine reflection system which is reduced, connected, symmetric, tame and which has unbroken root strings.*

(ii) *If* $F = \mathbb{R}$ *we can assume that* $(\cdot|\cdot)$ *is positive semidefinite.*

(b) *Let* (R, X) *be an affine reflection system with quotient root system* S *and extension datum* $\mathfrak{L}$. *Then* (R, X) *is an extended affine root system if and only if*

(i) S *is irreducible, hence* $\mathfrak{L} = (\Lambda_0, \Lambda_{\mathrm{sh}}, \Lambda_{\mathrm{lg}}, \Lambda_{\mathrm{div}})$,

(ii) $\Lambda_0 = \Lambda_{\mathrm{sh}} + \Lambda_{\mathrm{sh}}$, $\Lambda_{\mathrm{div}} \cap 2\Lambda_{\mathrm{sh}} = \emptyset$, *and*

(iii) (EARS7) *holds.*

(c) *For an extended affine root system* (R, X) *over* $F = \mathbb{R}$ *or* $F = \mathbb{C}$ *the following are equivalent:*

(i) R *is discrete;*

(ii) R^0 *is a discrete subset of* X*;*
(iii) $\mathrm{Span}_{\mathbb{Z}}(R^0)$ *is a discrete subgroup of* X*.*

In this case, all reflection subspaces Λ_ξ *of* (b) *are discrete too, and* $\mathrm{Span}_{\mathbb{Z}}(R^0)$ *is a free abelian group of finite rank.*

Proof (a) Obviously (AR1) = (EARS1) and (AR4) = (EARS3). The axiom (AR2), i.e., $s_\alpha(R) = R$, follows from (EARS2): $s_\alpha(\beta) = \beta + (u-d)\alpha$ and $-d \le u - d \le u$. Also (AR3) is immediate from (EARS2).

In any affine reflection system the root strings $R \cap (\beta + \mathbb{Z}\alpha)$ for $(\beta, \alpha) \in R \times R^{\mathrm{an}}$ are finite. This is Exercise 5.3.14(d) and an immediate consequence of the Structure Theorem 5.3.10. Also, a tame affine reflection system with unbroken roots strings is necessarily symmetric by Proposition 5.3.13. The characterization of an EARS in (i) is now clear. (ii) follows from Corollary 5.3.12.

(b) follows from Proposition 5.3.13 and Exercise 5.3.14, since for a connected R = irreducible S the formula (5.65) implies $\Lambda_{\mathrm{diff}} = \Lambda_{\mathrm{sh}} + \Lambda_{\mathrm{sh}}$.

(c) (i) $\Rightarrow$ (ii) is obvious. Suppose (ii) holds. We know from (b) that $R^0 = \Lambda_0 = \Lambda_{\mathrm{sh}} + \Lambda_{\mathrm{sh}}$. Since Λ_{sh} is a pointed reflection subspace, so is Λ_0 (Exercise 5.3.8). Hence $2 \cdot \mathrm{Span}_{\mathbb{Z}}(\Lambda_0) \subseteq \Lambda_0$ is discrete. But then so is $\mathrm{Span}_{\mathbb{Z}}(\Lambda_0)$. Thus (ii) $\Rightarrow$ (iii).

By (5.65) and (b.ii), $\Lambda_{\mathrm{div}} \subseteq \Lambda_{\mathrm{lg}} \subseteq \Lambda_{\mathrm{sh}} \subseteq \Lambda_0$. Hence, if (iii) holds, then all Λ_ξ are discrete subsets of X. But then so is is R, as a finite union of discrete subsets. This shows (iii) $\Rightarrow$ (i).

It is well-known fact that every discrete subgroup of a finite-dimensional real vector space is free of finite rank. □

A quick comparison of [1, Definition 2.1] and our Definition 5.3.20 together with Proposition 5.3.21(a) will convince the reader that an extended affine root system in the sense of [1] is the same as a discrete extended affine root system over $\mathbb{R}$ in our sense. The reason for the generalization and the change of name is the same as the one justifying our more general notion of extended affine Lie algebras: We are considering EALAs over arbitrary fields of characteristic 0 and the set of roots of an EALA will not be an extended affine root system in the sense of [1].

Finally, here is the result which brings us back to EALAs.

Theorem 5.3.22 *Let* (E, H) *be an EALA over* F *and let* $R \subseteq H^*$ *be its set of roots. Put* $X = \mathrm{Span}_F(R)$ *and let* $(\cdot|\cdot)_X$ *be the restriction of the bilinear form* (5.31) *to* X*. Then* $(R, X, (\cdot|\cdot)_X)$ *is an extended affine root system. If* $F = \mathbb{C}$*, then* (E, H) *is a discrete EALA if and only if* $(R, X, (\cdot|\cdot)_X)$ *is a discrete extended affine root system.*

Remarks 5.3.23 (a) For (E, H) a discrete EALA over $F = \mathbb{C}$, the theorem is proven in [1, I, Th. 2.16], using discreteness. The generalization to arbitrary EALAs is due to the author, see [186, Prop. 3]. It has been further generalized to other classes of Lie algebras, the so-called invariant affine reflection algebras, see [187, Th. 6.6 and Th. 6.8]. Special cases have also been proven in [16] and [176]. That R is symmetric, is an easy exercise, namely Exercise 5.2.1(c).

(b) In view of the theorem above, one can ask if every extended affine root system is the set of roots of some extended affine Lie algebra. This is however not the case, see [12, Th. 6.2] for a detailed discussion of this question.

As a first application of this theorem, we can now completely characterize EALAs of nullity 0.

Proposition 5.3.24 *The following are equivalent:*

(i) *(E, H) is an EALA of nullity* 0,

(ii) *(E, H) is an EALA with a finite-dimensional E,*

(iii) *E is a finite-dimensional split simple Lie algebra with splitting Cartan subalgebra H.*

In this case, E equals its core and the set of roots R coincides with the quotient root system S of R and is an irreducible reduced finite root system.

Proof We will show (i) $\Leftrightarrow$ (ii) $\Rightarrow$ (iii). The implication (iii) $\Rightarrow$ (i) has already been proven in (5.38).

(i) $\Rightarrow$ (ii): By definition of the nullity of (E, H), we have $R^0 = \{0\}$. Since R is an EARS by Theorem 5.3.22, it follows from Proposition 5.3.21 that $\Lambda_{\rm sh} \subseteq \Lambda_{\rm sh} + \Lambda_{\rm sh} = \Lambda_0 = R^0 = \{0\}$, so $\Lambda_{\rm sh} = \{0\}$ and then $\Lambda_{\rm div} \subseteq \Lambda_{\rm lg} \subseteq \{0\}$ by (5.65). Hence $R = S$ is a finite root system, which is reduced since R is reduced. Since $\dim H < \infty$ by the axiom (EA2) and $\dim E_\alpha = 1$ for $\alpha \in R^{\rm an} = R \setminus \{0\}$ by Proposition 5.2.2, it is now clear that E is finite-dimensional.

(ii) $\Rightarrow$ (i): We know $|R| < \infty$ since E is finite-dimensional. From $R^0 = \Lambda_0 = \Lambda_{\rm diff} = \mathbb{Z}\Lambda_{\rm diff}$ by Proposition 5.3.13, we can then conclude that $R^0 = \{0\}$. So (E, H) has nullity 0.

(i) and (ii) $\Rightarrow$ (iii): Let $\mathfrak{r}$ be the radical of the finite-dimensional Lie algebra E. As an ideal, it is invariant under the adjoint action of H. Hence $\mathfrak{r} = \bigoplus_{\alpha \in R}(\mathfrak{r} \cap E_\alpha)$. Suppose $0 \neq x_\alpha \in \mathfrak{r} \cap E_\alpha$ for some $\alpha \in R^{\rm an} = R \setminus \{0\}$. By Proposition 5.2.2, x_α embeds into an $\mathfrak{sl}_2$-triple $(x_\alpha, h_\alpha, y_\alpha) \in E_\alpha \times H \times E_{-\alpha}$. It then follows that $\mathfrak{sl}_2(F) \cong E_\alpha \times [E_\alpha, E_{-\alpha}] \oplus E_{-\alpha} \subseteq \mathfrak{r}$, contradiction. Therefore $\mathfrak{r} \subseteq E_0 = H$. We now get $[\mathfrak{r}, E] = \bigoplus_{\alpha \in R}[\mathfrak{r}, E_\alpha] = \{0\}$. Indeed, for $\alpha \neq 0$ we have $[\mathfrak{r}, E_\alpha] \subseteq \mathfrak{r} \cap E_\alpha = \{0\}$, while for $\alpha = 0$ we have $[\mathfrak{r}, E_0] = [\mathfrak{r}, H] = \{0\}$ since H is abelian. Thus $\mathfrak{r} = Z(E)$, and so E is reductive ([38, §6.4, Porp. 5]). This means $E = Z(E) \oplus [E, E]$ with $[E, E]$ semisimple. But $E_\alpha \subseteq [E, E]$ for $\alpha \in R^{\rm an} = R \setminus \{0\}$, and therefore the core E_c is contained in $[E, E]$ too. For the centralizer $C_E(E_c) = \{c \in E : [c, E_c] = 0\}$ of E_c in E we now get $Z(E) = C_E([E, E]) \subset C_E(E_c) \subset E_c$, where the last inclusion follows from the tameness axiom (EA5). Since then $Z(E) \subseteq [E, E]$ we have shown that $E = [E, E]$ is semisimple. The subalgebra H is a splitting Cartan subalgebra of E, and R is the root system of (E, H). As R is connected = irreducible, E is simple. □

5.4 The core and centreless core of an EALA

In the previous section we have studied affine reflection systems per se. The rationale for doing so became clear only in the end, when we saw in Theorem 5.3.22 that the set of roots R of an EALA (E, H) is an extended affine root system, a special type of an affine reflection system.

In this section we start by drawing consequences of the Structure Theorem 5.3.10 of affine reflection systems and the description of extended affine root systems in Proposition 5.3.21. The examples in Section 5.2.4 and Section 5.2.5 indicate that the core E_c and centreless core $E_{cc} = E_c/Z(E_c)$ of an extended affine Lie algebra (E, H) really are the "core" of the matter. We will show in Theorem 5.4.14 and in Corollary 5.4.16 that both are so-called Lie tori, a new class of Lie algebras which we will introduce in Section 5.4.1. We will present some basic properties of Lie tori in Section 5.4.2 and describe some examples in Section 5.4.4 and 5.4.5.

With some justification, this section could therefore also be entitled "On Lie tori". But the reader can be re-assured that we are not getting side-tracked too much: In the next section we will see that Lie tori are precisely what is needed to construct EALAs.

5.4.1 Lie tori: Definition. Lie tori are special objects in the following category of graded Lie algebras.

Definition 5.4.1 Let (S, Y) be a finite irreducible, but not necessarily reduced root system, as defined in Example 5.3.3. We denote by $\mathcal{Q}(S) = \mathrm{Span}_{\mathbb{Z}}(S) \subseteq Y$ the root lattice of S. To avoid some degeneracies we will always assume that $S \neq \{0\}$. Let Λ be an abelian group.

A $(\mathcal{Q}(S), \Lambda)$-*graded* Lie algebra is a Lie algebra L with compatible $\mathcal{Q}(S)$- and Λ-gradings. It is convenient (and helpful) to use subscripts for the $\mathcal{Q}(S)$-grading and superscripts for the Λ-grading. Thus,

$$L = \bigoplus\nolimits_{q \in \mathcal{Q}(S)} L_q = \bigoplus\nolimits_{\lambda \in \Lambda} L^\lambda$$

are $\mathcal{Q}(S)$- and Λ-gradings of L, and compatibility means

$$L = \bigoplus\nolimits_{q \in \mathcal{Q}(S),\, \lambda \in \Lambda} L_q^\lambda \quad \text{for } L_q^\lambda = L_q \cap L^\lambda.$$

Hence for $\lambda, \mu \in \Lambda$ and $p, q \in \mathcal{Q}(S)$

$$L^\lambda = \bigoplus\nolimits_{q \in \mathcal{Q}(S)} L_q^\lambda, \quad L_q = \bigoplus\nolimits_{\lambda \in \Lambda} L_q^\lambda \quad \text{and} \quad [L_q^\lambda, L_p^\mu] \subseteq L_{q+p}^{\lambda+\mu}.$$

Thus, L has three gradings, by $\mathcal{Q}(S)$, Λ and $\mathcal{Q}(S) \oplus \Lambda$ whose interplay will be crucial in the following. Corresponding to these three different gradings are three *support sets*:

$$\begin{aligned} \mathrm{supp}_{\mathcal{Q}(S)} L &= \{q \in \mathcal{Q}(S) : L_q \neq 0\}, \\ \mathrm{supp}_\Lambda L &= \{\lambda \in L : L^\lambda \neq 0\}, \text{ and} \\ \mathrm{supp}_{\mathcal{Q}(S) \oplus \Lambda} L &= \{(q, \lambda) \in (\mathcal{Q}(S), \Lambda) : L_q^\lambda \neq 0\}. \end{aligned}$$

Definition 5.4.2 We keep the notation of 5.4.1. A *Lie torus of type* (S, Λ) is a $(\mathcal{Q}(S), \Lambda)$-graded Lie algebra L over F, a field of characteristic 0, satisfying the axioms (LT1)–(LT3) below.

(LT1): $\mathrm{supp}_{\mathcal{Q}(S)} L \subseteq S$, hence $L = \bigoplus_{\xi \in S} L_\xi$.

(LT2): If $L_\xi^\lambda \neq 0$ and $\xi \neq 0$, then there exist $e_\xi^\lambda \in L_\xi^\lambda$ and $f_\xi^\lambda \in L_{-\xi}^{-\lambda}$ such that

$$L_\xi^\lambda = F e_\xi^\lambda, \quad L_{-\xi}^{-\lambda} = F f_\xi^\lambda, \tag{5.66}$$

and for $x_\tau \in L_\tau$ we have

$$[[e_\xi^\lambda, f_\xi^\lambda], x_\tau] = \langle \tau, \xi^\vee \rangle x_\tau. \tag{5.67}$$

(LT3): (a) $L_\xi^0 \neq 0$ if $\xi \in S_{\mathrm{ind}}^\times$, i.e., $0 \neq \xi \in S$ and $\xi/2 \notin S$.

(b) As a Lie algebra, L is generated by $\bigcup_{0 \neq \xi \in S} L_\xi$.

(c) $\Lambda = \mathrm{Span}_{\mathbb{Z}}(\mathrm{supp}_\Lambda L)$.

We will say that L is a *Lie torus* (without qualifiers) if L is a Lie torus of type (S, Λ) for some pair (S, Λ).

A Lie torus is called *invariant*, if L has an invariant nondegenerate symmetric bilinear form $(\cdot|\cdot)$ which is graded in the sense of (5.26):

$$(L_\xi^\lambda \mid L_\tau^\mu) = 0 \quad \text{if } \lambda + \mu \neq 0 \text{ or } \xi + \tau \neq 0.$$

Two Lie tori L and $\tilde{L}$, both of type (S, Λ), are called *graded-isomorphic* if there exists a Lie algebra isomorphism $f : L \to \tilde{L}$ such that $f(L_\xi^\lambda) = \tilde{L}_\xi^\lambda$ for all $(\xi, \lambda) \in S \times \Lambda$. Thus, a graded-isomorphism of Lie tori is an isomorphism in the category of graded Lie algebras. But we will nevertheless use the term "graded-isomorphism" to emphasize that Lie tori are graded algebras.

Remarks 5.4.3 (a) Let L be a Lie torus. Hence, by (LT1), $L = \bigoplus_{\xi \in S} L_\xi = \bigoplus_{\xi \in S,\, \lambda \in \Lambda} L_\xi^\lambda$. We will determine $\operatorname{supp}_{\mathcal{Q}(S)} L$ in Corollary 5.4.6 below. The axiom (LT2) implies that

$$\dim L_\xi^\lambda = 1 \quad \text{if } 0 \neq \xi \text{ and } L_\xi^\lambda \neq 0 \tag{5.68}$$

and that

$$(e_\xi^\lambda, h_\xi^\lambda, f_\xi^\lambda) \quad \text{with} \quad h_\xi^\lambda = [e_\xi^\lambda, f_\xi^\lambda] \in L_0^0 \tag{5.69}$$

is an $\mathfrak{sl}_2$-triple. The condition (LT3.a) together with (LT2) ensures that a Lie torus has enough $\mathfrak{sl}_2$-triples.

The other two conditions in (LT3) are not really serious; they just serve to normalize things: If (LT3.c) does not hold, one can simply replace Λ by $\operatorname{Span}_{\mathbb{Z}}(\operatorname{supp}_\Lambda L)$. Also,

$$\text{(LT3.b)} \quad \Longleftrightarrow \quad L_0^\lambda = \textstyle\sum_{0 \neq \xi \in S} \sum_{\mu \in \Lambda} [L_\xi^\mu, L_{-\xi}^{\lambda-\mu}] \tag{5.70}$$

for all $\lambda \in \Lambda$. If one has a Lie algebra for which all axioms except (5.70) hold, one can replace the subspaces L_0^λ by the right hand side of (5.70) and then gets a Lie torus. Observe that (5.70) for $\lambda = 0$ together with (LT2) yields

$$L_0^0 = \textstyle\sum F h_\xi^\lambda \tag{5.71}$$

where the sum in (5.71) is taken over all pairs (ξ, λ) for which h_ξ^λ exists, i.e., those with $L_\xi^\lambda \neq 0$ and $\xi \neq 0$.

(b) A Lie torus is a special type of a so-called division-(S, Λ)-graded Lie algebras, or more generally of a root-graded Lie algebra. This and also the different approaches to root-graded Lie algebras are discussed in [187, §5]. Lie tori were first defined by Yoshii in [224, 225], using the notion of a root-graded Lie algebra. The definition above is due to the author [185].

Viewing a Lie torus as a special type of a root-graded Lie algebra, is the approach used in the classification of Lie tori.

(c) Why was a Lie torus christened a "Lie torus"? The historically correct answer is: Because of pure analogy with already existing names like a quantum torus, defined in 5.4.25, or an alternative or Jordan torus. All of these are graded algebras, in which every non-zero homogeneous element is invertible. If one interprets the elements e and f of the $\mathfrak{sl}_2$-triple (5.69) as invertible elements of L, then a Lie torus is a graded Lie algebra in which most of the non-zero homogenous elements are invertible. It is certainly unusual to speak of "invertible elements" in a Lie algebra. But the examples below will provide some justification to that: We will see that the "invertible elements" of L are given by invertible elements of its coordinate algebra.

Besides the analogy with the already existing concepts of "tori" in categories of (non)associative algebras, the fact that a *toroidal* Lie algebra is a Lie *torus*, see Example 5.4.33, reinforces the choice of the name Lie torus.

5.4.2 Some basic properties of Lie tori. Throughout this section L is a Lie torus of type (S, Λ). We use the notation of Definition 5.4.2. We describe basic properties of Lie tori and prove some of them, in particular those for which there does not yet exist a published proof.

We first show that the homogeneous subspaces of the $\mathcal{Q}(S)$-grading of L are weight spaces for the ad-diagonalizable subalgebra

$$\mathfrak{h} = \operatorname{Span}_F\{h_\xi^0 : \xi \in S_{\text{ind}}^\times\}. \tag{5.72}$$

Lemma 5.4.4 *The subspaces L_τ, $\tau \in S$, are given by*

$$L_\tau = \{l \in L : [h_\xi^0, l] = \langle \tau, \xi^\vee \rangle l \text{ for all } \xi \in S_{\text{ind}}^\times\}. \tag{5.73}$$

Proof The inclusion from left to right holds by (5.67). For the proof of the other inclusion we write $l \in L$ as $l = \sum_\alpha l_\alpha$ with $l_\alpha \in L_\alpha$. Then l satisfies $[h_\xi^0, l] = \langle \tau, \xi^\vee \rangle l$ for all $\xi \in S_{\text{ind}}^\times$ if and only if for every $\alpha \in S$ we have $\langle \alpha - \tau, \xi^\vee \rangle l_\alpha = 0$ for all $\xi \in S_{\text{ind}}^\times$. Since $\operatorname{Span}_F(S_{\text{ind}}) = Y$ and the bilinear form on Y associated with the root system S is nondegenerate, there exists $\xi \in S_{\text{ind}}^\times$ with $\langle \alpha - \tau, \xi^\vee \rangle \neq 0$ for every pair $(\alpha, \tau) \in S^2$ with $\alpha \neq \tau$. Hence, any l belonging to the set on the right hand side of (5.73) has $l_\alpha = 0$ for $\alpha \neq \tau$, proving $l \in L_\tau$. □

Proposition 5.4.5 *For every $(\xi, \lambda) \in \operatorname{supp}_{\mathcal{Q}(S) \oplus \Lambda} L$ with $\xi \neq 0$ the map*

$$\varphi_\xi^\lambda = \exp\big(\operatorname{ad}(e_\xi^\lambda)\big) \exp\big(\operatorname{ad}(-f_\xi^\lambda)\big) \exp\big(\operatorname{ad}(e_\xi^\lambda)\big)$$

is a well-defined automorphism of the Lie algebra L with the property

$$\varphi_\xi^\lambda(L_\tau^\mu) = L_{s_\xi(\tau)}^{\mu - \langle \mu, \xi^\vee \rangle \lambda} \tag{5.74}$$

for all $\tau \in S$ and $\mu \in \Lambda$. Moreover, for every $w \in W(S)$, the Weyl group of S, there exists an automorphism φ_w of the Lie algebra L such that

$$\varphi_w(L_\tau^\mu) = L_{w(\tau)}^\mu$$

for all $\tau \in S$ and $\mu \in \Lambda$.

This proposition can be proven in the same way as [1, Prop. 1.27].

Corollary 5.4.6 ([5, Lemma 1.10]) *The $\mathcal{Q}(S)$-support of L satisfies*

$$\operatorname{supp}_{\mathcal{Q}(S)} L = \begin{cases} S & \text{if } S \text{ is reduced,} \\ S \text{ or } S_{\text{ind}} & \text{if } S \text{ is non-reduced.} \end{cases}$$

As a consequence of this corollary, $\operatorname{supp}_{\mathcal{Q}(S)} L$ is always a finite irreducible root system. It is immediate that L is also a Lie torus of type $(\operatorname{supp}_{\mathcal{Q}(S)} L, \Lambda)$. Without loss of generality we can therefore assume that $S = \operatorname{supp}_{\mathcal{Q}(S)} L$ if this is convenient.

Proposition 5.4.7 *For $\xi \in S$ define*

$$\Lambda_\xi = \{\lambda \in \Lambda : L_\xi^\lambda \neq 0\},$$

so that $\operatorname{supp}_\Lambda L = \bigcup_{\xi \in S} \Lambda_\xi$. Then the family $(\Lambda_\xi : \xi \in S)$ satisfies the axioms (ED1) *and* (ED2) *of* Definition 5.3.9,

$$\Lambda_\eta - \langle \eta, \xi^\vee \rangle \Lambda_\xi \subseteq \Lambda_{s_\xi(\eta)} \tag{ED1}$$

$$0 \in \Lambda_\xi \text{ for } \xi \in S_{\text{ind}} \text{ and } \Lambda_\xi \neq \emptyset \text{ for } \xi \in S_{\text{div}}^\times \tag{ED2}$$

Hence Λ_ξ *is a pointed reflection subspace for* $\xi \in S_{\rm ind}^\times$*, a symmetric reflection subspace for* $\xi \in S_{\rm div}^\times$ *and*

$$\Lambda_\xi = \Lambda_{w(\xi)} \quad \text{for all } w \in W(S). \tag{5.75}$$

Defining $\Lambda_{\rm sh}$, $\Lambda_{\rm lg}$ *and* $\Lambda_{\rm div}$ *as in* (5.63)*, we have*

$$\Lambda_{\rm sh} \supset \Lambda_{\rm lg} \supset \Lambda_{\rm div}, \tag{5.76}$$

$$\operatorname{supp}_\Lambda L = \Lambda_0 = \Lambda_{\rm sh} + \Lambda_{\rm sh}, \tag{5.77}$$

$$\emptyset = 2\Lambda_{\rm sh} \cap \Lambda_{\rm div}, \tag{5.78}$$

$$\Lambda = \operatorname{Span}_{\mathbb{Z}}(\Lambda_{\rm sh}). \tag{5.79}$$

The support families $(\Lambda_\xi : \xi \in S)$ *are the same for* L *and* $L/Z(L)$.

Proof (ED1) is a consequence of Proposition 5.4.5 and (ED2) of (LT3.a). It then follows as in Section 5.3 that Λ_ξ, $\xi \in S^\times$, are pointed respectively symmetric reflection subspaces such that (5.75) and (5.76) hold. (5.77) and (5.78) are proven in [224, Th. 5.1] and [5, Lemma 1.1.12]. (5.79) follows from (5.76) and (5.77). The last claim is also proven in [224, Th. 5.1]. □

Proposition 5.4.8 *Let* $\mathfrak{g}$ *be the subalgebra generated by* $\{L_\xi^0 : \xi \in S_{\rm ind}^\times\}$*, and define* $\mathfrak{h}$ *by* (5.72).

(a) *Then* $\mathfrak{g}$ *is a finite-dimensional split simple Lie algebra with splitting Cartan subalgebra* $\mathfrak{h}$.

(b) *The root system* $S_{\rm ind}$ *and the root system of* $(\mathfrak{g}, \mathfrak{h})$ *are canonically isomorphic. Namely, for every* $\xi \in S_{\rm ind}^\times$ *there exists a unique* $\tilde{\xi} \in \mathfrak{h}^*$*, defined by* $\tilde{\xi}(h_\eta^0) = \langle \xi, \eta^\vee \rangle$ *for* $\eta \in S_{\rm ind}^\times$*, such that the map* $\xi \mapsto \tilde{\xi}$ *extends to an isomorphism between the root system* $S_{\rm ind}$ *and the root system of* $(\mathfrak{g}, \mathfrak{h})$.

Proof This is a special case of a result for arbitrary root-graded Lie algebras, see [183, Remark 2 of §2.1] and [187, Prop. 5.9]. It is essentially a corollary to the Chevalley-Serre presentation of finite-dimensional split simple Lie algebras. For Lie tori it was announced in [185, §3]. The details of the proof are given in [5, Prop. 1.2.2]. □

The following Exercise 5.4.10 lists some more basic properties of Lie tori. You will need Exercise 5.4.9(a) in part (d) of 5.4.10.

Exercise 5.4.9 (a) Let K be a perfect Lie algebra. Then $K/Z(K)$ is a perfect and centreless Lie algebra.

(b) Let E be a Lie algebra with an invariant nondegenerate symmetric bilinear form $(\cdot|\cdot)$, and let K be an ideal of E with $K = [E, K]$. Then $\{z \in K : (z \mid K) = 0\} = Z(K)$.

Exercise 5.4.10 Let L be a Lie torus of type (S, Λ). Show:

(a) L_0^λ is given by the formula (5.70).

(b) L is perfect.

(c) The centre satisfies $Z(L) = \bigoplus_{\lambda \in \Lambda} Z(L)^\lambda$ for $Z(L)^\lambda = Z(L) \cap L_0^\lambda$.

(d) Let $Y = \bigoplus_{\lambda \in \Lambda} Y_0^\lambda$, $Y_0^\lambda = Y \cap L_0^\lambda$, be a graded subspace of $Z(L)$. Then L/Y is a Lie torus with respect to the subspaces $(L/Y)_\xi^\lambda = L_\xi^\lambda / Y_\xi^\lambda$, where for $\xi \neq 0$ we put $Y_\xi^\lambda = \{0\}$ and thus have $(L/Y)_\xi^\lambda \cong L_\xi^\lambda$ as vector spaces. In particular, $L/Z(L)$ is a centreless Lie torus.

(e) For $\lambda, \mu \in \Lambda_\xi$, $\xi \in S^\times$, we have $h_\xi^\lambda \equiv h_\xi^\mu \mod Z(L)$.

(f) $L^0 = \mathfrak{g} \oplus Z(L)^0$ and $L_0^0 = \mathfrak{h} \oplus Z(L)^0$.

(g) Let I be a Λ-graded ideal of L, whence $I = \bigoplus_{\lambda \in \Lambda} I^\lambda$ for $I^\lambda = I \cap L^\lambda$. Then either $I = L$ or $I \subseteq Z(L)$. In particular, a centreless Lie torus is graded-simple with respect to the Λ-grading of L.

Since a Lie torus is perfect by part (b) of the exercise above, it has a universal central extension (Theorem 5.1.10).

Theorem 5.4.11 ([185, §5], [188]) *Let $\mathfrak{u} : \mathfrak{uce}(L) \to L$ be a universal central extension of a Lie torus $L = \bigoplus_{\xi,\lambda} L_\xi^\lambda$ of type (S, Λ). Then $\mathfrak{uce}(L)$ is also a Lie torus of type (S, Λ), say $\mathfrak{uce}(L) = \bigoplus_{\xi,\lambda} \mathfrak{uce}(L)_\xi^\lambda$, and $\mathfrak{u}$ maps $\mathfrak{uce}(L)_\xi^\lambda$ onto L_ξ^λ.*

Remark 5.4.12 It follows from this theorem and the exercise above that in order to describe Lie tori up to graded isomorphism, one can proceed in two steps:

(A) Classify centreless Lie tori, up to graded isomorphism. We will discuss some examples of this classification in 5.4.4 and 5.4.5.

(B) Describe the universal central extension of the centreless Lie tori from (A). They are unique up to isomorphism. We will not say anything about this here. The reader can find some results in [29, 30, 186, 188] for Lie tori arising from EALAs and in [2, 3, 27, 188] for general root-graded Lie algebras.

Once (A) and (B) completed, an arbitrary Lie torus of type (S, Λ) is then obtained as $\mathfrak{uce}(L)/C$ where L is taken from the list in (A) and where C is a graded subspace of the centre of $\mathfrak{uce}(L)$.

The following results only holds for special types of Lie tori.

Theorem 5.4.13 ([185, Th. 5], proven in [188]) *Let $L = \bigoplus_{\xi \in S,\, \lambda \in \Lambda} L_\xi^\lambda$ be a Lie torus of type (S, Λ) where Λ is a finitely generated abelian group.*

(a) *Then L is finitely generated as Lie algebra and has bounded homogeneous dimension with respect to the $\mathcal{Q}(S) \oplus \Lambda$-grading of L.*

(b) *Moreover, the Lie algebra $\mathrm{Der}_F(L) = \mathrm{grDer}_F(L)$, where $\mathrm{grDer}_F(L)$ is naturally $\mathcal{Q}(S) \oplus \Lambda$-graded and has bounded homogeneous dimension with respect to this grading.*

(c) *If L is invariant, its universal central extension is isomorphic to the central extension $\mathrm{E}(L, D^{\mathrm{gr}\,*}, \psi_D)$ where D is any graded complement of $\mathrm{IDer}(L)$ in $\mathrm{SDer}_F(L)$.*

Part (c) of this theorem is an immediate corollary of parts (a) and (b) and of Theorem 5.1.25.

5.4.3 The core of an EALA. We will now connect extended affine Lie algebras and Lie tori, and first introduce some notation. Let (E, H) be an EALA with set of roots R. We have seen in Theorem 5.3.22 that R is an extended affine root system, hence an affine reflection system. We can therefore apply the Structure Theorem 5.3.10. Recall the following data describing the structure of R:

- $X = \mathrm{Span}_F(R) \subseteq H^*$, $X^0 = \{x \in X : (x \mid X) = 0\} = \{x \in X : (x \mid R) = 0\}$, $f : X \to X/X^0 = Y$ the canonical projection,
- $S = f(R)$ the quotient root system, a finite irreducible but possibly non-reduced root system, and $S_{\mathrm{ind}} = \{\alpha \in S : \alpha/2 \notin S\} \cup \{0\}$,
- $g : Y \to X$ a linear map satisfying $f \circ g = \mathrm{Id}_Y$ and $g(S_{\mathrm{ind}}) \subseteq R$,
- $(\Lambda_\xi : \xi \in S)$ the associated extension datum, defined by $R \cap f^{-1}(\xi) = g(\xi) \oplus \Lambda_\xi$ and $\Lambda_\xi \subseteq X^0$,

- $\Lambda = \mathrm{Span}_{\mathbb{Z}}\left(\bigcup_{\xi\in S}\Lambda_\xi\right)$, a free abelian group of finite rank (this is axiom (EARS7)).

Hence

$$\begin{aligned} R &= \textstyle\bigcup_{\xi\in S}\left(g(\xi)\oplus\Lambda_\xi\right)\subseteq g(Y)\oplus X^0,\\ R^{\mathrm{an}} &= \textstyle\bigcup_{\xi\in S^\times}\left(g(\xi)\oplus\Lambda_\xi\right),\\ R^0 &= 0\oplus\Lambda_0 = R\cap X^0. \end{aligned}$$

Theorem 5.4.14 ([12] for $F=\mathbb{C}$) *Let $K=E_c$ be the core of an EALA (E,H). We use the notation of above and define subspaces*

$$K_\xi^\lambda = K\cap E_{g(\xi)\oplus\lambda} = \begin{cases} E_{g(\xi)\oplus\lambda} & \xi\neq 0,\\ K\cap E_{0\oplus\lambda}, & \xi=0.\end{cases} \tag{5.80}$$

(a) *Then $K=\bigoplus_{\xi,\lambda}K_\xi^\lambda$ is a Lie torus of type (S,Λ), where Λ is free abelian of finite rank.*

(b) *K is a perfect ideal of E.*

(c) *Let $(\cdot|\cdot)$ be a nondegenerate invariant bilinear form on E, whose existence is guaranteed by the axiom* (EA1). *Then the radical of the restricted bilinear form $(\cdot|\cdot)|_{K\times K}$ equals the centre $Z(K)$, that is*

$$\{z\in K:(z\mid K)=0\} = Z(K) = \textstyle\bigoplus_{\lambda\in\Lambda}Z(K)\cap K_0^\lambda. \tag{5.81}$$

Remark 5.4.15 The subspaces K_ξ^λ in (5.80) and hence the Lie torus structure of K depend on the section g. A different choice of g leads to a so-called *isotope* of K, see [11] and [187, Prop. 6.4].

Corollary 5.4.16 *We use the notation of Theorem* 5.4.14, *and put $E_{cc} = K/Z(K)=L$, the* centreless core *of (E,H). Then L is an invariant centreless Lie torus of type (S,Λ) with respect to the homogeneous subspaces*

$$L_\xi^\lambda = K_\xi^\lambda/\left(Z(K)\cap K_\xi^\lambda\right)$$

and the bilinear form $(\cdot|\cdot)_L$ defined by

$$(\bar{x}\mid\bar{y})_L = (x\mid y)$$

where $x,y\in K$, $\bar{x}$ and $\bar{y}$ are the canonical images in L and $(\cdot|\cdot)$ is the bilinear form of 5.4.14(c).

Remark 5.4.17 Yoshii [225] has shown that any Lie torus of type (S,Λ) with Λ a torsion-free abelian group admits a non-zero graded invariant symmetric bilinear form. This implies the existence of a nondegenerate such form on E_{cc}. However, Yoshii's proof uses the existence of invariant nondegenerate symmetric bilinear forms on Jordan tori [191] and hence relies on the classification of Jordan tori.

We can now show that all root spaces of an EALA are finite-dimensional in a strong form.

Proposition 5.4.18 ([186, Prop. 3]) *An EALA has finite bounded dimension. The same is true for its core and centreless core.*

Proof Let $K = E_c$ be the core of the EALA (E, H). By Theorem 5.4.14, K is a Lie torus of type (S, Λ), where Λ is a free abelian group of finite rank. Hence, by Theorem 5.4.13, one knows that K has finite bounded dimension with respect to its double grading, say $\dim K_\xi^\lambda \le M_1$ for all pairs (ξ, λ). By the same reference, one also knows that the Lie algebra $\mathrm{Der}_F(K)$ of all F-linear derivations of K has a double grading by $\mathcal{Q}(S)$ and Λ,

$$\mathrm{Der}_F(K) = \bigoplus\nolimits_{\xi\in S,\, \lambda\in\Lambda} (\mathrm{Der}_F K)_\xi^\lambda,$$

where $(\mathrm{Der}_F K)_\xi^\lambda$ is the subspace of those derivations mapping K_τ^μ to $K_{\xi+\tau}^{\lambda+\mu}$, and that $\mathrm{Der}_F(K)$ has finite bounded dimension with respect to this grading, say $\dim_F(\mathrm{Der}_F K)_\xi^\lambda \le M_2$ for all pairs (ξ, λ).

Since K is an ideal, we have a Lie algebra homomorphism $\rho : E \to \mathrm{Der}_F(K)$, given by $\rho(e) = \mathrm{ad}\, e|_K$. It is homogenous of degree 0, i.e., $\rho(E_\alpha) \subseteq (\mathrm{Der}_F K)_\xi^\lambda$ for $\alpha = g(\xi) \oplus \lambda$ as in (5.80). Moreover, by the tameness axiom (EA5) for an EALA we know that $\ker \rho \subseteq K$ (whence $\ker\rho = Z(K)$, but we won't need this). It now follows that $\dim E_\alpha = \dim \ker(\rho|_{E_\alpha}) + \dim \rho(E_\alpha) \le \dim K_\xi^\lambda + \dim(\mathrm{Der}_F K)_\xi^\lambda \le M_1 + M_2$. □

5.4.4 Lie tori of type A_l, $l \ge 3$**.** As explained in Remark 5.4.12, in classifying Lie tori one can restrict one's attention to the case of centreless Lie tori, at least modulo the solution of problem (B) in 5.4.12. In this section we will describe centreless Lie tori of type A.

The reader will expect that this will have something to do with trace-0-matrices. This turns out to be correct, but only with the proper interpretation of "trace-0". It will not be sufficient to consider trace-0-matrices over F. We will see in Theorem 5.4.22 and the Example 5.4.29 that they only lead to nullity 0-examples. Rather, one must allow matrices with entries from a possibly non-commutative associative algebra. Things get even more general for Lie tori of type A_2 and A_1, see the Remark 5.4.28.

To avoid some degeneracies, in this section we let N be a natural number with $N \ge 3$. We start with an arbitrary associative unital F-algebra A. In particular, A need not be commutative, and hence

$$[A, A] = \mathrm{Span}_F\{a_1a_2 - a_2a_1 : a_1, a_2 \in A\}$$

is in general non-zero. As usual, $\mathfrak{gl}_N(A)$ is the Lie algebra of all $N \times N$ matrices with entries in A and Lie algebra product $[x, y] = xy - yx$, the usual commutator of the matrices x and y. We define the *special linear Lie algebra* $\mathfrak{sl}_N(A)$ as the derived algebra

$$\mathfrak{sl}_N(A) = [\mathfrak{gl}_N(A), \mathfrak{gl}_N(A)].$$

of $\mathfrak{gl}_N(A)$. In particular, $\mathfrak{sl}_N(A)$ is an ideal of $\mathfrak{gl}_N(A)$. To analyze the structure of $\mathfrak{sl}_N(A)$ we use the matrix units E_{ij}, i.e., the $N \times N$ matrices with 1 at the position (ij) and 0 at all other positions. They satisfy the basic multiplication rule

$$[aE_{ij}, bE_{mn}] = \delta_{jm}\, ab\, E_{in} - \delta_{ni}\, ba\, E_{mj} \tag{5.82}$$

where δ_* is the usual Kronecker delta. We put $E_N = \sum_{i=1}^N E_{ii}$. Some properties of the Lie algebra $\mathfrak{sl}_N(A)$ are listed in the following (very worthwhile) exercise.

Exercise 5.4.19 (a) $\mathfrak{sl}_N(A) = \{x \in \mathfrak{gl}_N(A) : \mathrm{tr}(x) \in [A, A]\}$.

(b) As a vector space, $\mathfrak{sl}_N(A)$ decomposes as

$$\begin{aligned} \mathfrak{sl}_N(A) &= \mathfrak{sl}_N(A)_0 \oplus \big(\textstyle\bigoplus_{i\neq j} AE_{ij}\big), \quad \text{where} \qquad (5.83)\\ \mathfrak{sl}_N(A)_0 &= \mathfrak{sl}_N(A) \cap \big(\textstyle\bigoplus_{i=1}^N AE_{ii}\big) \\ &= \textstyle\sum_{i\neq j}[AE_{ij}, AE_{ji}] = \sum_{i\neq j} \operatorname{Span}_F\{abE_{ii} - baE_{jj} : a,b \in A\} \\ &= \{cE_N : c \in [A,A]\} \oplus \big(\textstyle\bigoplus_{i=1}^{N-1}\{a(E_{ii} - E_{i+1,i+1}) : a \in A\}\big) \end{aligned}$$

(c) For a commutative A:

$$\mathfrak{sl}_N(A) = \{x \in \mathfrak{gl}_N(A) : \operatorname{tr}(x) = 0\} = \mathfrak{sl}_N(F) \otimes_F A.$$

(d) The centre of $\mathfrak{sl}_N(A)$ is $Z(\mathfrak{sl}_N(A)) = \{zE_N : z \in Z(A) \cap [A,A]\}$ where $Z(A) = \{z \in A : za = az$ for all $a \in A\}$ is the centre of A.

(e) For $a, b \in A$ and $i \neq j$,

$$(aE_{ij}, E_{ii} - E_{jj}, bE_{ji})$$

is an $\mathfrak{sl}_2$-triple if and only if a is invertible and $b = a^{-1}$.

The exercise shows that the structure of a general $\mathfrak{sl}_N(A)$ is quite similar to that of $\mathfrak{sl}_N(F)$. In particular, the decomposition (5.83) is a $\mathcal{Q}(\mathrm{A}_l)$-grading where

$$\mathrm{A}_l = \{\varepsilon_i - \varepsilon_j : 1 \leq i,j \leq N\}, \quad l = N-1.$$

is the root system of type A_l and

$$\mathfrak{sl}_N(A)_{\varepsilon_i - \varepsilon_j} = AE_{ij} \quad \text{for } i \neq j.$$

In fact, $\mathfrak{sl}_N(A)$ is the prototype of an A_l-graded Lie algebra (see [31]). At this level of generality we are far from the structure of a Lie torus. Most importantly, we are missing a compatible Λ-grading of $\mathfrak{sl}_N(A)$. We will use gradings of A, defined as follows.

Definition 5.4.20 Let $A = \bigoplus_{\lambda\in\Lambda} A^\lambda$ be a unital associative Λ-graded F-algebra. Then A is called an *associative torus of type* Λ if it satisfies (AT1)–(AT3) below.

(AT1): if every non-zero A^λ contains an invertible element,
(AT2): $\dim A^\lambda \leq 1$ for all $\lambda \in \Lambda$, and
(AT3): $\operatorname{Span}_{\mathbb{Z}}(\operatorname{supp}_\Lambda A) = \Lambda$.

One calls A simply an *associative torus* if A is an associative torus of type Λ for some abelian group Λ. See 5.4.23 for a short discussion of associative tori.

These definitions are justified by the following exercise, describing when $\mathfrak{sl}_N(A)$ is a Lie torus of type (A_l, Λ).

Exercise 5.4.21 (a) The Lie algebra $\mathfrak{sl}_N(A)$ has a Λ-grading compatible with the $\mathcal{Q}(\mathrm{A}_l)$-grading (5.83) if and only if A is Λ-graded. In this case, the compatible Λ-grading of $\mathfrak{sl}_N(A)$ is given by $\mathfrak{sl}_N(A) = \bigoplus_{\lambda\in\Lambda} \mathfrak{sl}_N(A)^\lambda$ where $\mathfrak{sl}_N(A)^\lambda$ consists of matrices in $\mathfrak{sl}_N(A)$, which have all their entries in A^λ.

(b) With respect to the compatible gradings of (a), the Lie algebra $\mathfrak{sl}_N(A)$ is a Lie torus of type $(\mathrm{A}_{N-1}, \Lambda)$ if and only if A is an associative torus of type Λ.

(c) The Lie torus $\mathfrak{sl}_N(A)$ is invariant with respect to the bilinear form $(\cdot|\cdot)_{\mathfrak{sl}}$ given by $(\sum_{i,j} x_{ij}E_{ij} \mid \sum_{p,q} y_{pq}E_{pq})_{\mathfrak{sl}} = \sum_{i,j}(x_{ij}y_{ji})_0$ where a_0 for $a \in A$ denotes the A^0-component of a.

But we not only have an example of a Lie torus of type (A_l, Λ), we actually have all centreless examples.

Theorem 5.4.22 *Let $l \geq 3$. A Lie algebra L is a centreless Lie torus of type (A_l, Λ) if and only if L is graded-isomorphic to $\mathfrak{sl}_{l+1}(A)$ for A an associative torus of type Λ. In this case, L is an invariant Lie torus.*

Proof This is a special case of the Coordinatization Theorem of A_l-graded Lie algebras ([31, Recognition Theorem 0.7]): A centreless Lie algebra L is A_l-graded ($l = N - 1$) if and only if L is $\mathcal{Q}(\mathrm{A}_l)$-graded-isomorphic to $\mathfrak{sl}_N(A)/Z(\mathfrak{sl}_N(A))$ for some associative F-algebra A. If L is a Lie torus, it follows as in the Exercise 5.4.21 above that A is an associative torus. But $Z(\mathfrak{sl}_N(A)) = \{0\}$ for an associative torus ([191, (3.3.2)] and Exercise 5.4.19). □

Besides [31], related results are proven in [29, Th. 2.65] (see Corollary 5.4.27 below), [108, 2.11 and 3.4] and [223, Prop. 2.13].

Review 5.4.23 (Associative tori versus twisted group algebras) In view of Theorem 5.4.22 it is of interest to know more about associative tori. First of all, the identity 1_A of an associative torus A satisfies $1_A \in A^0$. Hence $a^{-1} \in A^{-\lambda}$ for every invertible $a \in A^\lambda$. Moreover, since the product of two invertible elements in an associative algebra is again invertible, it follows that $\operatorname{supp}_\Lambda A$ is a subgroup of Λ, whence (AT3) is equivalent to

(AT3)′: $\operatorname{supp}_\Lambda A = \Lambda$.

Next, choose a family $(u_\lambda : \lambda \in \Lambda)$ of invertible elements $u_\lambda \in A^\lambda$. This is then in particular an F-basis of A so that the algebra structure of A is completely determined by the equations

$$u_\lambda u_\mu = c(\lambda, \mu) u_{\lambda+\mu} \tag{5.84}$$

for $\lambda, \mu \in \Lambda$ and suitable non-zero scalars $c(\lambda, \mu) \in F$. It is not necessary that $c(\lambda, \mu) = 1$, see for example Exercise 5.4.26. Rather, given an F-vector space with basis $(u_\lambda : \lambda \in \Lambda)$ one can define a multiplication on A by (5.84), and this multiplication is associative if and only if

$$c(\lambda, \mu)\, c(\lambda + \mu, \nu) = c(\mu, \nu)\, c(\lambda, \mu + \nu) \tag{5.85}$$

holds for all $\lambda, \mu, \nu \in \Lambda$. In this case, the algebra is an associative torus of type Λ. It is clear from the construction that, conversely, any associative torus is obtained in this way from a family $(c(\lambda, \mu))_{\lambda,\mu}$ of non-zero scalars. The algebras constructed in this way are called *twisted group algebras*. The reader with some knowledge in group cohomology will recognize that the families $(c(\lambda, \mu))$ satisfying (5.85) are precisely the 2-cocycles of Λ with values in $F \setminus \{0\}$. One can show that two families define graded-isomorphic tori if and only if their cohomology classes coincide.

Example 5.4.24 (Group algebra) Although, as we have pointed out, the $c(\lambda, \nu)$ need not equal 1 in general, the family for which all $c(\lambda, \mu) = 1$, satisfies (5.85) and so yields an example of a Λ-torus, called the *group algebra of* Λ and denoted $F[\Lambda]$. In particular, this implies that associative tori exist for all Λ, and hence Lie tori exist for all types (A_l, Λ), $l \geq 2$.

The Lie tori that arise as cores of an EALA have type (S, Λ) where Λ is a free abelian group of finite rank. This condition on Λ follows from the axiom (EA6) or, equivalently from the axiom (EARS7). We therefore discuss this special case now.

Definition 5.4.25 Let $\mathbf{q} = (q_{ij})$ be an $n \times n$ matrix such that the entries $q_{ij} \in F$ satisfy $q_{ii} = 1 = q_{ij}q_{ji}$ for all $1 \le i, j \le n$. The *quantum torus* associated to $\mathbf{q}$ is the associative algebra $\mathbb{F}_{\mathbf{q}}$ presented by the generators t_i, t_i^{-1}, $1 \le i \le n$ subject to the relations

$$t_i t_i^{-1} = t_i^{-1} t_i, \quad \text{and} \quad t_i t_j = q_{ij}\, t_j t_i \text{ for all } 1 \le i, j \le n.$$

For example, if all $q_{ij} = 1$, then $\mathbb{F}_{\mathbf{q}} = F[t_1^{\pm 1}, \dots, t_n^{\pm 1}]$ is the Laurent polynomial ring in n variables. Thus, a general $\mathbb{F}_{\mathbf{q}}$ is a non-commutative version of $F[t_1^{\pm 1}, \dots, t_n^{\pm 1}]$, the coordinate ring of the n-dimensional algebraic torus $(F \setminus \{0\})^n$, which explains the name "quantum torus".

Exercise 5.4.26 Let $\mathbb{F}_{\mathbf{q}}$ be a quantum torus. Show:
(a) $\mathbb{F}_{\mathbf{q}} = \bigoplus_{\lambda \in \mathbb{Z}^n} F t^\lambda$ for $t^\lambda = t_1^{\lambda_1} \cdots t_n^{\lambda_n}$.
(b) The t^λ satisfy the multiplication rule $t^\lambda t^\mu = c(\lambda, \mu) t^{\lambda + \mu}$ with

$$c(\lambda, \mu) = \textstyle\prod_{1 \le j < i \le n} q_{ij}^{\lambda_i \mu_j}.$$

(c) $\mathbb{F}_{\mathbf{q}}$ is an associative torus of type $\mathbb{Z}^n$.
(d) Every associative torus of type $\mathbb{Z}^n$ is graded-isomorphic to some quantum torus $\mathbb{F}_{\mathbf{q}}$.
(e) The centre $Z(\mathbb{F}_{\mathbf{q}}) = \{z \in \mathbb{F}_{\mathbf{q}} : [z, \mathbb{F}_{\mathbf{q}}] = 0\}$ of the associative algebra $\mathbb{F}_{\mathbf{q}}$ is a graded subspace of $\mathbb{F}_{\mathbf{q}}$, namely $Z(\mathbb{F}_{\mathbf{q}}) = \bigoplus_{\gamma \in \Gamma} F t^\gamma$, where Γ is a subgroup of $\mathbb{Z}^n$ given by

$$\begin{aligned} \Gamma &= \{\gamma \in \mathbb{Z}^n : c(\gamma, \mu) = c(\mu, \gamma) \text{ for all } \mu \in \mathbb{Z}^n\} \\ &= \{\gamma \in \mathbb{Z}^n : \textstyle\prod_{j=1}^n q_{ij}^{\gamma_j} = 1 \text{ for } 1 \le i \le n\}. \end{aligned}$$

Moreover, the following are equivalent:
(i) All q_{ij} are roots of unity.
(ii) $[\mathbb{Z}^n : \Gamma] < \infty$.
(iii) $\mathbb{F}_{\mathbf{q}}$ is finitely generated as a module over its centre $Z(\mathbb{F}_{\mathbf{q}})$.

Combining this exercise with Theorem 5.4.22 we obtain the classification of the cores of EALAs of type A:

Corollary 5.4.27 ([29, Th. 2.65]) *The Lie algebra $\mathfrak{sl}_N(\mathbb{F}_{\mathbf{q}})$ for a quantum torus $\mathbb{F}_{\mathbf{q}}$ is a centreless Lie torus of type $(\mathrm{A}_{N-1}, \mathbb{Z}^n)$. Conversely, any centreless Lie torus of type $(\mathrm{A}_{N-1}, \mathbb{Z}^n)$ with $N \ge 4$ is graded-isomorphic to some $\mathfrak{sl}_N(\mathbb{F}_{\mathbf{q}})$.*

Remark 5.4.28 The attentive reader will have noticed that we didn't say anything about Lie tori of type (A_1, Λ) and (A_2, Λ) in Theorem 5.4.22 and Corollary 5.4.27. Of course, $\mathfrak{sl}_3(A)$ of an associative Λ-torus will be centreless Lie tori of type (A_2, Λ). The limitation $N \ge 4$ in Theorem 5.4.22 is justified, since for $N = 3$ there are more examples: One needs coordinate algebras A which are no longer associative but only alternative. Moreover, one needs to replace the matrix algebra $\mathfrak{sl}_3(A)$ by something more general, a Tits-Kantor-Koecher algebra or an abstractly defined Lie algebra, see [30] for details. Analogous remarks apply for the A_1-case, in which the coordinates come from certain Jordan algebras (called *Jordan tori*) and in which $\mathfrak{sl}_2$ has to be replaced by a Tits-Kantor-Koecher algebra.

One has classification theorems for Lie tori of all types. For each type they are listed with references up to 2007 at the beginning of §7 in [11]. An additional recent reference is [190] for $S = \mathrm{B}_2$.

5.4.5 Some more easy examples of Lie tori. We describe some more easy examples, where easy means that they do not require some knowledge of non-associative algebras, like Jordan algebras, alternative or structurable algebras.

Example 5.4.29 ($\Lambda = \{0\}$) Let $\mathfrak{g}$ be a finite-dimensional split simple Lie algebra with splitting Cartan subalgebra $\mathfrak{h}$. Then $\mathfrak{g}$ has a root space decomposition $\mathfrak{g} = \bigoplus_{\xi \in S} \mathfrak{g}_\xi$ where $\mathfrak{g}_0 = \mathfrak{h}$ and S is the root system of $(\mathfrak{g}, \mathfrak{h})$, a finite reduced root system. Since $\mathfrak{g}$ is simple, S is also irreducible. Using standard properties of finite-dimensional split simple Lie algebras, it is easy to check that $\mathfrak{g} = \bigoplus_{\xi \in S} \mathfrak{g}_\xi$ *is a Lie torus of type* $(S, \{0\})$.

Conversely, if L is a Lie torus of type $(S, \{0\})$, then L is a finite-dimensional split simple Lie algebra. Indeed, $L = \mathfrak{g}$ in the notation of Proposition 5.4.8.

Note that this fits nicely into the picture of EALAs of nullity 0, which we have characterized in Proposition 5.3.24 as finite-dimensional split simple Lie algebras.

Example 5.4.30 As in the previous Example 5.4.29 let $\mathfrak{g}$ be a finite-dimensional split simple Lie algebra with splitting Cartan subalgebra $\mathfrak{h}$ and root system S. We would like to consider a Lie algebra of the form $\mathfrak{g} \otimes A$ where A is an associative algebra. For $\mathfrak{g}$ of type A we could take non-commutative "coordinates" A to get a Lie torus, see Section 5.4.4. However, for $\mathfrak{g}$ not of type A the algebra A must be commutative in order for $\mathfrak{g} \otimes A$ to be a Lie algebra.

Therefore, we let $A = \bigoplus_{\lambda \in \Lambda} A^\lambda$ be a *commutative* associative torus of type Λ and consider $\mathfrak{g} \otimes A$, which becomes a Lie algebra (over F) by defining

$$[u_1 \otimes a_1,\, u_2 \otimes a_2] = [u_1, u_2] \otimes a_1 a_2$$

for $u_i \in \mathfrak{g}$ and $a_i \in A$.

Exercise 5.4.31 Show: The Lie algebra $\mathfrak{g} \otimes A$ of Example 5.4.30 is centreless Lie torus of type (S, Λ) with respect to the homogeneous subspaces

$$(\mathfrak{g} \otimes A)_\xi^\lambda = \mathfrak{g}_\xi \otimes A^\lambda.$$

The support of the $\mathcal{Q}(S) \oplus \Lambda$-graded Lie algebra $\mathfrak{g} \otimes A$ is the set

$$\operatorname{supp}_{\mathcal{Q}(S) \oplus \mathbb{Z}^n} \mathfrak{g} \otimes A = S \times \Lambda.$$

Moreover, $\mathfrak{g} \otimes A$ is an invariant Lie torus with respect to the bilinear form

$$(x \otimes a^\lambda \mid y \otimes b^\mu) = \kappa(x, y)\, (a^\lambda b^\mu)_0$$

where κ is the Killing form of $\mathfrak{g}$ and c_0 for $c \in A$ is the 0-component of c.

The Example 5.4.30 works for any type of S. But it yields all examples only for special types of the root system S, as described in the following.

Theorem 5.4.32 *Any centreless Lie torus of type* (S, Λ) *for* S *of type* D_l, $l \geq 4$, *or* E_l, $l = 6, 7, 8$, *is graded-isomorphic to an example as in Example* 5.4.30 *for* $\mathfrak{g}$ *of the corresponding type and* A *a commutative associative torus of type* Λ.

Proof The proof is analogous to the proof of Theorem 5.4.22: One applies the Coordinatization Theorem of [31] to see that L has the form $\mathfrak{g} \otimes A$ for some commutative associative F-algebra A. One then has to discuss when such a Lie algebra is a Lie torus. This turns out to be the case exactly when A is a torus. □

Example 5.4.33 (Untwisted multiloop algebras) For EALAs it is of interest to describe the centreless Lie tori of type (S,Λ) with Λ a free abelian group of finite rank, say of rank n. Hence $\Lambda \cong \mathbb{Z}^n$. It is immediate that $\mathfrak{g}\otimes A$ is a Lie torus of type $(S,\mathbb{Z}^n)$ if and only if A is a commutative quantum torus, i.e., a Laurent polynomial ring in several, say n variables. In other words, these are the untwisted multiloop algebra of (5.11),

$$L(\mathfrak{g}) = \mathfrak{g}\otimes_F F[t_1^{\pm 1},\dots,t_n^{\pm 1}]$$

Hence by Theorem 5.4.11 the universal central extension of $L(\mathfrak{g})$, the toroidal Lie algebras of Section 5.1.2 are also Lie tori. Finally, Theorem 5.4.32 has the following corollary.

Corollary 5.4.34 ([29]) *Any centreless Lie torus of type $(S,\mathbb{Z}^n)$ with $S=\mathrm{D}_l$, $l\geq 4$ or $S=\mathrm{E}_l$, $l=6,7,8$, is graded-isomorphic to an untwisted multiloop algebra $L(\mathfrak{g})$ as in Example* 5.4.33.

Perhaps the reader now expects that the next example will be the general multiloop algebras $L(\mathfrak{g},\boldsymbol{\sigma})$ defined in (5.13). However, an arbitrary multiloop algebra is in general not a Lie torus, see [5, Th. 3.3.1] and [182, Th. 5.1.4] for a characterization of centreless Lie tori which are multiloop algebras. But this phenomenon does not occur in nullity 1.

Exercise 5.4.35 Verify that the loop algebra $L(\mathfrak{g},\sigma)$ of (5.3) is an invariant Lie torus of type $(S,\mathbb{Z})$ where S is the root system of Table 5.53.

5.5 The construction of all EALAs

Recall Theorem 5.4.14: If (E,H) is an EALA, its core E_c and its centreless core E_{cc} are Lie tori, the latter being an invariant Lie torus. Moreover, if (S,Λ) is the type of the Lie tori E_c and E_{cc}, then Λ is a free abelian group of finite rank. Thus:

core E_c (Lie torus) ⇜ EALA (E,H)
⇝ (downward)
centreless core E_{cc} (invariant Lie torus)

In this section we will reverse this diagram: We will construct an EALA starting from an invariant Lie torus.

To motivate the construction it is useful to look again at the construction of an affine Kac-Moody Lie algebra in Sections 5.1.1 and 5.2.4. It can be summarized as follows: We start with a twisted loop algebra $\mathcal{L}=L(\mathfrak{g},\sigma)$, which as we now know is an invariant Lie torus (Exercise 5.4.35). We then take a central extension $\tilde{\mathcal{L}}$, which in this example is the universal central extension and hence by Theorem 5.4.11 again a Lie torus (of course, in this example one can also verify this directly). Finally, we add some (not all) derivations to $\tilde{L}$ to get an affine Kac-Moody Lie algebra. Moreover and most importantly, all affine Kac-Moody Lie algebras are obtained in this way (Kac's Realization Theorem 5.1.3). To summarize, using the

EALA terminology:

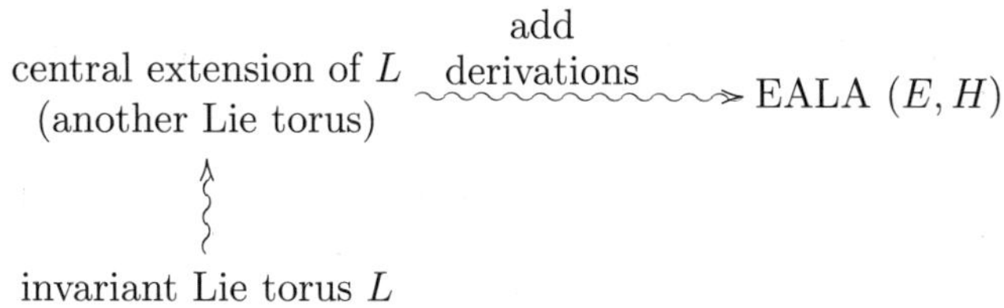

To do something like this in general, one faces the following two problems.

(A) An invariant Lie torus has in general many central extensions. For example, the untwisted multiloop algebra $L = \mathfrak{g} \otimes F[t_1^{\pm 1}, \dots, t_n^{\pm 1}]$ is an invariant Lie torus by Example 5.4.33. If $n \geq 2$, its universal central extension has an infinite-dimensional centre, a result we already mentioned in Section 5.1.2, see in particular Theorem 5.1.5. Hence, there are many possible central extensions. Should we only consider the universal central extension?

(B) Which derivations should we add? Already in the affine case we did not add all derivations, as follows for example from Exercise 5.1.7!

It turns out that the two problems are closely related, and we will solve both at the same time. Rather than taking a 2-step approach, we will take one big step, by taking what one may call an *affine extension* (after all, the result will be an extended affine Lie algebra). In fact, affine extensions are a special case of so-called *double extensions*, see for example [37]. Thus, the goal is

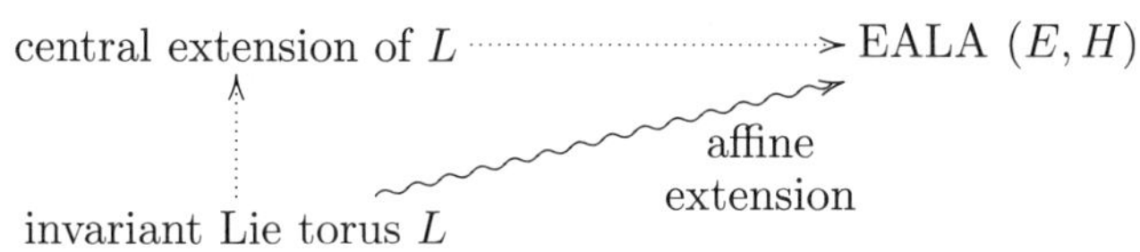

(5.86)

The key idea is based on the construction of a 2-cocyle in Example 5.1.19: Any subspace D of skew-symmetric derivations will give rise to a 2-cocycle and hence to a central extension. But not only do we get examples of central extensions. By Theorem 5.1.25 and Exercise 5.1.27, up to isomorphism all central coverings of L are of the form $\mathrm{E}(L, D, \psi_D)$ for some graded subspace D of $\mathrm{SDer}(L)$ with $D \cap \mathrm{IDer}(L) = \{0\}$. Observe that we can indeed apply this theorem: The invariant Lie torus L

(i) is perfect by Exercise 5.4.10 and is finitely generated as Lie algebra by Theorem 5.4.13 (recall that Λ is free of finite rank, where (S, Λ) is the type of L),

(ii) has finite homogeneous dimension, even bounded homogeneous dimension also by Theorem 5.4.13, and

(iii) has an invariant nondegenerate Λ-graded symmetric bilinear form, by definition of an invariant Lie torus.

But we need more than just a central covering. For example, the axiom (EA2) requires that we construct an ad-diagonalizable subalgebra H for which the subspaces L_ξ^λ, $\xi \neq 0$, are root spaces, as can be seen from Theorem 5.4.14. By Lemma 5.4.4 we can realize the subspaces L_ξ as root spaces of some natural subalgebra $\mathfrak{h} \subseteq L$. But we do not have a result, which describes the subspaces L^λ in a similar fashion, i.e., as root spaces of some toral subalgebra. There is in fact no natural choice of a subalgebra to do so. Rather, we will distinguish these subspaces "externally", i.e., via an action of some non-inner derivation algebra. The required formalism to do this, is described in the next section. This has nothing to do with Lie algebras. Rather, it is a topic in the theory of graded vector spaces, and we will therefore describe it in this setting.

5.5.1 Degree maps. In this section, V is vector space over a field F, which could be of arbitrary characteristic until Proposition 5.5.3(b). Also, Λ denotes an arbitrary abelian group. We recall that a Λ-grading of V is simply a direct vector space decomposition of V by a family $(V^\lambda : \lambda \in \Lambda)$ of subspaces $V^\lambda \subseteq V$. Our goal in this section is to present a method describing the homogeneous subspaces V^λ of a given Λ-grading of V as the joint eigenspaces of a subspace of diagonalizable endomorphisms.

To motivate the construction, let us first look at the converse, namely inducing a grading of V via the action of endomorphisms. We will say that a subspace $T \subseteq \operatorname{End}_F(V)$ is a *subspace of simultaneously diagonalizable endomorphisms*, if

$$\begin{aligned} V &= \textstyle\bigoplus_{\lambda \in T^*} V^\lambda \quad \text{where} \\ V^\lambda &= \{v \in V : t(v) = \lambda(t)v \text{ for all } t \in T\}. \end{aligned} \tag{5.87}$$

In this case, T obviously consists of pairwise commuting diagonalizable endomorphisms. Conversely, it is well-known that a finite-dimensional subspace of pairwise commuting diagonalizable endomorphisms is a subspace of simultaneously diagonalizable endomorphisms (this is no longer true if T is infinite-dimensional). Observe that the decomposition (5.87) is a grading of V by the group $\operatorname{Span}_{\mathbb{Z}}(\operatorname{supp}_{T^*} V)$ where $\operatorname{supp}_{T^*} V = \{\lambda \in T^* : V^\lambda \neq 0\}$. Our goal is to realize a given grading of V in this way.

To do so, we will use the F-vector space

$$\mathrm{D}(\Lambda) = \operatorname{Hom}_{\mathbb{Z}}(\Lambda, F)$$

consisting of all maps $\theta : \Lambda \to F$ which are $\mathbb{Z}$-linear: $\theta(\lambda_1 + \lambda_2) = \theta(\lambda_1) + \theta(\lambda_2)$ for all $\lambda_i \in \Lambda$. This is an F-vector space by defining for $\theta, \theta_i \in \mathrm{D}(\Lambda)$ and $s \in F$ the sum $\theta_1 + \theta_2$ and the scalar multiplication $s\theta$ by $(\theta_1 + \theta_2)(\lambda) = \theta_1(\lambda) + \theta_2(\lambda)$ and $(s\theta)(\lambda) = s(\theta(\lambda))$.

Exercise 5.5.1 (a) Show $\mathrm{D}(\Lambda) \cong \operatorname{Hom}_F(\Lambda \otimes_{\mathbb{Z}} F, F) = (\Lambda \otimes_F F)^*$. Thus $\mathrm{D}(\Lambda)$ is naturally a dual vector space.

(b) If Λ is free of rank n, say with $\mathbb{Z}$-basis $\varepsilon_1, \dots, \varepsilon_n$, then $\mathrm{D}(\Lambda) = F\partial_1 \oplus \cdots \oplus F\partial_n$ where $\partial_i \in \mathrm{D}(\Lambda)$ is defined by $\partial_i(\sum_j m_j \varepsilon_j) = m_i$. In particular, $\dim_F \mathrm{D}(\Lambda) = n$.

We now suppose that $V = \bigoplus_{\lambda \in \Lambda} V^\lambda$ is a Λ-grading of the vector space V. Any $\theta \in \mathrm{D}(\Lambda)$ defines an endomorphism $\partial_\theta \in \operatorname{End}_F(V)$ by

$$\partial_\theta(v^\lambda) = \theta(\lambda)\, v^\lambda \quad \text{for } v^\lambda \in V^\lambda.$$

We put

$$\mathcal{D}(V) = \{\partial_\theta : \theta \in \mathrm{D}(\Lambda)\}$$

and call the elements of $\mathcal{D}(V)$ *degree maps*. If $A = \bigoplus_{\lambda \in \Lambda} A^\lambda$ is a Λ-grading of an algebra A, the maps ∂_θ are derivations and $\mathcal{D}(A)$ is then called the space of *degree derivations*.

The map $\partial : \mathrm{D}(\Lambda) \to \mathcal{D}(V)$ is clearly F-linear and surjective by definition. Its kernel is $\{\theta \in \mathrm{D}(\Lambda) : \theta(\operatorname{supp}_\Lambda V) = 0\}$. To make ∂ an isomorphism we will

$$\textit{from now on assume } \operatorname{Span}_{\mathbb{Z}}(\operatorname{supp}_\Lambda V) = \Lambda. \tag{5.88}$$

As we have pointed out at previous occasions, this is not a serious assumptions since one can always replace Λ by $\operatorname{Span}_{\mathbb{Z}}(\operatorname{supp}_\Lambda V)$ without changing the given grading. Since now ∂ is an isomorphism, we can define a linear form $\mathrm{ev}_\lambda \in \mathcal{D}(V)^*$ for every $\lambda \in \Lambda$:

$$\mathrm{ev}_\lambda(\partial_\theta) = \theta(\lambda)$$

The F-linear map

$$\mathrm{ev} : \Lambda \to \mathcal{D}(V)^*, \quad \lambda \mapsto \mathrm{ev}_\lambda$$

is called the *evaluation map*. By construction,

$$V^\lambda \subseteq \{v \in V : d(v) = \mathrm{ev}_\lambda(d)v \text{ for all } d \in \mathcal{D}(V)\} \tag{5.89}$$

since for $d = \partial_\theta$ and $v \in V^\lambda$ we have $\partial_\theta(v^\lambda) = \theta(\lambda)v^\lambda = \mathrm{ev}_\lambda(\partial_\theta)v^\lambda$.

Definition 5.5.2 In the setting of above, i.e., $V = \bigoplus_{\lambda\in\Lambda} V^\lambda$ is Λ-graded and (5.88) holds, we will say that a subspace $T \subseteq \mathcal{D}(V)$ *induces the Λ-grading of V* if

$$V^\lambda = \{v \in V : t(v) = \mathrm{ev}_\lambda(t)v \text{ for all } t \in T\}$$

holds for all $\lambda \in \Lambda$.

Proposition 5.5.3 *Let $V = \bigoplus_{\lambda\in\Lambda} V^\lambda$ be a Λ-grading of the vector space V such that* (5.88) *holds.*

(a) *A subspace $T \subseteq \mathcal{D}(V)$ induces the Λ-grading of V if the restricted evaluation map*

$$\mathrm{ev}_T : \Lambda \to T^*, \quad \mathrm{ev}_T(\lambda) = \mathrm{ev}_\lambda|_T$$

is injective.

(b) *Suppose F has characteristic* 0 *and Λ is torsion-free, i.e., $n\lambda = 0$ for some $n \in \mathbb{Z}$ implies $\lambda = 0$. Then Λ embeds into the F-vector space $U = \Lambda \otimes_{\mathbb{Z}} F$ and for every subspace $S \subseteq \mathrm{D}(\Lambda)$ separating the points of Λ in U the corresponding subspace $T = \partial(S) \subseteq \mathcal{D}(V)$ induces the Λ-grading of V. In particular, $\mathcal{D}(V)$ induces the Λ-grading of V.*

5.5.2 The centroid of Lie algebras, in particular of Lie tori. After the intermezzo on how to induce gradings of vector spaces in the previous Section 5.5.1 we now come back to Lie algebras, but not immediately to Lie tori and EALAs. Of course, the topic of this section is motivated by the over-all goal of this section: The construction of EALAs from Lie tori using certain subspaces of derivations. The derivations, which in Section 5.5.4 will be used in the general construction, are products of degree maps, studied in Section 5.5.1, and so-called centroidal transformations, to which this section is devoted.

The basic idea of the centroid of a Lie algebra (or of any algebra for that matter) is that it identifies the largest ring over which the given algebra can be considered as an algebra. For example, if one studies the real Lie algebra L which is $\mathfrak{sl}_n(\mathbb{C})$ considered as a real Lie algebra by restricting the scalars to $\mathbb{R}$, the centroid will be $\cong \mathbb{C}$ and will thus indicate that L can also be considered as a complex Lie algebra.

In general, the centroid will not be a field but only a (commutative) ring. Hence, considering a Lie algebra as algebra over its centroid, necessitates that in the following definition and in Lemma 5.5.6 we will deviate from our standard assumption and consider Lie algebras over rings. The definition of a Lie algebra L defined over a ring, say k, is not surprising: L is a k-module with a k-bilinear map $[.,.] : L \times L \to L$ which is alternating, i.e., $[l,l] = 0$ for all $l \in L$, and satisfies the Jacobi identity.

Definition 5.5.4 ([128, Ch. X]) The *centroid* $\mathrm{Cent}_k(L)$ of a Lie algebra L defined over a ring k is defined as

$$\mathrm{Cent}_k(L) = \{\chi \in \mathrm{End}_k(L) : \chi([l_1, l_2]) = [l_1, \chi(l_2)] \text{ for all } l_1, l_2 \in L\}.$$

Of course, $\chi \in \mathrm{Cent}_k(L) \Leftrightarrow \chi([l_1, l_2]) = [\chi(l_1), l_2]$ for all $l_1, l_2 \in L$. It is important to indicate k in the notation $\mathrm{Cent}_k(L)$ since the centroid depends on the base ring k.

We have $k\,\mathrm{Id}_L \subseteq \mathrm{Cent}_k(L)$ for every L. One calls L *central* if the map $k \to \mathrm{Cent}_F(L)$, $s \mapsto s\,\mathrm{Id}_L$, is an isomorphism, and one says that L is *central-simple* if L is just that, namely central and simple.

Let $L = \bigoplus_{\lambda \in \Lambda} L^\lambda$ be a Lie algebra graded by an abelian group Λ. We can then also define the Λ-*graded centroid* as

$$\mathrm{grCent}_k(L) = \mathrm{grEnd}_k(L) \cap \mathrm{Cent}_k(L) = \textstyle\bigoplus_{\lambda \in \Lambda} \mathrm{Cent}_k(L)^\lambda$$

where $\mathrm{Cent}_k(L)^\lambda$ consists of the centroidal transformations which have degree λ, i.e., $\chi(L^\mu) \subseteq L^{\lambda+\mu}$ for all $\mu \in \Lambda$.

Example 5.5.5 As an immediate example we calculate the centroid of the Lie algebra $L = \mathfrak{sl}_2(\mathbb{C})$, considered as real Lie algebra by restricting the scalars to $\mathbb{R}$. We let (e,h,f) be an $\mathfrak{sl}_2$-triple in $\mathfrak{sl}_2(\mathbb{C})$. Then the relations $[h, ce] = 2ce$ and $[h, cf] = -2cf$ for $c \in \mathbb{C}$ show that $\chi(ce)$ and $\chi(cf)$ are uniquely determined by $\chi(h)$. For example, $2\chi(ce) = \chi([h, ce]) = [\chi(h), ce]$. Moreover, $\chi(h) \in \mathbb{C}h$ because $[\chi(h), ch] = \chi([h, ch]) = 0$. Hence $\dim_{\mathbb{R}} \mathrm{Cent}_{\mathbb{R}}(L) \le 2$. On the other side, $\mathbb{C}\,\mathrm{Id}_L \subseteq \mathrm{Cent}_{\mathbb{R}}(L)$ is clear, whence $\mathbb{C}\,\mathrm{Id}_L = \mathrm{Cent}_{\mathbb{R}}(L)$.

We leave it to the reader to show $\mathrm{Cent}_{\mathbb{R}}(L) = \mathbb{C}\,\mathrm{Id}_L$ for $L = \mathfrak{sl}_n(\mathbb{C})$ considered as real Lie algebra, without using any of the results mentioned below!

The following lemma gives a mathematical meaning to the claims made before Definition 5.5.4 and lists the most important properties of the centroid of arbitrary Lie algebras, not only of Lie tori.

Lemma 5.5.6 (Folklore) *Let L be a Lie algebra defined over a ring k.*

(a) *The centroid of L is always a unital associative subalgebra of the endomorphism algebra $\mathrm{End}_k(L)$ of L. Hence $\mathrm{Cent}_k(L)$ is a k-algebra and L becomes a $\mathrm{Cent}_k(L)$-module by defining the action of $\mathrm{Cent}_k(L)$ on L by $\chi \cdot l = \chi(l)$ for $\chi \in \mathrm{Cent}_k(L)$ and $l \in L$.*

(b) *If the centroid of L is commutative, then with respect to the action of $\mathrm{Cent}_k(L)$ on L defined in* (a), *L is a Lie algebra over the ring $\mathrm{Cent}_k(L)$. Moreover, L is central as a Lie algebra over its centroid.*

(c) *If L is perfect, its centroid is commutative and does not depend on the base ring k: $\mathrm{Cent}_k(L) = \mathrm{Cent}_{\mathbb{Z}}({}_{\mathbb{Z}}L)$ where ${}_{\mathbb{Z}}L$ is the Lie algebra L with scalars restricted to $\mathbb{Z}$.*

(d) *If L is simple, its centroid is a field and L as a Lie algebra over the field $\mathrm{Cent}_F(L)$ is central-simple. In particular:*

(i) *a finite-dimensional simple Lie algebra over an algebraically closed field F is central-simple, and*

(ii) *the centroid of a simple real Lie algebra L is either* $\cong \mathbb{R}\,\mathrm{Id}$, *in which case L is central-simple, or is* $\cong \mathbb{C}\,\mathrm{Id}$, *in which case L is a simple complex Lie algebra, considered as a real Lie algebra.*

(e) *Suppose L is Λ-graded. Then* $\mathrm{grCent}_k(L)$ *is a Λ-graded subalgebra of the full centroid* $\mathrm{Cent}_k(L)$. *Moreover,* $\mathrm{grCent}_k(L) = \mathrm{Cent}_k(L)$ *if L is finitely generated as an ideal, i.e., there exist $l_1, \ldots, l_n \in L$ such that the ideal generated by $l_1, \ldots, l_n$ is all of L.*

(f) *If $\chi \in \mathrm{Cent}_k(L)$ and $d \in \mathrm{Der}_k(L)$, then $\chi \circ d \in \mathrm{Der}_k(L)$. With respect to this operation,* $\mathrm{Der}_k(L)$ *is a* $\mathrm{Cent}_k(L)$*-module and* $\mathrm{IDer}(L)$ *is a submodule of the* $\mathrm{Cent}_k(L)$*-module* $\mathrm{Der}_k(L)$.

The proof of this lemma is a straightforward exercise, which the reader will be asked to do now. The exercise also lists some interesting additional facts on the centroid.

Exercise 5.5.7 (a) For any $\chi \in \mathrm{Cent}_k(L)$ the kernel $\ker \chi$ and the image $\mathrm{Im}\,\chi$ are ideals of L satisfying $[\ker \chi, \mathrm{Im}\,\chi] = 0$.

(b) Prove Lemma 5.5.6. For part (c) of the Lemma use (a) above.

(c) If L is perfect, any $\chi \in \mathrm{Cent}_k(L)$ is symmetric with respect to any invariant bilinear form on L.

Here is the result, which describes the centroid of the Lie algebras of interest in this section. We will use the notion of an associative torus and a twisted group algebra, introduced in Definition 5.4.20 and further discussed in 5.4.23–5.4.26.

Proposition 5.5.8 ([26, Prop. 3.13]) *Let $L = \bigoplus_{\xi \in S,\, \lambda \in \Lambda} L_\xi^\lambda$ be a centreless Lie torus of type (S, Λ).*

(a) *With respect to the $(\mathcal{Q}(S) \oplus \Lambda)$-grading of L we have*

$$\mathrm{Cent}_F(L) = \bigoplus\nolimits_{\lambda \in \Lambda} \mathrm{Cent}_F(L)_0^\lambda = \mathrm{grCent}_F(L). \tag{5.90}$$

In particular, $\chi(L_\xi) \subseteq L_\xi$ for any $\chi \in \mathrm{Cent}_k(L)$ and $\xi \in S$.

(b) *Moreover, with respect to the decomposition* (5.90) *the centroid* $\mathrm{Cent}_F(L)$ *is an associative commutative torus of type Γ, where $\Gamma = \mathrm{supp}_\Lambda \mathrm{Cent}_F(L)$ is a subgroup of Λ. Hence* $\mathrm{Cent}_F(L)$ *is a twisted group algebra over Γ.*

(c) *In particular, if Λ is free abelian of finite rank n, the centroid* $\mathrm{Cent}_F(L)$ *is graded-isomorphic to $F[\Gamma]$, the group algebra of Γ as defined in* (5.4.24), *and is thus isomorphic to a Laurent polynomial ring in ν variables, $0 \le \nu \le n$. Moreover, L is a free module over its centroid.*

Proof Parts (a) and (b) of this proposition are proven in [26, Prop. 3.13]. Part (c) is ([185, Th. 7]). The first part of (c) follows from (b): A twisted group algebra over a free group is a group algebra. The second part is a special case of a general fact: Any graded module over an associative torus is free. □

Example 5.5.9 Let $L = \mathfrak{sl}_N(A)$ for A an associative F-algebra, see Section 5.4.4, and let $Z(A) = \{z \in A : [z, A] = 0\}$ be the centre of the associative algebra A. Any $z \in Z(A)$ induces a centroidal transformation χ_z defined by mapping $x = (x_{ij}) \in \mathfrak{sl}_N(A)$ to $\chi_z(x) = (zx_{ij})$. It is easily seen ([187, 7.9]) that

$$Z(A) \to \mathrm{Cent}_F(\mathfrak{sl}_N(A)), \quad z \mapsto \chi_z$$

is an isomorphism of F-algebras (the only non-obvious part is surjectivity).

Let us now specialize to the case of a Lie torus $\mathfrak{sl}_N(A)$ of type $(\mathrm{A}_{N-1}, \mathbb{Z}^n)$. Thus, by Corollary5.4.27, $A = \mathbb{F}_\mathbf{q}$ is a quantum torus. A description of the centre $Z(\mathbb{F}_\mathbf{q})$ is given in Exercise 5.4.26(e) (see [29, Prop. 2.44] for a proof): $Z(\mathbb{F}_\mathbf{q}) = \bigoplus_{\gamma\in\Gamma} Ft^\gamma$ where Γ is the subgroup

$$\Gamma = \{\gamma \in \mathbb{Z}^n : \textstyle\prod_{j=1}^n q_{ij}^{\gamma_j} = 1 \text{ for } 1 \le i \le n\}$$

of $\mathbb{Z}^n$. The centre of $\mathbb{F}_\mathbf{q}$ is therefore isomorphic to a Laurent polynomial ring in, say, ν variables, as claimed in Proposition 5.5.8(c). To see that the inequalities $0 \le \nu \le n$ stated there are sharp, we consider the quantum torus associated to the matrix

$$\mathbf{q} = \begin{bmatrix} 1 & q \\ q^{-1} & 1 \end{bmatrix}.$$

Specializing the description of Γ above we get

$$\Gamma = \begin{cases} \{0\}, & q \text{ not a root of unity,} \\ m\mathbb{Z} \oplus m\mathbb{Z}, & q \text{ an } m\text{th root of unity.} \end{cases}$$

Hence $\nu = 0$ in the first case and $\nu = 2 = n$ in the second case.

However, the following result says it can only happen in type A that the centroidal grading group Γ has smaller rank than Λ.

Theorem 5.5.10 ([185, Th. 7]) *Let L be a centreless Lie torus of type $(S, \mathbb{Z}^n)$ with S not of type* A. *Then $[\mathbb{Z}^n : \Gamma] < \infty$ and L is a free* $\mathrm{Cent}_F(L)$*-module of finite rank.*

This result, together with the Realization Theorem of [4] implies that an invariant Lie torus of type $(S, \mathbb{Z}^n)$, $S \ne \mathrm{A}_l$, is graded-isomorphic to a multiloop algebra as defined in (5.13). A characterization of which multiloop algebras are Lie tori is the main result of [5]. A more general approach to realizing Lie tori as multiloop algebras is developed in [182].

It is easy to verify Theorem 5.5.10 in case L is a Lie torus of type $(S, \mathbb{Z}^n)$ and S of type D or E. As we have seen in Theorem 5.4.32, in this case $L = \mathfrak{g} \otimes F[t_1^{\pm 1}, \dots, t_n^{\pm 1}]$. The centroids of these types of Lie algebras are described in the next example.

Example 5.5.11 Let $\mathfrak{g}$ be a finite-dimensional central-simple Lie algebra. For example, by [26, Remark 3.6] any finite-dimensional split simple Lie algebra is central and thus central-simple. (Over algebraically closed fields, this also follows from Lemma 5.5.6(d).) Also, let A be an associative commutative F-algebra.

A straightforward verification shows that for $s \in F$ and $a \in A$ the map $\chi_{s,a}$, defined by $u \otimes b \mapsto su \otimes ab$, is a centroidal transformation of the Lie algebra $\mathfrak{g} \otimes A$. It follows from [9, Lemma 2.3(a)] or [17, Lemma 1.2] or [26, Cor. 2.23] that these are all the maps in the centroid of $\mathfrak{g} \otimes A$:

$$F\,\mathrm{Id}_\mathfrak{g} \otimes A \cong \mathrm{Cent}_F(\mathfrak{g} \otimes A), \quad \text{via } s \otimes a \mapsto \chi_{s,a}.$$

Although this will not be needed in the following, we mention that the centroid of an EALA is known too.

Proposition 5.5.12 *Let E be an EALA, let $K = E_c$ be its core and put $D = E/K$. Then K is a central Lie algebra, and*

$$\mathrm{Cent}_F(E) = F\,\mathrm{Id}_E \oplus \mathcal{V}(K), \quad \mathcal{V}(K) = \{\chi \in \mathrm{Cent}_F(E) : \chi(K) = 0\}.$$

As a vector space, the ideal $\mathcal{V}(K)$ of $\mathrm{Cent}_F(E)$ is canonically isomorphic to the D-module homomorphisms $D \to Z(K)$:

$$\mathcal{V}(K) \cong \mathrm{Hom}_D(D, Z(K)).$$

This is proven in [26, Cor. 4.13]. Observe that the reference to [186, Th.6] in the proof of [26] can now be replaced by the combination of Theorem 5.4.13(c) and Exercise 5.1.27.

5.5.3 Centroidal derivations of Lie tori. In this section L is a centreless Lie torus of type (S, Λ). Regarding L as a Λ-graded Lie algebra, the results of Section 5.5.1 apply and provide us with the subspace

$$\mathcal{D} = \mathcal{D}(L) = \{\partial_\theta : \theta \in \mathrm{D}(\Lambda)\}$$

of degree derivations of L. Moreover, we can apply Lemma 5.5.6(f) and get that $\chi \circ \partial_\theta \equiv \chi\partial_\theta$ is a derivation for any $\chi \in \mathrm{Cent}_F(L)$. We call the elements of

$$\mathrm{CDer}_F(L) = \mathrm{Cent}_F(L)\,\mathcal{D}$$

centroidal derivations. (A notion of centroidal derivations for arbitrary Λ-graded Lie algebras is developed in [187, 4.9].) Recall from Proposition 5.5.8 that $\mathrm{Cent}_F(L) = \bigoplus_{\gamma\in\Gamma} \mathrm{Cent}_F(L)^\gamma$ is a commutative associative torus of type Γ, where Γ is a subgroup of Λ. Since $\mathcal{D}$ consist of degree 0 endomorphisms, $\mathrm{CDer}(L)$ is Γ-graded,

$$\begin{aligned} \mathrm{CDer}_F(L) &= \textstyle\bigoplus_{\gamma\in\Gamma} \mathrm{CDer}_F(L)^\gamma \quad \text{for} \\ \mathrm{CDer}_F(L)^\gamma &= \mathrm{Cent}_F(L)^\gamma\,\mathcal{D} = \mathrm{Cent}_F(L) \cap \mathrm{End}_F(L)^\gamma. \end{aligned} \tag{5.91}$$

It is then easily seen that $\mathrm{CDer}_F(L)$ is a Γ-graded subalgebra of $\mathrm{Der}_F(L)$. For $\chi^\gamma \in \mathrm{Cent}_F(L)^\gamma$, $\chi^\delta \in \mathrm{Cent}_F(L)^\delta$ and $\theta, \psi \in \mathrm{D}(\Lambda)$ the Lie algebra product of $\mathrm{CDer}_F(L)$ is given by the formula

$$[\chi^\gamma\partial_\theta,\, \chi^\delta\partial_\psi] = \chi^\gamma\chi^\delta\,\big(\theta(\delta)\,\partial_\psi - \psi(\gamma)\,\partial_\theta\big). \tag{5.92}$$

Thus, $\mathrm{CDer}_F(L)$ is a generalized Witt algebra, see for example [191, 1.9].

Suppose now that L is an invariant Lie torus, say with respect to the invariant bilinear from $(\cdot|\cdot)$. We can then consider the *skew centroidal derivations*

$$\mathrm{SCDer}_F(L) = \mathrm{SDer}_F(L) \cap \mathrm{CDer}_F(L),$$

defined as the centroidal derivations which are skew-symmetric with respect to $(\cdot|\cdot)$. This is a Γ-graded subalgebra of $\mathrm{CDer}_F(L)$ whose homogenous components are given by

$$\mathrm{SCDer}_F(L)^\gamma = \{\chi^\gamma\partial_\theta : \chi^\gamma \in \mathrm{Cent}_F(L)^\gamma, \theta(\gamma) = 0\}. \tag{5.93}$$

In particular,

$$\mathrm{SCDer}_F(L)^0 = \mathcal{D}$$

is a toral subalgebra of $\mathrm{SCDer}_F(L)$ since $[\partial_\theta, \chi^\delta\partial_\psi] = \theta(\delta)\partial_\psi$ by (5.92). It is also of interest to point out that $[\mathrm{SCDer}_F(L)^\gamma, \mathrm{SCDer}_F(L)^{-\gamma}] = 0$, which implies that $\mathrm{SCDer}_F(L)$ is the semidirect product of the toral subalgebra $\mathcal{D}$ and the ideal spanned by the homogeneous subspaces of non-zero degree,

$$\mathrm{SCDer}_F(L) = \mathcal{D} \ltimes \big(\textstyle\bigoplus_{\gamma\neq 0} \mathrm{SCDer}_F(L)^\gamma\big).$$

For the construction of EALAs, the following theorem is fundamental.

Theorem 5.5.13 ([185, Th. 9]) *Let L be an invariant Lie torus of type (S, Λ) with Λ free of finite rank. Then $\mathrm{Der}_F(L)$ is a semi-direct product,*

$$\mathrm{Der}_F(L) = \mathrm{IDer}_F(L) \rtimes \mathrm{CDer}_F(L), \quad \textit{hence} \qquad (5.94)$$
$$\mathrm{SDer}_F(L) = \mathrm{IDer}_F(L) \rtimes \mathrm{SCDer}_F(L),$$

where $\mathrm{IDer}_F(L)$ denotes the ideal of all inner derivations, see (5.24).

Some remarks on the proof of this theorem follow. By Proposition 5.5.8 the centroid of L is a Laurent polynomial ring. Let K be its field of fractions, a field of rational functions. As a $\mathrm{Cent}_F(L)$-module, L is torsion-free and hence L embeds into the Lie K-algebra

$$\tilde{L} = L \otimes_{\mathrm{Cent}_F(L)} K,$$

the so-called *central closure of* L. If the $\mathrm{Cent}_F(L)$-module L is finitely generated, its central closure is a finite-dimensional central-simple Lie algebra. Hence, in this case $\mathrm{Der}_K(\tilde{L}) = \mathrm{IDer}(\tilde{L})$, from which the theorem easily follows. If however L is not finitely generated as a $\mathrm{Cent}_F(L)$-module, then we know from Theorem 5.5.10 that L is a Lie torus of type A. More precisely, as a consequence of the results in [222] and [191] for type A_1, [30] for type A_2 and Corollary 5.4.27 for type A_l, $l \geq 3$, such a Lie torus is graded-isomorphic to $\mathfrak{sl}_n(\mathbb{F}_\mathfrak{q})$. But in this case the result follows from [29, 2.17, 2.53], [30, Th. 1.40] and [191, Th. 4.11]. We will discuss the special case $L = \mathfrak{g} \otimes F[t_1^{\pm 1}, \dots, t_n^{\pm n}]$ in Example 5.5.14. To avoid any confusion, we note that the splitting (5.94) is not the one proven in [24, Th. 3.12] for arbitrary root-graded Lie algebras.

The importance of the theorem stems from Theorem 5.1.25: It identifies a natural complement of $\mathrm{IDer}(L)$ in $\mathrm{SDer}_F(L)$. Hence, up to graded-isomorphism, any graded covering of L has the form $\mathrm{E}(L, D^{\mathrm{gr}\,*}, \psi_D)$ for a graded subspace $D \subseteq \mathrm{SCDer}_F(L)$. Moreover, since $\mathfrak{D} \subseteq \mathrm{SCDer}_F(L)$, we can require that $D^0 \subseteq \mathfrak{D}$ be not too small and use it to distinguish the homogeneous spaces L^λ by applying Proposition 5.5.3. This will be our approach in Section 5.5.4. But first some examples.

Example 5.5.14 Let $L = \mathfrak{g} \otimes A$ where $\mathfrak{g}$ is a split simple finite-dimensional Lie algebra with root system S and where $A = F[t_1^{\pm 1}, \dots, t_n^{\pm 1}]$ is a Laurent polynomial ring in n variables. This is an invariant Lie torus of type $(S, \mathbb{Z}^n)$, see the Examples 5.4.30 and 5.4.33. We have seen in (5.18) that

$$\mathrm{Der}_F(\mathfrak{g} \otimes A) = \mathrm{IDer}(\mathfrak{g} \otimes A) \oplus (\mathrm{Id}_\mathfrak{g} \otimes \mathrm{Der}_F(A)).$$

The reader has (or should have) determined $\mathrm{Der}_F(A)$ in Exercise 5.1.7: $\mathrm{Der}_F(A) = A\,\mathfrak{D}$ where $\mathfrak{D} = \mathrm{Span}_\mathbb{Z}(\{\partial_i : 1 \leq i \leq n\})$ in the notation of the quoted exercise. But by Exercise 5.5.1, $\mathfrak{D} = \mathfrak{D}(A)$ is also the space of degree derivations of A. Since the Λ-grading of $L = \mathfrak{g} \otimes A$ is concentrated in the factor A, it follows that $\mathrm{Id} \otimes \mathfrak{D}$ is the space of degree derivations of L, whence, by Example 5.5.11, we have

$$\mathrm{CDer}_F(\mathfrak{g} \otimes A) = \mathrm{Id}_\mathfrak{g} \otimes A\,\mathfrak{D} = \mathrm{Id}_\mathfrak{g} \otimes \mathrm{Der}_F(A).$$

Thus, for the invariant Lie torus $\mathfrak{g} \otimes A$ the decomposition (5.18) is the same as the decomposition (5.94)!

We have seen in Example 5.4.30 that L is an invariant Lie torus with respect to the tensor product form $(\cdot|\cdot) = \kappa \otimes \beta$ where κ is the Killing form of $\mathfrak{g}$ and where β is the bilinear form on A defined by $\beta(t^\lambda, t^\mu) = \delta_{\lambda,-\mu}$. It is then easy to identify $\mathrm{SCDer}_F(L)$ using (5.93). In particular, for $n = 1$ we see that $\mathrm{SCDer}(\mathfrak{g} \otimes F[t^{\pm 1}]) =$

Fd, where d is the degree derivation of (5.8). In particular, this together with Theorem 5.1.25 gives a new proof of the theorem, mentioned in Section 5.1.1, that the Lie algebra $\tilde{\mathcal{L}}(\mathfrak{g},\sigma)$ of (5.7) is the universal central extension of the twisted loop algebra $\mathcal{L}(\mathfrak{g},\sigma)$.

5.5.4 The general construction. Finally, we can describe the ingredients (L, D, τ) of the general construction:

- $L = \bigoplus_{\xi\in S,\lambda\in\Lambda} L_\xi^\lambda$ is an invariant Lie torus of type (S,Λ) with Λ a free abelian group of finite rank; we put $\Gamma = \operatorname{supp}_\Lambda \operatorname{Cent}(L)$, see Proposition 5.5.8.

- $D = \bigoplus_{\gamma\in\Gamma} D^\gamma \subseteq \operatorname{SCDer}_F(L)$ is a graded subalgebra such that the evaluation map
$$\operatorname{ev}_{D^0} : \Lambda \to D^{0\,*}, \quad \lambda \to \operatorname{ev}_\lambda|_{D^0} \text{ is injective.} \tag{5.95}$$
- $\tau : D\times D \to D^{\mathrm{gr}\,*}$ is an *affine cocycle*, i.e., τ is a bilinear map satisfying for all $d, d_i \in D$
$$\tau(d,d)=0 \quad\text{and}\quad \textstyle\sum_{\circlearrowleft} d_1\cdot\tau(d_2,d_3) = \sum_{\circlearrowleft}\tau([d_1,d_2],d_3), \tag{5.96}$$
$$\tau(D^0,D)=0, \quad\text{and}\quad \tau(d_1,d_2)(d_3) = \tau(d_2,d_3)(d_1) \tag{5.97}$$

Let us discuss the meaning of the conditions above. Recall from Proposition 5.5.3 that the condition (5.95) implies that D^0 induces the Λ-grading of L, i.e.,
$$L^\lambda = \{l\in L : d^0(l) = \operatorname{ev}_\lambda(d^0)l, \text{ for all } d^0\in D^0\}. \tag{5.98}$$
For example, (5.95) holds for $D = \mathfrak{D} = \operatorname{SCDer}_F(L)^0$ or D any graded subalgebra with $D^0 = \mathfrak{D}$. In (5.96), the symbol $\sum_{\circlearrowleft}$ denotes the cyclic sum: $\sum_{\circlearrowleft} d_1\cdot\tau(d_2,d_3) = d_1\cdot\tau(d_2,d_3)+d_2\cdot\tau(d_3,d_1)+d_3\cdot\tau(d_1,d_2)$ and analogously for $\sum_{\circlearrowleft}\tau([d_1,d_2],d_3)$. Moreover, $d\cdot c$ for $c\in D^{\mathrm{gr}\,*}$ is the contragredient action of D on the graded dual space $D^{\mathrm{gr}\,*}$. The condition (5.96) says that τ is an *abelian* 2*-cocycle*, meaning that $D^{\mathrm{gr}\,*}\oplus D$ is a Lie algebra with respect to the product formula
$$[c_1\oplus d_1,\, c_2\oplus d_2] = \big(d_1\cdot c_2 - d_2\cdot c_1 + \tau(d_1,d_2)\big)\oplus[d_1,d_2] \tag{5.99}$$
for $c_i\in D^{\mathrm{gr}\,*}$ and $d_i\in D$. Thus,
$$0 \longrightarrow D^{\mathrm{gr}\,*} \xrightarrow{\text{inc}} D^{\mathrm{gr}\,*}\oplus D \xrightarrow{\operatorname{pr}_D} D \longrightarrow 0$$
is an abelian extension: $D^{\mathrm{gr}\,*}$ is an abelian ideal, not necessarily contained in the centre. The conditions in (5.97) will allow us to define a toral subalgebra H and an invariant bilinear form $(\cdot|\cdot)$ below. We note that an affine cocycle is necessarily graded of degree 0:
$$\tau(D^\gamma, D^\delta)\subseteq (D^{\mathrm{gr}\,*})^{\gamma+\delta}$$
for $\gamma,\delta\in\Gamma$. Moreover, there do exist non-trivial affine cocycles, see [29, Rem. 3.71] and [91].

To data (L,D,τ) as above we associate a Lie algebra
$$E = L\oplus D^{\mathrm{gr}\,*}\oplus D$$
with product ($l_i\in L$, $c_i\in D^{\mathrm{gr}\,*}$ and $d_i\in D$)
$$\begin{aligned}[l_1\oplus c_1\oplus d_1,\, l_2\oplus c_2\oplus d_2] &= \big([l_1,l_2]_L + d_1(l_2) - d_2(l_1)\big)\\ &\quad\big(\psi_D(l_1,l_2) + d_1\cdot c_2 - d_2\cdot c_1 + \tau(d_1,d_2)\big)\oplus[d_1,d_2].\end{aligned} \tag{5.100}$$

Here $[.,.]_L$ is the Lie algebra product of L, $d_i(l_j)$ is the natural action of D on L, and ψ_D is the central 2-cocycle of (5.25). It is immediate from the product formula that

(i) $L \oplus D^{\mathrm{gr}\,*}$ is an ideal of E, and the canonical projection $L \oplus D^{\mathrm{gr}\,*} \to L$ is a central extension.

(ii) The Lie algebra $D^{\mathrm{gr}\,*} \oplus D$ of (5.99) is a subalgebra of E.

The Lie algebra E has a a subalgebra

$$H = \mathfrak{h} \oplus D^{0\,*} \oplus D^0$$

where $\mathfrak{h} = \mathrm{Span}_F\{h_\xi^\lambda : \xi \in S^\times, \lambda \in \Lambda\} = \mathrm{Span}_F\{h_\xi^0 : 0 \neq \xi \in S_{\mathrm{ind}}\}$ as in (5.72). We embed S into the dual space $\mathfrak{h}^*$, using the evaluation map of (5.95), and extend $\xi \in S \subseteq \mathfrak{h}^*$ to a linear form of H by $\xi(D^{0\,*} \oplus D^0) = 0$. We embed $\Lambda \subseteq D^{0\,*}$, using the evaluation map of Proposition 5.5.3, and then extend $\lambda \in \Lambda \subseteq D^{0\,*}$ to a linear form of H by putting $\lambda(\mathfrak{h} \oplus D^{0\,*}) = 0$. Then H is a toral subalgebra of E with root spaces

$$E_{\xi\oplus\lambda} = \begin{cases} L_\xi^\lambda, & \xi \neq 0, \\ L_0^\lambda \oplus (D^{-\lambda})^* \oplus D^\lambda, & \xi = 0. \end{cases}$$

Observe $H = E_0$ since $\mathfrak{h} = L_0^0$ by Exercise 5.4.10. The symmetric bilinear form $(\cdot|\cdot)$ on E, defined by

$$(l_1 \oplus c_1 \oplus d_1 \mid l_2 \oplus c_2 \oplus d_2) = (l_1 \mid l_2)_L + c_1(d_2) + c_2(d_1),$$

is nondegenerate and invariant. Here $(\cdot|\cdot)_L$ is of course the given bilinear form of the invariant Lie torus L. With respect to this bilinear form the set of roots of (E, H) is $R = R^0 \cup R^{\mathrm{an}}$ where

$$R^0 = \{\lambda \in \Lambda \subseteq H^* : L_0^\lambda \neq 0\} \quad \text{and}$$
$$R^{\mathrm{an}} = \{\xi \oplus \lambda : \xi \neq 0 \text{ and } L_\xi^\lambda \neq 0\}.$$

We have now indicated that the axioms (EA1) and (EA2) of an extended affine Lie algebra holds for the pair (E, H). The verification of the remaining axioms can be easily be done by the reader, or can be looked up in [182, Prop. 5.2.4]. This then shows part (a) of the following theorem.

Theorem 5.5.15 ([186, Th. 6]) (a) *The pair (E, H) constructed above is an extended affine Lie algebra, denoted* $\mathrm{E} = \mathrm{E}(L, D, \tau)$. *Its core is $L \oplus D^{\mathrm{gr}\,*}$ and its centreless core is L.*

(b) *Conversely, let (E, H) be an extended affine Lie algebra, and let $L = E_c/Z(E_c)$ be its centreless core, which by Corollary* 5.4.16 *is an invariant Lie torus, say of type (S, Λ), with Λ free of finite rank.*

Then there exists a subalgebra $D \subseteq \mathrm{SCDer}_F(L)$ and an abelian 2-cocycle τ satisfying the conditions (5.95)–(5.97) *on (D, τ) such that $E \cong \mathrm{E}(L, D, \tau)$.*

We have defined discrete EALAs in Section 5.2.1 as a special class of EALAs over the base field $F = \mathbb{C}$. They can now be characterized as follows.

Corollary 5.5.16 ([186, Th. 8]) *Let $F = \mathbb{C}$.* (a) *Let L be an invariant Lie torus of type (S, Λ) with Λ free of finite rank and let $D \subseteq \mathrm{SCDer}_{\mathbb{C}}(L)$ be a graded subalgebra such that the evaluation map* $\mathrm{ev} : \Lambda \to D^{0\,*}$ *is injective with discrete image. Then,* *for any affine 2-cocycle τ the extended affine Lie algebra* $\mathrm{E}(L, D, \tau)$ *is a discrete EALA. Conversely, any discrete EALA arises in this way.*

Erhard Neher
Department of Mathematics and Statistics
University of Ottawa
Ottawa, ON, Canada K1N 6N5
neher@uottawa.ca

Fields Institute Communications
Volume **59**, 2011

CHAPTER 6

Representations of Affine and Toroidal Lie Algebras

VYJAYANTHI CHARI

Abstract. We discuss the category $\mathcal{I}$ of level zero integrable representations of loop algebras and their generalizations. The category is not semisimple and so one is interested in its homological properties. We begin by looking at some approaches which are used in the study of other well–known non–semisimple categories in the representation theory of Lie algebras. This is done with a view to seeing if and how far these approaches can be made to work for $\mathcal{I}$. In the later sections we focus first on understanding the irreducible level zero modules and later on certain universal modules, the local and global Weyl modules which in many ways play a role similar to the Verma modules in the BGG–category $\mathcal{O}$. In the last section, we discuss the connections with the representation theory of finite–dimensional associative algebras and on some recent work with J. Greenstein.

Introduction

In these notes we discuss aspects of the representation theory of affine Kac–Moody Lie algebras. There is a vast literature on this subject and our focus, for the most part, will be on the level zero integrable representations (which include the finite–dimensional representations) of the loop algebra. At the moment, it appears that this direction is the one most likely to generalize to the case of multi–loop algebras and perhaps to extended affine Lie algebras.

A finite–dimensional representation of the loop algebra need not be completely reducible and thus one encounters problems of the type which usually appear in the representation theory of algebraic groups in positive characteristic or in the BGG–category $\mathcal{O}$ for semisimple Lie algebras. However, many of the methods used in the study of the latter subjects are not available for affine Lie algebras. We have tried to illustrate some of the similarities and the difficulties throughout the notes. Thus, we begin with a brief discussion of some of the basic ideas, tools and important results in the representation theory of semisimple Lie algebras and we restrict our attention to ideas that we will use in the later lectures. We then discuss the integrable representations and the BGG–category $\hat{\mathcal{O}}$ for affine algebras, the integer form of the universal enveloping algebra of the affine Lie algebra and the positive level integrable modules. The remaining sections are devoted to integrable level zero modules, the finite–dimensional representations of the loop algebra and generalizations of these results to multiloop algebras. We conclude the notes with a section on the connections between the category of graded representations of a maximal parabolic subalgebra of the affine Lie algebra and representations of associative finite–dimensional algebras, highest weight categories and quivers.

We have tried to maintain some level of informality, reflecting the fact that these are notes of a summer school. This allows for some repetition and topics sometimes appear before they should since they serve as motivation for future lectures. One is also guided by one's own interests and necessarily, these notes do not include many important directions in the study of representations of affine Lie algebras: as, for instance, the connections with mathematical physics, number theory, vertex algebras and the relations with the monster group. The interested reader will find, however, that the list of references includes a selection of books and papers dealing with these aspects of the subject.

Acknowledgments. The organizers of the summer school did a superb job of identifying and attracting excellent graduate students and postdoctoral scholars, and I thank them for giving me the opportunity to be a part of the school. Partial support from the NSF grant DMS-0901253 is also acknowledged.

6.1 Simple Lie algebras

Much of the material in this section can be found in introductory graduate text books [38]–[40], [53], [124] on Lie algebras and representation theory. Further references to the literature may be found throughout the section although these are far from comprehensive.

6.1.1 Let $\mathfrak{g}$ be a finite–dimensional complex simple Lie algebra of rank n with a fixed Cartan subalgebra $\mathfrak{h}$ and let Φ be the corresponding root system and let $\Pi = \{\alpha_i : i \in I\}$ (where $I = \{1, \cdots, n\}$) be a set of simple roots for Φ. The root lattice Q is the $\mathbb{Z}$–span of the simple roots while Q^+ is the $\mathbb{N}$–span of the simple roots, and the set $\Phi^+ = \Phi \cap Q^+$ is the set of positive roots in Φ. The restriction of the Killing form $\kappa : \mathfrak{g} \times \mathfrak{g} \to \mathbb{C}$ to $\mathfrak{h} \times \mathfrak{h}$ induces a nondegenerate bilinear form $(\ ,\ )$ on $\mathfrak{h}^*$ and we let $\{\omega_1, \cdots, \omega_n\} \subseteq \mathfrak{h}^*$ be the fundamental weights given by $2(\omega_j, \alpha_i) = \delta_{i,j}(\alpha_i, \alpha_i)$. Let P (resp. P^+) be the $\mathbb{Z}$ (resp. $\mathbb{N}$) span of the $\{\omega_i : 1 \le i \le n\}$ and note that $Q \subseteq P$. Given $\lambda, \mu \in P$ we say that $\mu \le \lambda$ iff $\lambda - \mu \in Q^+$. Clearly $\le$ is a partial order on P. The set Φ^+ has a unique maximal element with respect to this order which is denoted by θ and is called the highest root of Φ^+. Set

$$\rho = \sum_{i=1}^{n} \omega_i = \frac{1}{2} \sum_{\alpha \in \Phi^+} \alpha.$$

Let W be the Weyl group of $\mathfrak{g}$ generated by simple refections $\{s_i : i \in I\}$ and let $\ell : W \to \mathbb{N}$ be the length function which assigns to an element $w \in W$, the length of a reduced expression for w as a product of simple reflections.

6.1.2 Given $\alpha \in \Phi$, let $\mathfrak{g}_\alpha$ be the corresponding root space and set $\mathfrak{n}^\pm = \oplus_{\alpha \in \Phi^+} \mathfrak{g}_{\pm\alpha}$. Then $\mathfrak{n}^\pm$ are subalgebras of $\mathfrak{g}$ and we have an isomorphism of vector spaces

$$\mathfrak{g} = \mathfrak{n}^- \oplus \mathfrak{h} \oplus \mathfrak{n}^+. \tag{6.1}$$

For $\alpha \in \Phi^+$, fix elements $x_\alpha^\pm \in \mathfrak{g}_{\pm\alpha}$ and $h_\alpha \in \mathfrak{h}$ such that they span a Lie subalgebra of $\mathfrak{g}$ which is isomorphic to $\mathfrak{sl}_2$, i.e., we have

$$[h_\alpha, x_\alpha^\pm] = \pm 2x_\alpha^\pm, \quad [x_\alpha^+, x_\alpha^-] = h_\alpha,$$

and more generally, assume that the set

$$\{x_\alpha^\pm :, \alpha \in \Phi^+\} \cup \{h_{\alpha_i} : i \in I\},$$

is a Chevalley basis for $\mathfrak{g}$.

6.1.3 For any Lie algebra $\mathfrak{a}$, we let $\mathbf{U}(\mathfrak{a})$ be the universal enveloping algebra of $\mathfrak{a}$. Representations of $\mathfrak{a}$ are the same as the representations of $\mathbf{U}(\mathfrak{a})$ and since $\mathbf{U}(\mathfrak{a})$ is a Hopf algebra, one can define the notions of the trivial representation, the tensor product of representations and the dual of a representation. If $\dim \mathfrak{a} < \infty$, then the algebra $\mathbf{U}(\mathfrak{a})$ is *Noetherian* and this fact plays an important role in the representation theory of the complex semisimple Lie algebras. The Poincare–Birkhoff–Witt theorem implies that we have an isomorphism of vector spaces

$$\mathbf{U}(\mathfrak{g}) \cong \mathbf{U}(\mathfrak{n}^-) \otimes \mathbf{U}(\mathfrak{h}) \otimes \mathbf{U}(\mathfrak{n}^+). \tag{6.2}$$

Another fact that plays a major role in the representation theory of $\mathfrak{g}$ is that the center $\mathcal{Z}$ of $\mathbf{U}(\mathfrak{g})$ is large and one has a good theory of central characters, as we shall see below. We recall here the definition of the *Casimir element* Ω in $\mathcal{Z}$ and we do this in a way that is adapted for use in the affine case. Let $h_1', \cdots, h_n'$ be a basis of $\mathfrak{h}$ which is dual to the basis $h_{\alpha_1}, \cdots, h_{\alpha_n}$ with respect to the Killing form of $\mathfrak{g}$ and set

$$\Omega = \sum_{i=1}^n h_i h_i' + h_\rho + 2 \sum_{\alpha \in \Phi^+} x_\alpha^- x_\alpha^+, \tag{6.3}$$

where $h_\rho \in \mathfrak{h}$ is defined by requiring $\rho(h) = \kappa(h, h_\rho)$.

6.1.4 Recall that a $\mathfrak{g}$–module V is said to be a *weight module* if

$$V = \bigoplus_{\mu \in \mathfrak{h}^*} V_\mu,$$

where $V_\mu = \{v \in V : hv = \mu(h)v \quad h \in \mathfrak{h}\}$. An element $\mu \in \mathfrak{h}^*$ is a *weight* of V if $V_\mu \neq 0$ and we set $\mathrm{wt}(V) = \{\mu \in \mathfrak{h}^* : V_\mu \neq 0\}$.

The *BGG–category* $\mathcal{O}$ is the full subcategory of $\mathfrak{g}$–modules M satisfying the condition that M is a finitely generated weight module which is $\mathfrak{n}^+$–locally finite (for all $m \in M$ we have $\dim \mathbf{U}(\mathfrak{n}^+)m < \infty$) as a $\mathfrak{n}^+$–module. The morphisms in $\mathcal{O}$ are just the $\mathfrak{g}$–module maps. The following is straightforward.

Lemma *The category $\mathcal{O}$ is abelian and all objects of $\mathcal{O}$ are Noetherian modules for $\mathbf{U}(\mathfrak{g})$. Moreover if $M \in \mathcal{O}$ then*

$$\dim M_\mu < \infty \quad \text{for all} \quad \mu \in \mathfrak{h}^*,$$

and there exist finitely many elements $\mu_1, \cdots, \mu_k \in \mathfrak{h}^*$ (depending on M) such that

$$\mathrm{wt}\, M \subseteq \bigcup_{s=1}^k (\mu_s - Q^+).$$

The formal character $\mathrm{ch}(V)$ of a weight module V, encodes the information on the dimensions of the weight spaces, i.e., it is the function $\mathfrak{h}^* \to \mathbb{N}$ which sends λ to $\dim V_\lambda$. This is usually written as follows. Given $\lambda \in \mathfrak{h}^*$ let $e(\lambda) : \mathfrak{h}^* \to \mathbb{Z}$ be defined by $e(\lambda)\lambda = 1$, $e(\lambda)\mu = 0$ if $\mu \neq \lambda$, then

$$\mathrm{ch}(V) = \sum_{\mu \in \mathfrak{h}^*} \dim V_\mu \, e(\mu). \tag{6.4}$$

Formal characters behave well with respect to direct sums and tensor products.

6.1.5 An important family of objects in $\mathcal{O}$ are the highest weight modules. A module M is said to be of *highest weight* λ with *highest weight vector* m if $M = \mathbf{U}(\mathfrak{g})m$ and

$$hm = \lambda(h)m, \qquad x_\alpha^+ m = 0, \qquad h \in \mathfrak{h}, \quad \alpha \in \Phi^+. \tag{6.5}$$

For $\lambda \in \mathfrak{h}^*$ let $M(\lambda) = \mathbf{U}(\mathfrak{g})m_\lambda$ be the universal highest weight module (also called the *Verma module*) with highest weight λ and highest weight vector m_λ, in other words the equation (6.5) gives the defining relations of m_λ. Using (6.2), it is straightforward to show that

$$\mathrm{wt}\, M(\lambda) = \lambda - Q^+, \quad \dim M(\lambda)_\lambda = 1, \quad \dim M(\lambda)_\mu < \infty, \qquad \mu \in \mathfrak{h}^*.$$

In particular, it follows that $M(\lambda)$ has a unique irreducible quotient $V(\lambda)$ and using Lemma 6.1.4, we get

Lemma *Any irreducible object in $\mathcal{O}$ is isomorphic to $V(\lambda)$ for some $\lambda \in \mathfrak{h}^*$.*

6.1.6 The most obvious family of $\mathfrak{g}$–modules which are in $\mathcal{O}$ are of course the finite–dimensional ones and the following result completely describes the corresponding full subcategory of $\mathcal{O}$.

Theorem

(i) *Any finite–dimensional $\mathfrak{g}$–module is isomorphic to a direct sum of irreducible modules, i.e., the full subcategory of $\mathcal{O}$ consisting of finite–dimensional modules is semisimple.*

(ii) *For $\lambda \in \mathfrak{h}^*$, the irreducible module $V(\lambda)$ is finite–dimensional iff $\lambda \in P^+$ and the character of $V(\lambda)$ is given by the Weyl–character formula,*

$$\left(\sum_{w \in W} (-1)^{\ell(w)} e(w\rho)\right) ch(V(\lambda)) = \sum_{w \in W} (-1)^{\ell(w)} e(w(\lambda + \rho)). \tag{6.6}$$

(iii) *For $\lambda \in P^+$, the module $V(\lambda)$ is generated by an element v_λ with defining relations:*

$$hv_\lambda = \lambda(h)v_\lambda, \quad x_{\alpha_i}^+ v_\lambda = 0, \quad (x_{\alpha_i}^-)^{\lambda(h_{\alpha_i})+1} v_\lambda = 0,$$

where $h \in \mathfrak{h}$ and $1 \leq i \leq n$.

We shall say that a $\mathfrak{g}$–module V is *locally finite–dimensional* if it isomorphic to a sum of finite–dimensional modules. The preceding theorem implies that V is isomorphic to a direct sum of modules $V(\lambda)$, $\lambda \in P^+$. If in addition, we have

$$\dim \mathrm{Hom}_{\mathfrak{g}}(V(\lambda), V) < \infty \quad \lambda \in P^+,$$

then we can define a function $P^+ \to \mathbb{N}$ which maps λ to $\dim \mathrm{Hom}_{\mathfrak{g}}(V(\lambda), V)$. We denote this function by $\mathrm{ch}_{\mathfrak{g}}(V)$ and observe that it encodes the multiplicity of an irreducible representation in V (whereas $\mathrm{ch}(V)$ encodes the dimension of a weight space). Clearly, we can write,

$$\mathrm{ch}_{\mathfrak{g}}(V) = \sum_{\lambda \in P^+} \dim \mathrm{Hom}_{\mathfrak{g}}(V(\lambda), V) e(\lambda), \tag{6.7}$$

where $e(\lambda)$ is the function (or rather its restriction to P^+) defined in Section 6.1.4.

6.1.7 The category $\mathcal{O}$ itself is not semisimple, for instance the Verma–modules are indecomposable but not always irreducible and we give an example of this, in the case of $\mathfrak{sl}_2$. Let x, y, h be the standard basis of $\mathfrak{sl}_2$. Since $\mathfrak{h}$ is one–dimensional, we identify the set P^+ with $\mathbb{N}$ and consider the Verma–module $M(0)$ and its highest weight vector m_0. It is straightforward to see that the element $ym_0 \in M(0)$ generates a highest weight submodule isomorphic to $M(-2)$ and one has the following non–split short exact sequence of $\mathfrak{sl}_2$–modules:

$$0 \to M(-2) \to M(0) \to \mathbb{C} \to 0. \tag{6.8}$$

This lack of semi–simplicity has resulted in the development of many approaches designed to understand the structure of $\mathcal{O}$ and we briefly sketch a few directions. The interested reader could refer to [127] and to the references in that book. In the later sections of these notes, we shall see that some of these approaches can or may be adapted to work for affine Lie algebras while others will clearly have no parallel for infinite–dimensional Lie algebras.

6.1.8 We start with the theory of central characters as an example of something that is very powerful in the study of $\mathcal{O}$ but is essentially not available for affine Lie algebras. A *central character* is a homomorphism of algebras $\chi : \mathcal{Z} \to \mathbb{C}$ (where $\mathcal{Z}$ is the center of $\mathbf{U}(\mathfrak{g})$) and a $\mathfrak{g}$–module M is said to *admit a central character* χ if:

$$zm = \chi(z)m, \quad z \in \mathcal{Z}, \quad m \in M.$$

If $z \in \mathcal{Z}$, one can prove without too much difficulty that there exists a unique element $\beta(z) \in \mathbf{U}(\mathfrak{h})$ such that $z - \beta(z) \in \mathfrak{n}^- \mathbf{U}(\mathfrak{g}) \mathfrak{n}^+$. The corresponding map $\beta : \mathcal{Z} \to \mathbf{U}(\mathfrak{h})$ is an algebra homomorphism and is called the *Harish–Chandra homomorphism*. Suppose now that $\lambda \in \mathfrak{h}^*$ and let $m \in M(\lambda)$, say $m = gm_\lambda$ for some $g \in \mathbf{U}(\mathfrak{g})$. One has

$$zm = gzm_\lambda = \lambda(\beta(z))gm_\lambda,$$

where $\lambda \in \mathfrak{h}^*$ is extended to an algebra homomorphism also denoted $\lambda : \mathbf{U}(\mathfrak{h}) \to \mathbb{C}$. It is clear that the composite map $\chi_\lambda : \mathcal{Z} \to \mathbb{C}$ sending $z \to \lambda(\beta(z))$ is an algebra homomorphism and hence $M(\lambda)$ admits a central character.

It is now natural to ask if the χ_λ are all distinct and if any homomorphism from $\mathcal{Z} \to \mathbb{C}$ is of the form χ_λ for some $\lambda \in \mathfrak{h}^*$. The answer to the first question is quite clearly no. Consider the example given in (6.8). Since $M(0)$ admits a central character any submodule or quotient of $M(0)$ also has the same central character and hence we get in this case that $\chi_0 = \chi_{-2}$. It is a theorem of Harish–Chandra, that for $\lambda, \mu \in \mathfrak{h}^*$:

$$\chi_\lambda = \chi_\mu \iff \lambda + \rho = w(\mu + \rho) \text{ for some } w \in W, \tag{6.9}$$

and also that the answer to the second question above is yes. As a consequence of this theorem one can show that:

Theorem *Any module $M \in \mathrm{Ob}\,\mathcal{O}$ is Artinian. In particular, M has a Jordan–Hölder series and can be written uniquely (up to isomorphism and re-indexing) as a direct sum of indecomposable modules.*

For $M \in \mathcal{O}$ and $\mu \in \mathfrak{h}^*$ denote by $[M : V(\mu)]$ the multiplicity of $V(\mu)$ in the Jordan–Hölder series of M. The problem of determining the multiplicities of the composition factors of the Verma module is a very difficult one and the answer is given by the famous Kazhdan–Lusztig conjecture [146] in terms of the Kazhdan–Lusztig (KL) polynomials. The conjecture is now a theorem due to Beilinson–Bernstein [21] and independently Brylinski–Kashiwara [35]. The KL–polynomials are defined recursively and explicit computations using the definition are formidable. There is extensive literature devoted to understanding the combinatorics of the Kazhdan–Lusztig polynomials and to developing programs to compute the polynomials.

6.1.9 It is natural to consider the homological properties of $\mathcal{O}$ by asking questions such as: does $\mathcal{O}$ contain enough projectives? In fact it does and one can determine all indecomposable projective objects in $\mathcal{O}$. These are again indexed by elements of $\mathfrak{h}^*$ and we denote by $P(\lambda)$ the corresponding indecomposable projective module. The module $P(\lambda)$ has a Verma flag which is a filtration in which the subsequent quotients are Verma–modules. As in the case of Jordan–Hölder series of a module, the length of any two filtrations of $P(\lambda)$ by Verma modules is the same and so is the multiplicity of a Verma module $M(\mu)$ in any two filtrations.

And this allows us now to state the fundamental result proved by Bernstein–Gelfand and Gelfand:

Theorem *For $\lambda, \mu \in \mathfrak{h}^*$ we have*

$$[P(\lambda) : M(\mu)] = [M(\mu) : V(\lambda)],$$

where $[P(\lambda) : M(\mu)]$ is the multiplicity of $M(\mu)$ in a Verma flag of $P(\lambda)$ and $[M(\mu) : V(\lambda)]$ is the multiplicity of $V(\mu)$ in the Jordan–Hölder series of $M(\mu)$.

Much of the work on $\mathcal{O}$ including the Kazhdan–Lusztig conjecture was stimulated by this result.

6.1.10 One could try to understand $\mathcal{O}$ by looking at smaller, more manageable subcategories of $\mathcal{O}$ and this is done as follows. It is not true that the center always acts on objects (even indecomposable ones) of $\mathcal{O}$ via a central character. However, since $\mathcal{Z}$ is commutative and we are working over the complex numbers it is true that any $M \in \mathrm{Ob}\,\mathcal{O}$ can be written as a direct sum of generalized eigenspaces for the action of $\mathcal{Z}$. Hence if M is indecomposable then there exists an algebra homomorphism $\chi : \mathcal{Z} \to \mathbb{C}$ such that

$$M = M^{\chi} = \{m \in M : (z - \chi(z))^k m = 0 \text{ for some } k = k(m) \in \mathbb{N}\}.$$

Moreover in this case, if $V(\mu)$ occurs in a Jordan–Hölder series for M then $\chi = \chi_\mu$. This motivates the following definition. Let $\mathcal{O}_\chi$ be the full subcategory of M consisting of $M \in \mathrm{Ob}\,\mathcal{O}$ such that $M = M^{\chi}$. Using (6.9) we see that $\mathcal{O}_\chi$ has only finitely many simple objects $V(\mu)$ with $\chi_\mu = \chi$ and it suffices to understand $\mathcal{O}_\chi$. The endomorphism algebra of the projective generator of $\mathcal{O}_\chi$ is a finite–dimensional associative algebra $\mathcal{A}_\chi$ and its (left)–module category is equivalent to $\mathcal{O}_\chi$. The category $\mathcal{O}_\chi$ fits into the axiomatic framework of highest weight categories developed by Cline, Parshall and Scott, [79] and $\mathcal{A}_\chi$ is quasi–hereditary. The algebra $\mathcal{A}_\chi$ is given by a quiver with relations although it is in general very hard to compute these relations, see however [212] where an algorithm to compute these relations is given. Also, under suitable conditions [22], [211], the algebra $\mathcal{A}_\chi$ has a Koszul grading and hence the homological properties of $\mathcal{O}_\chi$ can also be understood from this viewpoint.

6.1.11 Before we leave the realm of finite–dimensional simple Lie algebras, we discuss the Kostant $\mathbb{Z}$–form of $\mathbf{U}(\mathfrak{g})$ and the passage to positive characteristic. Given any associative algebra A over $\mathbb{C}$ and elements $a \in A$ and $r \in \mathbb{N}$, the rth–divided power of a is the element $a^{(r)} = a^r/r!$. Let $\mathbf{U}_\mathbb{Z}(\mathfrak{g})$ be the $\mathbb{Z}$–subalgebra of $\mathbf{U}(\mathfrak{g})$ generated by the elements $(x_\alpha^\pm)^{(r)}$, $\alpha \in \Phi^+$, $r \in \mathbb{N}$ and define $\mathbf{U}_\mathbb{Z}(\mathfrak{n}^\pm)$ in the obvious way. To define the analog of the Cartan subalgebra, one does not do the obvious and take the divided powers of the h_i, $1 \le i \le n$. Instead, one is guided by the following formula which rewrites the product $(x_\alpha^+)^{(s)}(x_\alpha^-)^{(r)}$ in Poincare–Birkhoff–Witt order as in (6.2),

$$(x_\alpha^+)^{(s)}(x_\alpha^-)^{(r)} = \sum_{m=0}^{\min(r,s)} (x_\alpha^-)^{(r-m)} \binom{h_\alpha - r - s + 2m}{m} (x_\alpha^+)^{(s-m)}, \tag{6.10}$$

where for any $m \in \mathbb{N}$ and $h \in \mathfrak{h}$, we set

$$\binom{h}{m} = \frac{h(h-1)\cdots(h-m+1)}{m!}.$$

We then define $\mathbf{U}_\mathbb{Z}(\mathfrak{h})$ to be the $\mathbb{Z}$–subalgebra generated by the elements $\binom{h_{\alpha_i}}{m}$, $1 \le i \le n$, $m \in \mathbb{N}$. It is a theorem of Kostant that $\mathbf{U}_\mathbb{Z}(\mathfrak{g})$ has a PBW–basis, by which we mean a $\mathbb{Z}$–basis of $\mathbf{U}_\mathbb{Z}(\mathfrak{h})$ consisting of all ordered monomials from the set

$$\{(x_\alpha^\pm)^{(r)}, \binom{h_i}{r} : \alpha \in \Phi^+, \quad i \in I, r \in \mathbb{N}\}.$$

This means that $\mathbf{U}_\mathbb{Z}(\mathfrak{g})$ is a $\mathbb{Z}$–lattice in $\mathbf{U}(\mathfrak{g})$ and moreover one can prove that if $\lambda \in P^+$, and we set

$$V_\mathbb{Z}(\lambda) = \mathbf{U}_\mathbb{Z}(\mathfrak{g})v_\lambda,$$

then $V_\mathbb{Z}(\lambda)$ is also a $\mathbb{Z}$–lattice in $V(\lambda)$. Let $\mathbb{F}$ be a field of characteristic p and regard $\mathbb{F}$ as a left module for $\mathbb{Z}$. Set

$$\mathbf{U}_\mathbb{F}(\mathfrak{g}) = \mathbf{U}_\mathbb{Z}(\mathfrak{g}) \otimes_\mathbb{Z} \mathbb{F}, \qquad V_\mathbb{F}(\lambda) = V_\mathbb{Z}(\lambda) \otimes_\mathbb{Z} \mathbb{F}.$$

Clearly, $\mathbf{U}_\mathbb{F}(\mathfrak{g})$ is an associative algebra over $\mathbb{F}$ (called the *hyperalgebra* of $\mathfrak{g}$ over $\mathbb{F}$) and $V_\mathbb{F}(\lambda)$ is an indecomposable (usually reducible) module for this algebra, and is called the Weyl module of weight λ. There is extensive literature on the study of finite–dimensional representations of the hyperalgebra. One can talk about weight modules (whose definition requires a natural modification) and characters of these modules and we summarize the relevant points for our next sections in the following theorem.

Theorem (i) *Given $\lambda \in P^+$, the module $V_\mathbb{F}(\lambda)$ has a unique irreducible quotient and any irreducible finite–dimensional module of $\mathbf{U}_\mathbb{F}(\mathfrak{g})$ is isomorphic to such a quotient for some $\lambda \in P^+$.*

(ii) *The character of $V_\mathbb{F}(\lambda)$ is given by the Weyl character formula.*

(iii) *The module $V_{\mathbb{F}}(\lambda)$ is generated by the element $v_\lambda \otimes 1$ and the defining relations are:*

$$(x_\alpha^+)^{(r)}(v_\lambda \otimes 1) = 0, \quad \binom{h_i}{r}(v_\lambda \otimes 1) = \binom{\lambda(h_i)}{r}(v_\lambda \otimes 1), \quad (x_\alpha^-)^{(s)}(v_\lambda \otimes 1) = 0,$$

for all $\alpha \in \Phi^+$, $i \in I$, $r, s \in \mathbb{N}$ with $s \geq \lambda(h_i) + 1$.

Note that the fact that the modules $V_{\mathbb{F}}(\lambda)$ are indecomposable but not irreducible implies that the category of finite–dimensional representations of the hyperalgebra is also not semisimple and many of the problems and methods discussed for the category $\mathcal{O}$ are also studied and used in this case.

6.2 Affine Lie algebras

We now turn our attention to the case of the best understood Kac–Moody Lie algebras: the affine Lie algebras. There are two definitions available for these algebras: one is via a finite set of generators and relations and the other is very explicit as we shall see below. It is the interplay between these two definitions that makes the study of these algebras and their representations tractable. The explicit realization is what we need for our purposes and so we refer the reader to [53], [136] for the presentation via Chevalley generators and relations. Most of the results discussed in this section can also be found in those two books. We use freely the notation of the first section.

6.2.1 Let t be an indeterminate and $\mathbb{C}[t, t^{-1}]$ the algebra of Laurent polynomials in t with the coefficients in $\mathbb{C}$. For any complex Lie algebra $\mathfrak{a}$, the loop algebra $L(\mathfrak{a})$ is defined by,

$$L(\mathfrak{a}) = \mathfrak{a} \otimes \mathbb{C}[t, t^{-1}]$$

with Lie bracket,

$$[a \otimes f(t), b \otimes g(t)] = [a, b] \otimes f(t)g(t), \quad a, b \in \mathfrak{a}, \quad f, g \in \mathbb{C}[t, t^{-1}].$$

The algebra $\mathfrak{a}$ is identified with the subalgebra $\mathfrak{a} \otimes 1$ and we shall assume this from now on without further mention. Given an automorphism σ of $\mathfrak{a}$ of order m and ζ a fixed primitive m^{th} root of unity, we can write,

$$\mathfrak{a} = \bigoplus_{s=0}^{m-1} \mathfrak{a}_s, \qquad \mathfrak{a}_s = \{a \in \mathfrak{a} : \sigma(a) = \zeta^s a\},$$

and we set

$$L(\mathfrak{a}, \sigma, m) = \bigoplus_{s=0}^{m-1} \mathfrak{a}_s \otimes t^s \mathbb{C}[t^m, t^{-m}].$$

It is a simple matter to check that $L(\mathfrak{a}, \sigma, m)$ is a Lie subalgebra of $L(\mathfrak{a})$ and it is a proper subalgebra if and only if σ is not the identity automorphism of $\mathfrak{a}$, and in this case, we call $L(\mathfrak{a}, \sigma, m)$ a twisted loop algebra.

6.2.2 The Lie algebra $L(\mathfrak{g})$ admits a unique (up to an isomorphism) non–trivial central extension which we denote by $\tilde{L}(\mathfrak{g})$. It can be constructed as follows: set

$$\tilde{L}(\mathfrak{g}) = L(\mathfrak{g}) \oplus \mathbb{C}c$$

and define the Lie bracket by requiring c to be central and by setting

$$[x \otimes t^p, y \otimes t^r] = [x, y] \otimes t^{p+r} + p\delta_{p+r,0}\kappa(x, y)c, \quad x, y \in \mathfrak{g}, \quad p, r \in \mathbb{Z},$$

where $\delta_{p+r,0}$ is the Kronecker δ symbol and recall that κ is the Killing form of $\mathfrak{g}$. The derivation $d = td/dt$ of $\mathbb{C}[t, t^{-1}]$ acts on $L(\mathfrak{g})$ and on $\tilde{L}(\mathfrak{g})$ by

$$d(x \otimes t^r) = rx \otimes t^r, \quad [d, c] = 0.$$

The semi–direct product Lie algebra

$$\hat{L}(\mathfrak{g}) = \tilde{L}(\mathfrak{g}) \oplus \mathbb{C}d,$$

is called the untwisted affine Lie algebra and $\mathfrak{g}$ is naturally isomorphic to a subalgebra of $\hat{L}(\mathfrak{g})$. If σ is a Dynkin diagram automorphism of $\mathfrak{g}$, then $\tilde{L}(\mathfrak{g}, \sigma, m)$ and $\hat{L}(\mathfrak{g}, \sigma, m)$ are defined in the obvious way and are Lie subalgebras of $\tilde{L}(\mathfrak{g})$ and $\hat{L}(\mathfrak{g})$ respectively. The algebra $\hat{L}(\mathfrak{g}, \sigma, m)$ is called a *twisted affine Lie algebra*.

For the most part, we will deal only with the untwisted affine Lie algebras and we will refer to these simply as affine Lie algebras. We will usually indicate if the results we discuss are known for the twisted affine algebras.

6.2.3 The affine Lie algebra comes equipped with a naturally defined root system and a set of simple roots. Set $\hat{\mathfrak{h}} = \mathfrak{h} \oplus \mathbb{C}c \oplus \mathbb{C}d$ and define $\delta \in \hat{\mathfrak{h}}^*$ by $\delta(d) = 1$, $\delta(\mathfrak{h} \oplus \mathbb{C}c) = 0$. Then $\hat{\mathfrak{h}}$ is abelian and given $\mu \in \mathfrak{h}^*$ we extend μ to an element of $\hat{\mathfrak{h}}^*$ by requiring

$$\mu(d) = 0, \qquad \mu(c) = 0,$$

and by abuse of notation, we continue to denote this element by μ. The adjoint action of $\hat{\mathfrak{h}}$ on $\hat{L}(\mathfrak{g})$ is semisimple and the non–zero eigenvalues (the set of roots) are,

$$\hat{\Phi} = \{\alpha + r\delta : \alpha \in \Phi, r \in \mathbb{Z}\} \bigsqcup \{r\delta : r \in \mathbb{Z}, \quad r \neq 0\}.$$

The corresponding eigenspaces (root spaces) are $\mathfrak{g}_\alpha \otimes t^s$, $\mathfrak{h} \otimes t^r$ where $\alpha \in \Phi$, $r, s \in \mathbb{Z}$ and $r \neq 0$. Notice that the eigenspaces corresponding to $r\delta$ are of dimension $\dim \mathfrak{h}$ and hence bigger than one in general. Setting

$$\hat{\Pi} = \{\alpha_i : 0 \leq i \leq n\}, \quad \alpha_0 = -\theta + \delta,$$

we find that any element of $\hat{\Phi}$ can be written uniquely as an integer linear combination of elements of $\hat{\Pi}$ where the coefficients are all either non–negative or non–positive. Elements of the set $\hat{\Pi}$ are called simple roots and we have

$$\hat{\Phi}^+ = \{\pm\alpha + r\delta : r \in \mathbb{N}_+, \quad \alpha \in \Phi\} \quad \sqcup \Phi^+ \quad \sqcup \{r\delta : \ r \in \mathbb{N}_+\}.$$

The subalgebras $\hat{\mathfrak{n}}^\pm$ are defined in the natural way and one has triangular decompositions similar to the ones given in (6.1) and (6.2) for $\mathfrak{g}$. Set

$$x_\theta^\mp \otimes t^{\pm 1} = x_{\alpha_0}^\pm, \qquad h_0 = c - h_\theta.$$

The elements $x_{\alpha_i}^\pm$, h_{α_i} $0 \leq i \leq n$ are called the Chevalley generators of $\hat{L}(\mathfrak{g})$ and an abstract presentation of the Lie algebra in terms of these generators can be given. The fundamental weights Λ_i, $0 \leq i \leq n$ are elements of $\hat{\mathfrak{h}}^*$ defined by

$\Lambda_i(h_{\alpha_j}) = \delta_{i,j}$. The notion of root lattice $\hat{Q}$, weight lattice $\hat{P}$, weight modules, weight spaces, highest weight modules and so on, have obvious analogs in the affine case, one just replaces $\mathfrak{h}$ by $\hat{\mathfrak{h}}$ and so on. The Killing form of $\hat{L}(\mathfrak{g})$ cannot be defined in the usual way since $\hat{L}(\mathfrak{g})$ is infinite–dimensional. However, the formulae

$$< x \otimes t^r, y \otimes t^s >= \kappa(x,y)\delta_{r+s,0}, \quad < x \otimes t^r, \mathbb{C}c \oplus \mathbb{C}d >= 0,$$
$$< c, c >=< d, d >= 0, \quad < c, d >= 1,$$

define an invariant symmetric nondegenerate bilinear form on $\hat{L}(\mathfrak{g})$. The affine Weyl group $\hat{W}$ is the subgroup of $\mathrm{Aut}(\mathfrak{h}^* \oplus \mathbb{C}\delta)$ generated by the simple reflections $\{s_i : i = 0, \cdots, n\}$ and is an infinite Coxeter group. As for finite–dimensional simple Lie algebras, the length of an element w in $\hat{W}$ is just the number of positive roots which are turned negative by w: in particular (unlike in the situation for finite–dimensional simple Lie algebras) there does not exist an element in $\hat{W}$ which maps $\hat{\Phi}^+$ to $\hat{\Phi}^-$.

The many parallels with the structure of $\mathfrak{g}$ suggest that the study of the representation theory of $\hat{L}(\mathfrak{g})$ should proceed in the same way as the study of representations of $\mathfrak{g}$. However, the differences noted above, together with the observation that $\mathbf{U}(\hat{L}(\mathfrak{g}))$ is not Noetherian means that there are significant differences and difficulties in the representation theory of $\hat{L}(\mathfrak{g})$ and we shall see some of these in the rest of the notes.

6.2.4 Before defining the category $\hat{\mathcal{O}}$, we note the following result proved in [64]:

Proposition *The center of* $\mathbf{U}(\hat{L}(\mathfrak{g}))$ *is the polynomial algebra generated by the center* $\mathbb{C}c$ *of* $\hat{L}(\mathfrak{g})$.

This proposition means that unlike Section 6.2, there is no good theory of central characters available for the study of representations of affine Lie algebras. However, many of the results discussed for $\mathcal{O}$ can be proved for $\hat{\mathcal{O}}$ and this is essentially because it is possible to define the analog of the Casimir operator. Recall that the definition given in (6.3) of the Casimir, only used dual bases with respect to the Killing form of $\mathfrak{g}$. This formula can be used verbatim for $\hat{L}(\mathfrak{g})$ by using the symmetric form defined above. However the sum is infinite and the Casimir operator lives in some completion of $\mathbf{U}(\hat{L}(\mathfrak{g}))$ and this means that it does not operate on all $\hat{L}(\mathfrak{g})$–modules. But it does act on modules in the category $\hat{\mathcal{O}}$ whose definition we now recall.

6.2.5 We say that an $\hat{L}(\mathfrak{g})$–module M is an object of $\hat{\mathcal{O}}$ if it is a weight module, i.e., $\hat{\mathfrak{h}}$ acts semi–simply, $M = \oplus_{\mu \in \hat{\mathfrak{h}}^*} M_\mu$, and

$$\dim M_\mu < \infty, \quad \mathrm{wt}\, M \subseteq \bigcup_{s=1}^{p} \mu_s - \hat{Q}^+, \tag{6.11}$$

for some $p \in \mathbb{N}$ and elements $\mu_1, \cdots, \mu_p \in \hat{\mathfrak{h}}^*$. This definition differs from that of $\mathcal{O}$, in two ways. The first is that one does not require the module to be finitely generated. This is because one wants $\mathcal{O}$ to be closed under taking submodules and quotients and since $\mathbf{U}(\hat{L}(\mathfrak{g}))$ is not Noetherian a submodule of a finitely generated module need not be finitely generated. The condition that $\hat{\mathfrak{n}}^+$ acts locally finitely on M does not have all the implications it does in the category $\mathcal{O}$ (see Lemma 6.1.4) and hence is replaced by the stronger conditions in (6.11). Moreover, notice also that the second condition in (6.11) implies that if $\mu \in \mathrm{wt}\, M$, then $\mu + \beta \in \mathrm{wt}\, M$ only for finitely many $\beta \in \hat{\Phi}^+$. It is now immediate that the Casimir operator acts in a well–defined way on every $M \in \hat{\mathcal{O}}$ and moreover one can prove that it commutes with the action of $\hat{L}(\mathfrak{g})$.

6.2.6 For $\lambda \in \hat{\mathfrak{h}}^*$ we let $\hat{M}(\lambda)$ be the Verma module with highest weight λ defined (with obvious modifications) in the same way as the Verma modules for simple Lie algebras given in Section 6.1.5. It has a unique irreducible quotient $\hat{V}(\lambda)$ and any irreducible module in $\hat{\mathcal{O}}$ is isomorphic to one of these and these results are proved in the same way as those for simple Lie algebras.

The following is a fairly simple exercise.

Lemma *Let* $M \in \hat{\mathcal{O}}$ *and assume that* $\dim M < \infty$. *Then* $(L(\mathfrak{g}) \oplus \mathbb{C}c)M = 0$.

In other words, there are no interesting finite dimensional modules in $\hat{\mathcal{O}}$ and in view of Theorem 6.1.6 this raises the natural question, are irreducible modules associated to $\lambda \in \hat{P}^+$ distinguished in some way? The answer is yes and this leads us to the definition of an integrable module for an affine Lie algebra.

6.2.7 A weight module $\hat{V}$ for $\hat{L}(\mathfrak{g})$ is said to be *integrable* if the elements $x^{\pm}_{\alpha_i}$, $0 \le i \le n$ act locally nilpotently on $\hat{V}$. Integrable modules share many of the properties of finite–dimensional $\mathfrak{g}$ modules. For instance a standard $\mathfrak{sl}_2$–argument shows that if V is integrable then $\mathrm{wt}\, V \subseteq \hat{P}$ and also that $\mathrm{wt}\, V$ is $\hat{W}$–invariant. (Of course, one could have also defined integrable modules for $\mathfrak{g}$ but the definition is not interesting since one can prove that any integrable $\mathfrak{g}$–module is a sum of finite–dimensional $\mathfrak{g}$–modules). Let $\mathrm{ch}\hat{V}(\lambda)$ be the formal character of $\hat{V}(\lambda)$ as in (6.4). The following theorem explains the analogy between finite–dimensional modules for $\mathfrak{g}$ and integrable modules in $\hat{\mathcal{O}}$.

Theorem *For* $\lambda \in \hat{P}^+$, *the module* $\hat{V}(\lambda)$ *is integrable and the full subcategory of* $\hat{\mathcal{O}}$ *consisting of integrable modules is completely reducible. Finally,*

$$\left(\sum_{w \in \hat{W}} (-1)^{\ell(w)} e(w\rho)\right) ch(\hat{V}(\lambda)) = \sum_{w \in \hat{W}} (-1)^{\ell(w)} e(w(\lambda + \rho)). \tag{6.12}$$

The formula (6.12) is called the Weyl–Kac character formula and the Casimir operator $\hat{\Omega}$ is an adequate substitute for the role played by the center in proving the Weyl character formula for a simple Lie algebra. The sums that appear in the Weyl–Kac character formula are infinite. This formula can be specialized in many ways and leads to interesting number–theoretic identities, including combinatorial identities of Macdonald and the classical Jacobi triple product identity.

6.2.8 Suppose that V is a weight module for $\hat{L}(\mathfrak{g})$, in which case we have

$$V = \oplus_{a \in \mathbb{C}} V^a, \quad V^a = \{v \in V : cv = av\},$$

and V^a is a $\hat{L}(\mathfrak{g})$–submodule. We say that V^a is a module of *level* a.

Lemma *For* $\lambda \in \hat{\mathfrak{h}}^*$, *the module* $\hat{V}(\lambda)$ *has level* $\lambda(h_\theta + h_0)$. *Hence every irreducible integrable module in* $\hat{\mathcal{O}}$ *has nonnegative integer level. The only level zero irreducible modules in* $\hat{\mathcal{O}}$ *are one-dimensional and correspond to taking* λ *to be a scalar multiple of* δ.

Thus all the affine Lie algebras have a canonical integrable representation of level one corresponding to the weight Λ_0 and this is sometimes called the basic representation of $\hat{L}(\mathfrak{g})$. It is the simplest integrable representation in $\hat{\mathcal{O}}$ and the character formula given in Section 6.2.7 can be made explicit. The theory of vertex algebras, the relationship with physics and the connections with the monster group (see [36], [101], [102], [136], for instance) all have their roots in the effort to understand and explicitly construct the basic representation of the affine Lie algebra.

To conclude this section, we note that many of the problems studied for $\mathcal{O}$ have also been studied for $\hat{\mathcal{O}}$ and one does have a block decomposition [83] and a Kazhdan–Lusztig theory (see [127] for references to the literature on these topics). But the fact that the category $\hat{\mathcal{O}}$ is neither Noetherian nor Artinian does make things much more complicated. In the remaining sections our focus will be only on the integrable representations of affine Lie algebras. Finally, note that the results of this section are also known for the twisted affine algebras.

6.3 Affine Lie algebras integrable representations and integral forms

The reader has noticed by now that there are integrable representations of $\hat{L}(\mathfrak{g})$ which are not in $\hat{\mathcal{O}}$. The most obvious representation of a Lie algebra is the adjoint representation, and it is a simple matter to see from Section 6.2.2 that the adjoint representation of $\hat{L}(\mathfrak{g})$ is integrable but not in $\hat{\mathcal{O}}$. This raises the problem of classifying the irreducible integrable representations of affine Lie algebras, rather than just the ones in $\hat{\mathcal{O}}$. We address this problem in the first part of this section.

The adjoint representation is an example of a level zero representation which is indecomposable and reducible, since the center of $\hat{L}(\mathfrak{g})$ is a proper non–split submodule under the adjoint action. This shows that the category of level zero integrable representations of $\hat{L}(\mathfrak{g})$ is not semisimple and hence should have interesting homological properties. To study these properties and to pursue the directions of study which have been fruitful in the case of simple Lie algebras, it is helpful to understand (following H. Garland) the integral form of the universal enveloping algebra of $\hat{L}(\mathfrak{g})$. Garland also proved that the integrable modules in $\hat{\mathcal{O}}$ have an integral form, but relatively little is known about the corresponding representations of the hyperalgebra, except in the case of the basic representation which we discuss briefly.

6.3.1 We begin by making some comments about duality. Let $w_0 \in W$ be the longest element of the Weyl group of $\mathfrak{g}$. It is well–known that the dual of an irreducible finite–dimensional representation of $\mathfrak{g}$ with highest weight λ is a highest weight representation with highest weight $-w_0\lambda$. Let us now consider the situation of irreducible integrable modules in $\hat{\mathcal{O}}$. The first difficulty one encounters, is that the module $\hat{V}(\lambda)$ is infinite–dimensional and the dual space is too big. The weight spaces, however, are finite–dimensional and so one works instead with the restricted dual

$$\hat{V}(\lambda)^{\#} = \bigoplus_{\mu \in \hat{P}^{+}} V(\lambda)^{*}_{\mu},$$

which is an integrable $\hat{L}(\mathfrak{g})$–module. Since $\hat{W}$ is an infinite group, one does not have the analog of the longest element: there does not exist $w \in \hat{W}$ such that $w\hat{\Pi} = -\hat{\Pi}$ and this means that $\hat{V}(\lambda)^{\#}$ is not a highest weight module. However, it is easily seen that it is a lowest weight module, which is defined in the obvious way, by replacing $\hat{\mathfrak{n}}^+$ by $\hat{\mathfrak{n}}^-$. The lowest weight space is $V(\lambda)^*_\lambda$ and the lowest weight is $-\lambda$. But apart from this modification, it is clear that understanding highest weight modules is the same as understanding lowest weight modules.

6.3.2 The category of all integrable modules for $\hat{L}(\mathfrak{g})$ is very big, even if we restrict to a fixed level. To make the study more manageable we let $\mathcal{I}$ be the category whose objects are integrable $\hat{L}(\mathfrak{g})$–modules with finite–dimensional weight spaces. Our goal is to classify the irreducible objects in $\mathcal{I}$, and in the previous section we saw how to construct irreducible objects of $\mathcal{I}$ which are also in $\hat{\mathcal{O}}$. But so far we have only encountered one interesting example (see Lemma 6.2.8) of a level zero representation, the adjoint, and that was reducible. Our first task then, is to construct natural examples of irreducible representations of level zero.

Given a representation V of $\mathfrak{g}$, set $L(V) = V \otimes \mathbb{C}[t, t^{-1}]$. For $a \in \mathbb{C}^\times$, one can define the structure of a $\hat{L}(\mathfrak{g})$–module on $L(V)$ by:

$$(x \otimes t^r)(v \otimes t^s) = a^r xv \otimes t^{r+s}, \quad c(L(V)) = 0, \quad d(v \otimes t^r) = rv \otimes t^r. \tag{6.13}$$

We denote this module by $L(\mathrm{ev}_a^* V)$ and we will say more about this notation later. This construction can be generalized further as follows. Suppose that we are given representations $\{V_s : 1 \le s \le k\}$ of $\mathfrak{g}$ and k non–zero complex numbers $a_1, \cdots, a_k$. The space $L(V_1 \otimes \cdots \otimes V_k)$ has a $\hat{L}(\mathfrak{g})$–module structure defined by

$$\begin{aligned}
&(x \otimes t^r)(v_1 \otimes \cdots \otimes v_k \otimes t^m) \\
&\quad = \sum_{s=1}^{k} (v_1 \otimes \cdots \otimes v_{s-1} \otimes xv_s \otimes v_{s+1} \otimes \cdots \otimes v_k \otimes a_s^r t^{r+m}),
\end{aligned}$$

$$cL(V_1 \otimes \cdots \otimes V_k) = 0,$$

$$d(v_1 \otimes \cdots \otimes v_k \otimes t^m) = m(v_1 \otimes \cdots \otimes v_k \otimes t^m),$$

and we denote this module by $L(\mathrm{ev}_{a_1}^* V_1 \otimes \cdots \otimes \mathrm{ev}_{a_k}^* V_k)$. Modules of these kind are sometimes called *loop modules*. The following result was proved in [71].

Proposition *Let $k \in \mathbb{N}$, and for $1 \le s \le k$ let $\lambda_s \in P^+$ and $a_s \in \mathbb{C}^\times$ and assume that $a_s \ne a_m$ if $s \ne m$. There exists $r \ge 1$ such that*

$$\sum_{s=1}^{k} \lambda_s(h) a_s^j = 0, \quad 1 \le j < r, \text{ for all } h \in \mathfrak{h}.$$

For $1 \le j \le r$, the submodule

$$L^j(\mathrm{ev}_{a_1}^* V(\lambda_1) \otimes \cdots \otimes \mathrm{ev}_{a_k}^* V(\lambda_k)) = \mathbf{U}(\hat{L}(\mathfrak{g}))(v_{\lambda_1} \otimes \cdots \otimes v_{\lambda_k} \otimes t^j)$$

is irreducible and we have an isomorphism of $\hat{L}(\mathfrak{g})$–modules,

$$L(\mathrm{ev}_{a_1}^* V(\lambda_1) \otimes \cdots \otimes \mathrm{ev}_{a_k}^* V(\lambda_k)) \cong \bigoplus_{s=1}^{r} L^s(\mathrm{ev}_{a_1}^* V(\lambda_1) \otimes \cdots \otimes \mathrm{ev}_{a_k}^* V(\lambda_k)).$$

The statement in [71] is more precise and gives the value of r for a fixed set of λ_s, a_s, $1 \le s \le k$. As an example, one has $r = 2$ when $\lambda_1 = \lambda_2$ and $a_1 = -a_2$.

6.3.3 The following is an amalgamation of results proved in [56] and [71] and achieves our goal of classifying the irreducible objects of $\mathcal{I}$. Recall that if V is integrable, then $\operatorname{wt} V \subseteq \hat{P}$ and hence the center $c = h_0 + h_\theta$ acts on V with integer eigenvalues.

Theorem *Let $V \in \operatorname{Ob}\mathcal{I}$ be irreducible and let $k \in \mathbb{Z}$ be such that $cv = kv$ for all $v \in V$.*

(i) *If $k > 0$ then $V \in \hat{\mathcal{O}}$ and hence $V \cong \hat{V}(\lambda)$ for some $\lambda \in \hat{P}^+$.*
(ii) *If $k < 0$ then $V \cong \hat{V}(\lambda)^\#$ for some $\lambda \in \hat{P}^+$.*
(iii) *If $k = 0$, then*

$$V \cong L^s(\mathrm{ev}_{a_1}^* V(\lambda_1) \otimes \cdots \otimes \mathrm{ev}_{a_k}^* V(\lambda_k)),$$

for some $\lambda_1, \cdots, \lambda_k \in P^+$, $a_1, \cdots, a_k \in \mathbb{C}^\times$ and s as in Proposition 6.3.2.

The twisted version of this theorem is known and may be found in [72].

We now make some remarks which will allow us to limit our focus in the next sections on finite–dimensional representations of $L(\mathfrak{g})$. The first observation is that to produce a representation of $\hat{L}(\mathfrak{g})$ from a representation of $\mathfrak{g}$ one has to tensor with $\mathbb{C}[t, t^{-1}]$ only to keep track of the grading that d defines on $\hat{L}(\mathfrak{g})$. But one can get around this problem by defining a functor $\mathbf{L}$ from the category of $L(\mathfrak{g})$–modules to $\hat{L}(\mathfrak{g})$–modules, which is given on objects by,

$$\mathbf{L}(V) = V \otimes \mathbb{C}[t, t^{-1}],$$

$$(x \otimes t^r)(v \otimes t^s) = ((x \otimes t^r)v) \otimes t^{r+s}, \quad c\mathbf{L}(V) = 0, \quad d(v \otimes t^r) = rv \otimes t^r.$$

Clearly $\mathbf{L}$ takes finite–dimensional $L(\mathfrak{g})$–modules to integrable $\hat{L}(\mathfrak{g})$–modules. It will become clear from the results of the next section, that the functor $\mathbf{L}$ in general sends irreducible finite–dimensional modules to completely reducible integrable ones. In [60] it is shown that $\mathbf{L}$ preserves enough of the homological properties of $L(\mathfrak{g})$–modules to make it worthwhile to restrict one's attention to modules for $L(\mathfrak{g})$.

6.3.4 We digress, briefly, from our study of $\mathcal{I}$ to define the imaginary integral root vectors and the integral form of $\mathbf{U}(\hat{L}(\mathfrak{g}))$. These were introduced by Garland in [109] and allow us to study lattices in integrable modules and hence also the representations of the hyperalgebra of $\hat{L}(\mathfrak{g})$. It is also worthwhile to note that the integral imaginary root vectors play a major role in the vertex operator construction [100] of the basic representation which was defined in Section 6.2.8.

As in the case of $\mathfrak{g}$ we let $\mathbf{U}_\mathbb{Z}(\hat{L}(\mathfrak{g}))$ be the $\mathbb{Z}$–subalgebra of $\mathbf{U}(\hat{L}(\mathfrak{g}))$ generated by the elements $(x_\alpha^\pm \otimes t^s)^{(r)}$, where $\alpha \in \Phi^+$, $s \in \mathbb{Z}$ and $r \in \mathbb{N}$. We know from Section 6.1.11 that one has to be more careful with the Cartan type elements $(h_i \otimes t^s)$, $s \in \mathbb{Z}$. For $s = 0$, $r \in \mathbb{N}$ and $1 \le i \le n$, the elements $\binom{h_i}{r}$ are in $\mathbf{U}_\mathbb{Z}(\mathfrak{g})$ and hence in $\mathbf{U}_\mathbb{Z}(\hat{L}(\mathfrak{g}))$.

To define the analogs of the divided powers for the imaginary root vectors $h_i \otimes t^s$, $i \in I$, $s \in \mathbb{Z}$, $s \neq 0$ one tries to modify suitably the principle we used in Section 6.1.11 (see equation (6.10)). Consider the triangular decomposition

$$\mathbf{U}(\hat{L}(\mathfrak{g})) = \mathbf{U}(\hat{\mathfrak{n}}^-)\mathbf{U}(\hat{\mathfrak{h}})\mathbf{U}(\hat{\mathfrak{n}}^+),$$

and note that for all $\alpha \in \Phi^+$, $r, m \in \mathbb{N}$ the element $(x_\alpha^+ \otimes t)^{(r)}(x_\alpha^-)^{(m)}$, is not in the order prescribed on the right hand side. In [109], Garland proved a remarkable formula which rewrites this element in the correct order. We will not reproduce his entire formula here, but content ourselves with writing down the correct analog (which emerges from his formula) of the divided power of the imaginary root vectors. For each $i \in I$ we define formal power series $P_i^{\pm}(u)$ in an indeterminate u with values in the commutative algebra $\mathbf{U}(\mathfrak{h} \otimes t^{\pm 1}\mathbb{C}[t^{\pm 1}])$ by

$$P_i^{\pm}(u) = \sum_{s=0}^{\infty} P_{i,s}^{\pm} u^s = \exp\left(-\sum_{k=1}^{\infty} \frac{h_i \otimes t^{\pm k}}{k}\right), \tag{6.14}$$

and we can now state Garland's theorem.

Theorem *Fix an order on the set*

$$\{(x_{\alpha}^{\pm} \otimes t^s)^{(r)} : \alpha \in \Phi^+, s \in \mathbb{Z}, r \in \mathbb{N}\} \cup \{P_{i,s}^{\pm} : 1 \le i \le n, s \in \mathbb{N}\} \cup \{\binom{h_i}{r} : i \in I, r \in \mathbb{N}\}.$$

The ordered monomials from this set are a $\mathbb{Z}$–basis of $\mathbf{U}_{\mathbb{Z}}(\hat{L}(\mathfrak{g}))$ and a $\mathbb{C}$–basis of $\mathbf{U}(\hat{L}(\mathfrak{g}))$. Moreover, if $\lambda \in \hat{P}^+$ and $\hat{v}_{\lambda} \in \hat{V}(\lambda)_{\lambda}$, then $\mathbf{U}_{\mathbb{Z}}(\hat{L}(\mathfrak{g}))\hat{v}_{\lambda}$ is a $\mathbb{Z}$–lattice in $\hat{V}(\lambda)$.

The twisted version of this theorem may be found in [172].

6.3.5 In view of this theorem it would be natural to pass to a field $\mathbb{F}$ of characteristic p as we did in Section 6.1.11 and study the representations $\mathbf{U}(\hat{L}(\mathfrak{g}))_{\mathbb{F}}$ and in particular the modules $\hat{V}(\lambda)_{\mathbb{F}}$. However, this direction does not seem to have been pursued and except for the result we now discuss, virtually nothing is known about the irreducible modules of positive level in characteristic p. In [65], a quantum analog of this problem was studied and it was shown there that the basic representation of the quantum affine algebra remained irreducible at an mth root of unity, where m is odd and coprime to the determinant of the Cartan matrix of $\mathfrak{g}$. This was proved essentially by constructing the basic representation at a root of unity and showing that it had the same character. The proof given in that paper works verbatim in the characteristic p case as long as p is odd and coprime to the determinant of the Cartan matrix of $\mathfrak{g}$. The case of level zero modules in characteristic p has been studied by Jakelic and Moura in [129], [130], [131] and we shall discuss this in the next sections.

6.4 Finite–dimensional modules for loop algebras and their generalizations

In this section, we discuss the category of finite–dimensional representations of loop algebras, study extensions between irreducible modules and describe the blocks of this category.

6.4.1 For $a \in \mathbb{C}^{\times}$ let $\mathrm{ev}_a : L(\mathfrak{g}) \to \mathfrak{g}$ be the homomorphism of Lie algebras given by

$$\mathrm{ev}_a(x \otimes f) = f(a)x.$$

If V is a $\mathfrak{g}$–module, we denote the corresponding $L(\mathfrak{g})$-module by $\mathrm{ev}_a^* V$. Together with the discussion in Section 6.3.3, this explains the notation $L(\mathrm{ev}_a^* V)$ used in Section 6.3.2. The following result classifies irreducible finite–dimensional representations of $L(\mathfrak{g})$.

Theorem *An irreducible finite–dimensional representation of $L(\mathfrak{g})$ is isomorphic to a tensor product $\mathrm{ev}_{a_1}^* V(\lambda_1) \otimes \cdots \otimes \mathrm{ev}_{a_k}^* V(\lambda_k)$, where $\lambda_s \in P^+$, $a_s \in \mathbb{C}^{\times}$ with $a_s \ne a_r$ for $1 \le s, r \le k$. Conversely any such tensor product is irreducible.*

This result can be deduced easily from Proposition 6.3.2 and Theorem 6.3.3(iii). A more direct proof was given by Rao in [200] along the following lines. If $\rho : L(\mathfrak{g}) \to \mathrm{End}V$ is a finite–dimensional representation, then the kernel of ρ is an ideal of finite–codimension in $L(\mathfrak{g})$. It is not too hard to show that any ideal in $L(\mathfrak{g})$ must be of the form $\mathfrak{g} \otimes I(V)$ where $I(V)$ is an ideal in $\mathbb{C}[t, t^{-1}]$. If V is irreducible then one proves that $I(V)$ is a product of distinct maximal ideals and hence is generated by an element $(t - a_1) \cdots (t - a_k)$ for some $k \in \mathbb{N}$ and distinct elements $a_1, \cdots, a_k$ in $\mathbb{C}^\times$ Moreover $\mathbb{C}[t, t^{-1}]/I(V)$ is a vector space of dimension k and it follows now that $L(\mathfrak{g})/(\mathfrak{g} \otimes I(V))$ is isomorphic to a direct sum of k–copies of $\mathfrak{g}$ and hence is a semisimple Lie algebra. The theorem now follows by using the representation theory of semisimple Lie algebras.

For the twisted affine Lie algebras the corresponding result was proved in [57]. To describe it, note that since $L(\mathfrak{g}, \sigma, m)$ is a subalgebra of $L(\mathfrak{g})$, one can regard the tensor product $V = \mathrm{ev}_{a_1}^* V(\lambda_1) \otimes \cdots \otimes \mathrm{ev}_{a_k}^* V(\lambda_k)$ as a module for the twisted algebra. In general it is not irreducible as a module for the twisted algebra, but if one imposes the additional condition that $a_s^m \neq a_r^m$ for all $1 \leq s, r \leq k$, where m is the order of σ, then V is an irreducible module for $L(\mathfrak{g}, \sigma, m)$. Moreover, one can also prove that these are exactly (up to isomorphism) all the finite–dimensional irreducible modules.

6.4.2 An element $a \in \mathbb{C}^\times$ determines a maximal ideal $(t - a)$ of $\mathbb{C}[t, t^{-1}]$ and another way to formulate Theorem 6.4.1 is to say that the irreducible finite–dimensional representations are parametrized by finitely–supported functions from the maximal spectrum of $\mathbb{C}[t, t^{-1}]$ to P^+, where by finitely supported, we mean $\chi(M) = 0$ for all but finitely many maximal ideals. We generalize the theorem to the case of $\mathfrak{g} \otimes A$, where A is a finitely generated commutative associative algebra algebra over $\mathbb{C}$. The space $\mathfrak{g} \otimes A$ is a Lie algebra as usual: $[x \otimes a, y \otimes b] = [x, y] \otimes ab$. Let $\max A$ be the set of maximal ideals in A and denote by $\Xi(\max A, P^+)$ the set of finitely supported functions from $\max A$ to P^+, and set

$$\mathrm{supp}\,\psi = \{M \in \max A : \psi(M) \neq 0\}, \quad \psi \in \Xi(\max A, P^+).$$

For $M \in \max A$, let $\mathrm{ev}_M : \mathfrak{g} \otimes A \to \mathfrak{g}$ be the map of Lie algebras which is induced by the algebra homomorphism $A \to A/M \cong \mathbb{C}$ and we denote by $\mathrm{ev}_M^* V(\lambda)$ the representation of $\mathfrak{g} \otimes A$ obtained by pulling back $V(\lambda)$ through ev_M.

Theorem *Let A be a finitely generated commutative and associative algebra over $\mathbb{C}$. For $\chi \in \Xi(\max A, P^+)$, the $\mathfrak{g} \otimes A$–module $\bigotimes_{M \in \mathrm{supp}\,\chi} \mathrm{ev}_M^* V(\chi(M))$ is irreducible and conversely any finite dimensional irreducible representation is isomorphic to one of these.*

The theorem has been proved by many people in various cases: in the case when A is the polynomial ring in r variables the result was first proved by Rao [200], the general case given above appears in [58]. In [153], Lau also obtains a proof of this statement for the Laurent polynomial ring in r variables and he is also able prove the result in the twisted case. It is important to note here, that unlike in the case when $A = \mathbb{C}[t, t^{-1}]$, the twisted algebras are not determined by a Dynkin diagram automorphism and one has many non–isomorphic twisted algebras and Lau works in this generality. The methods of all these papers are algebraic and similar to the proof sketched above for the loop algebra. A more geometric approach is developed in [189] and their methods allow them to develop a uniform approach which works for twisted and untwisted algebras of the form $\mathfrak{a} \otimes A$ where $\mathfrak{a}$

is an arbitrary Lie algebra. In this generality the irreducible representations are not always tensor products of evaluations, but they are very nearly so, up to tensoring with a one–dimensional representation.

6.4.3 An obvious and not very difficult question that one should ask is what kind of module one gets if one allows $a_r = a_s$ for some pair $1 \leq r, s, \leq k$ in the statement of Theorem 6.4.1. The following proposition proved in [59] shows that the module is completely reducible and more generally gives a necessary and sufficient condition for a finite–dimensional module to be completely reducible. Recall that for any $L(\mathfrak{g})$–module V we have defined in Section 6.4.1 an ideal $I(V)$ of $\mathbb{C}[t, t^{-1}]$ which is maximal with the property that $(\mathfrak{g} \otimes I(V))V = 0$. More precisely, $\mathfrak{g} \otimes I(V)$ is the annihilating ideal of V in $L(\mathfrak{g})$.

Proposition *Let $\lambda, \mu \in P^+$ and $a \in \mathbb{C}^\times$. Then $\mathrm{ev}_a^* V(\lambda) \otimes \mathrm{ev}_a^* V(\mu)$ is isomorphic to a direct sum of submodules of the form $\mathrm{ev}_a^* V(\nu)$, $\nu \in P^+$ and*

$$\mathrm{Hom}_{L(\mathfrak{g})}(\mathrm{ev}_a^* V(\nu), \mathrm{ev}_a^* V(\lambda) \otimes \mathrm{ev}_a^* V(\mu)) \cong \mathrm{Hom}_{\mathfrak{g}}(V(\nu), V(\lambda) \otimes V(\mu)).$$

More generally, a finite–dimensional module V is completely reducible iff the ideal $I(V)$ is a product of distinct maximal ideals. Analogous statements hold for algebras of type $\mathfrak{g} \otimes A$.

6.4.4 We now describe extensions between irreducible $L(\mathfrak{g})$–modules. We begin with a simple case and compute $\mathrm{Ext}^1_{L(\mathfrak{g})}(\mathrm{ev}_a^* V(\lambda), \mathrm{ev}_a^* V(\mu))$ for $a \in \mathbb{C}^\times$.

Let

$$0 \to \mathrm{ev}_a^* V(\mu) \to U \to \mathrm{ev}_a^* V(\lambda) \to 0 \tag{6.15}$$

be a short exact sequence of $L(\mathfrak{g})$–modules. Regarded as a short exact sequence of $\mathfrak{g}$–modules this sequence is split and we can pick $u_\lambda \in U_\lambda$ with $\mathbf{U}(\mathfrak{g})u_\lambda \cong V(\lambda)$. If (6.15) is not split as a short exact sequence of $L(\mathfrak{g})$–modules, there must exist $r \in \mathbb{Z}$ such that the $\mathfrak{g}$–module map,

$$\mathfrak{g} \otimes \mathbf{U}(\mathfrak{g})u_\lambda \to U, \qquad (xt^r \otimes gu_\lambda) \to xt^r gu_\lambda,$$

has a non–zero projection (as a $\mathfrak{g}$–module) onto the image of $\mathrm{ev}_a^* V(\mu)$, i.e.,

$$\mathrm{Ext}^1_{L(\mathfrak{g})}(\mathrm{ev}_a^* V(\lambda), \mathrm{ev}_a^* V(\mu)) \neq 0 \implies \mathrm{Hom}_{\mathfrak{g}}(\mathfrak{g} \otimes V(\lambda), V(\mu)) \neq 0.$$

Conversely, given $\pi : \mathfrak{g} \otimes V(\lambda) \to V(\mu)$ a map of $\mathfrak{g}$–modules it is proved in [69] that the formula

$$\begin{aligned} xt^r(v, w) &= (a^r xv, a^r xw + ra^{r-1}\pi(xv)), \\ & \quad v \in V(\lambda), w \in V(\mu), \quad a \in \mathbb{C}^\times, \quad r \in \mathbb{Z}, \quad x \in \mathfrak{g}, \end{aligned}$$

defines an indecomposable $L(\mathfrak{g})$–module denoted $V(\lambda, \mu, a)$ on $V(\lambda) \oplus V(\mu)$ and hence one can conclude that

$$\mathrm{Ext}^1_{L(\mathfrak{g})}(\mathrm{ev}_a^* V(\lambda), \mathrm{ev}_a^* V(\mu)) \cong \mathrm{Hom}_{\mathfrak{g}}(\mathfrak{g} \otimes V(\lambda), V(\mu)),$$

which can be shown to be equivalent to

$$\mathrm{Ext}^1_{L(\mathfrak{g})}(\mathrm{ev}_a^* V(\lambda), \mathrm{ev}_a^* V(\mu)) \cong \mathrm{Hom}_{L(\mathfrak{g})}(\mathrm{ev}_a^* \mathfrak{g} \otimes \mathrm{ev}_a^* V(\lambda), \quad \mathrm{ev}_a^* V(\mu)).$$

The general statement proved in [59] (for $\mathfrak{g} \otimes \mathbb{C}[t]$ but which has an obvious modification to $L(\mathfrak{g})$) is:

Theorem *Let V and V' be irreducible finite–dimensional $L(\mathfrak{g})$–modules. Then*

$$\dim \operatorname{Ext}^1_{L(\mathfrak{g})}(V, V') = \sum_{a \in \mathbb{C}^\times} \dim \operatorname{Hom}_{L(\mathfrak{g})}(\operatorname{ev}_a^* \mathfrak{g} \otimes V, V').$$

This condition can be made quite explicit using the results stated above, namely if we write V and V' as a tensor product of evaluation representations as in Theorem 6.4.1 one can compute the dimension of $\operatorname{Ext}^1_{L(\mathfrak{g})}(V, V')$ as the sum of multiplicities of the adjoint representation of $\mathfrak{g}$ in a tensor product of the form $V(\lambda) \otimes V(\mu)^*$ for suitable λ, μ.

6.4.5 The analogous result has not been established as yet for the twisted Lie algebras. It has been solved in [148] for the more general Lie algebra $\mathfrak{g} \otimes A$ and in fact the main theorem in this paper is written in a very explicit way. There is one new feature though which is not seen in the case of $\mathbb{C}[t, t^{-1}]$ which we explain again in a simple case. For $\lambda, \mu \in P^+$ and $M \in \max A$, we have

$$\operatorname{Ext}^1_{\mathfrak{g}\otimes A}(\operatorname{ev}_M^* V(\lambda), \operatorname{ev}_M^* V(\mu)) \cong \operatorname{Hom}_{\mathfrak{g}}(\mathfrak{g} \otimes V(\lambda), V(\mu)) \otimes \operatorname{Der}(A, A/M),$$

where $\operatorname{Der}(A, A/M)$ is the space of all linear maps $A \to A/M$ satisfying $D(ab) = aD(b) + bD(a)$. In the case when $A = \mathbb{C}[t, t^{-1}]$ this space is one–dimensional and so is invisible.

6.4.6 To continue our study of finite–dimensional representations, we look back at Section 6.2 and notice that one of the natural things to do is to understand the blocks of the category. As we already noted in Section 6.4, the theory of central characters is not available for affine or loop algebras. However, it is still possible to describe the blocks and to do this one uses a family of universal indecomposable finite–dimensional modules introduced in [75] and called the Weyl modules for affine Lie algebras. These are not the same modules as the ones discussed in Section 6.2, but are in some sense obtained in the same way. They can be regarded as the classical limit of irreducible representations of quantum affine algebras and we shall say more about this later. The first step is to organize the irreducible representations in some nice way and for this, we recall from [69] the notion of the spectral character of a $L(\mathfrak{g})$–module.

Definition Given an irreducible representation $V = \otimes_{s=1}^k \operatorname{ev}_{a_s}^* V(\lambda_s)$ of $L(\mathfrak{g})$ the *spectral character* of V is defined to be the function $\chi_V : \mathbb{C}^\times \to P/Q$ given by

$$\chi_V(z) = \begin{cases} 0, & z \notin \{a_1, \cdots, a_k\}, \\ \lambda_s + Q, & z = a_s, \quad 1 \le s \le k. \end{cases}$$

A finite dimensional $L(\mathfrak{g})$–module is said to *have spectral character* if all the irreducible modules occurring in a Jordan–Hölder series have the same spectral character.

Note that it is very easy for two modules to have the same spectral character, for instance if $\lambda, \mu \in P^+ \cap Q$, then

$$\chi_{\operatorname{ev}_a^* V(\lambda)} = \chi_{\operatorname{ev}_b^* V(\mu)}, \quad \text{for all } \; a, b \in \mathbb{C}^\times.$$

6.4.7 We recall the definition of the blocks of a category $\mathcal{C}$.

Definition Say that two indecomposable objects $U, V \in \mathrm{Ob}\,\mathcal{C}$ are *linked* if there do not exist abelian subcategories $\mathcal{C}_k$, $k = 1, 2$ such that $\mathcal{C} = \mathcal{C}_1 \oplus \mathcal{C}_2$ with $U \in \mathcal{C}_1$ and $V \in \mathcal{C}_2$. If U and V are decomposable then we say that they are *linked* if every indecomposable summand of U is linked to every indecomposable summand of V. This defines an equivalence relation on $\mathcal{C}$ and the equivalence classes are called the *blocks* of $\mathcal{C}$.

The main result proved in [69] is the following theorem which describes the blocks of the category of finite–dimensional representations.

Theorem *Let V be a finite–dimensional $L(\mathfrak{g})$ module. Then V is isomorphic to a direct sum of finite–dimensional modules V_r, $1 \le r \le k$ each of which admits a spectral character. Moreover, any two modules with the same spectral character are linked.*

As a consequence of this theorem, we see that the blocks are very large subcategories, for instance when $\mathfrak{g}$ is of type E_8 there is only one block, i.e., any two finite–dimensional modules are linked. A similar result was proved in [206] for the twisted algebras $L(\mathfrak{g}, \sigma, m)$ and in [148] for the Lie algebras $\mathfrak{g} \otimes A$. All the proofs are similar and involve two ingredients. The first is the module $V(\lambda, \mu, a)$ constructed in Section 6.4.4 and the second are the Weyl modules which we discuss in the next section.

6.4.8 Most of the results of this section have analogs in positive characteristic for the hyperalgebras $\mathbf{U}(L(\mathfrak{g}))_{\mathbb{F}}$ and these are studied in [129], [130]. The one exception is the result on extensions between irreducible modules. This is because the proof given in characteristic zero relies on the fact that a finite–dimensional representation of a complex simple Lie algebras is completely reducible, which as we remarked in Section 6.1.11 is false in positive characteristic.

6.5 Weyl modules, restricted Kirillov–Reshetikhin and beyond

We begin this section by discussing briefly the input coming from the representation theory of quantum affine algebras, which has helped to identify interesting families of finite–dimensional representations of loop algebras. The interested reader could consult [63] and the references therein for further details on representations of quantum affine algebras.

6.5.1 In the 1980's Drinfeld and Jimbo introduced the quantized enveloping algebras $\mathbf{U}_q(\mathfrak{a})$ of a Kac–Moody Lie algebra $\mathfrak{a}$. These algebras depend on a parameter q which can be either a complex number, in which case one regards $\mathbf{U}_q(\mathfrak{a})$ as an algebra over $\mathbb{C}$ or a formal variable in which case one regards it as an algebra over the function field $\mathbb{C}(q)$, and informally speaking if we put $q = 1$ we get the usual enveloping algebra of $\mathfrak{a}$. In the case when $\mathfrak{g}$ is simple, Lusztig [164] proved that for generic q, the irreducible finite–dimensional representations of $\mathbf{U}_q(\mathfrak{g})$ are given by elements of P^+ and moreover that the module associated to $\lambda \in P^+$ has the same character (suitably understood) as the module in the $q = 1$ case, namely it is the character of the $\mathfrak{g}$–module $V(\lambda)$. An analogous statement was also proved for the positive level integrable modules of the quantized affine algebras. In the case when q is a primitive m^{th}–root of unity, the picture resembles the case of positive characteristic and the irreducible modules are in general smaller.

6.5.2 In the case of finite–dimensional modules for the quantized affine algebras the difference between generic q and the roots of unity case is already visible when $q = 1$. In other words the action of $L(\mathfrak{g})$ on evaluation representations and their tensor products given in Theorem 6.4.1 cannot be deformed to give an action of the quantum affine algebra on the same space. Or, to put it in yet another way, the $q = 1$ limit of irreducible representations of quantum affine algebras generally give rise to a reducible indecomposable representation of $L(\mathfrak{g})$. This was the motivation for introducing in [75] the definition of Weyl modules for affine Lie algebras. We should remark here that this phenomenon was first noted in [89] in connection with the representation theory of the closely related Yangians, which are deformations of the current algebra $\mathfrak{g} \otimes \mathbb{C}[t]$.

6.5.3 The classification of irreducible finite–dimensional representations for quantum affine algebras was obtained in [73] and is similar to the one obtained by Drinfeld for the Yangians. It was also very similar to the abstract classification (rather than the more explicit one described in these notes) of loop modules given in [56]. However, the fact that there is no analog of the evaluation homomorphism from the quantum loop algebra to the quantum simple algebra has meant that describing the irreducible modules in a concrete way has been challenging and a variety of methods have been developed to understand these modules better. One of these methods is to understand the multiplicity with which an irreducible module for the quantum simple algebra occurs in an irreducible module for the quantum loop algebra. This can be turned into a problem of the non–quantum case for the following reason. It was proved in [75] that under natural conditions the irreducible representations of the quantum affine algebra admitted a $\mathbb{C}[t, t^{-1}]$–form and hence could be specialized to $q = 1$ and gives a representation of $L(\mathfrak{g})$. A result of Lusztig implies that understanding the $\mathfrak{g}$–module decomposition of this module is the same as answering the question for the quantum case. With this motivation out of the way, we now return our focus to representations of $L(\mathfrak{g})$.

6.5.4 For $\lambda \in P^+$, let $W(\lambda)$ be the $L(\mathfrak{g})$–module generated by an element w_λ with relations

$$hw_\lambda = \lambda(h)w_\lambda, \quad L(\mathfrak{n}^+)w_\lambda = 0, \quad (x^-_{\alpha_i})^{\lambda(h_{\alpha_i})+1}w_\lambda = 0.$$

We call $W(\lambda)$ the *global Weyl module*. The following is elementary.

Lemma *The module $W(\lambda)$ is integrable for all $\lambda \in P^+$. Moreover if V is any integrable module, then*

$$\operatorname{Hom}_{L(\mathfrak{g})}(W(\lambda), V) \cong V_\lambda^+, \qquad V_\lambda^+ = \{v \in V_\lambda : L(\mathfrak{n}^+)v = 0\}.$$

In particular, if $\mathrm{ev}_{a_1}^* V(\lambda_1) \otimes \cdots \otimes \mathrm{ev}_{a_k}^* V(\lambda_k)$ is an irreducible finite–dimensional $L(\mathfrak{g})$–module with $\sum_{s=1}^k \lambda_s = \lambda$, then

$$\dim \operatorname{Hom}_{L(\mathfrak{g})}(W(\lambda), \mathrm{ev}_{a_1}^* V(\lambda_1) \otimes \cdots \otimes \mathrm{ev}_{a_k}^* V(\lambda_k)) = 1.$$

The lemma shows immediately that the module $W(\lambda)$ is infinite–dimensional if $\lambda \neq 0$ and in fact that $W(\lambda)_\lambda$ is infinite–dimensional. For otherwise, the action of the algebra $L(\mathfrak{h})$ on $W(\lambda)_\lambda$ would be a direct sum of a finite number of generalized eigenspaces. On the other hand the lemma says that $W(\lambda)$ has infinitely many irreducible quotients $\mathrm{ev}_{a_1}^* V(\lambda_1) \otimes \cdots \otimes \mathrm{ev}_{a_k}^* V(\lambda_k)$ and a simple calculation shows that the eigenvalues of $L(\mathfrak{h})$ on the image of w_λ in each of these quotients is different.

6.5.5 Let us look at a further consequence of this lemma. Recall the modules $V(\lambda, \mu, a)$ defined in Section 6.4.4. If $\lambda - \mu \in Q^+$, then with a little computation, one can prove that $V(\lambda, \mu, a)$ is generated as a $L(\mathfrak{g})$–module by the element $v_\lambda \in V(\lambda)$ and that $L(\mathfrak{n}^+)v_\lambda = 0$ and $\dim V(\lambda, \mu, a)_\lambda = 1$, if $\lambda \neq \mu$. Hence $\mathrm{Hom}_{L(\mathfrak{g})}(W(\lambda), V(\lambda, \mu, a)) \neq 0$ and so $W(\lambda)$ also has cyclic reducible (indecomposable) finite–dimensional quotients. It is reasonable therefore, to ask if there are a family of maximal or universal finite dimensional quotients of $W(\lambda)$: a family of finite–dimensional quotients V which are cyclic and have $\dim V_\lambda = 1$ and are such that any other such finite–dimensional quotient of $W(\lambda)$ is a quotient of one of these. The answer is yes and these are called the *local Weyl modules* and we shall spend some time discussing these modules. We remark that it is these local Weyl modules which appear as the $q = 1$ limits of representations of quantum affine algebras.

6.5.6 The module $W(\lambda)$ admits a right $L(\mathfrak{h})$–module structure, given by

$$(yw_\lambda)u = yuw_\lambda, \quad y \in \mathbf{U}(L(\mathfrak{g})), \quad u \in L(\mathfrak{h}),$$

and hence $W(\lambda)$ is a $(L(\mathfrak{g}), L(\mathfrak{h}))$–bimodule. Given $\chi : L(\mathfrak{h}) \to \mathbb{C}$, set

$$W(\lambda, \chi) = W(\lambda) \otimes_{L(\mathfrak{h})} \mathbb{C}_\chi,$$

where $\mathbb{C}_\chi$ is the one–dimensional $L(\mathfrak{h})$–module defined by χ. The following was proved in [75].

Theorem *Let $\lambda \in P^+$ and $\chi \in L(\mathfrak{h})^*$. Then the module $W(\lambda, \chi)$ is finite–dimensional and is non–zero iff there exist $k \in \mathbb{N}$, $\lambda_1, \cdots, \lambda_k \in P^+$ and $a_1, \cdots, a_k \in \mathbb{C}^\times$ such that*

$$\chi(h \otimes t^s) = \sum_{j=1}^{k} \lambda_j(h) a_j^s.$$

Moreover if (λ, χ) and (λ', χ') are such that $W(\lambda, \chi)$ and $W(\lambda', \chi')$ are non–zero, then

$$W(\lambda, \chi) \cong W(\lambda', \chi') \iff \lambda = \lambda', \quad \chi = \chi'.$$

The notation used here, although different from that used in [75], is consistent with the notation of the previous sections of these notes.

6.5.7 The local Weyl modules have a tensor product factorization proved in [75] which is similar to that given in Theorem 6.4.1. For $\lambda \in P^+$ and $a \in \mathbb{C}^\times$, let $\chi_{\lambda,a} \in L(\mathfrak{h})^*$ be given by $\chi_{\lambda,a}(h \otimes t^r) = a^r \lambda(h)$.

Proposition *Any non–zero local Weyl module is isomorphic to a tensor product of the form $W(\lambda_1, \chi_{\lambda_1,a_1}) \otimes \cdots \otimes W(\lambda_k, \chi_{\lambda_k,a_k})$ for some $\lambda_s \in P^+$, $a_s \in \mathbb{C}^\times$, $1 \le s \le k$ with $a_j \neq a_m$ if $j \neq m$.*

As a result of the proposition, to understand the $\mathfrak{g}$–module structure or the dimension of the local Weyl modules it suffices to study the modules $W(\lambda, \chi_{\lambda,a})$ for $(\lambda, a) \in P^+ \times \mathbb{C}^\times$. These questions are addressed in [75, 68, 96], for $\mathfrak{g}$ of type A_1, A_n and A, D, E respectively and in these cases the local Weyl modules can be identified with Demazure modules in positive level representations of affine Lie algebras. This is definitely false for non–simply laced algebras where the local Weyl modules are bigger than the Demazure modules. In general the dimension of the local Weyl modules can be deduced from the work of [20]. As a consequence one knows that the dimension of $W(\lambda, \chi_{\lambda,a})$ is independent of the choice of $a \in \mathbb{C}^\times$ which in turn can be used to prove that $W(\lambda)$ is free as a right module for a suitably defined polynomial ring (the quotient of $\mathbf{U}(L(\mathfrak{h}))$ by the annihilator of w_λ).

6.5.8 The Weyl modules for $\mathfrak{g} \otimes A$ have been defined and studied in [58] for an arbitrary associative, commutative algebra A. The results of that paper show that there are many important differences between the study of Weyl modules for the loop algebra and the more general case. As in the case for irreducible modules, one works with maximal ideals, and the local Weyl modules for a fixed λ are parametrized by a maximal ideal of A. The dimension of the local Weyl modules even in the simplest cases does depend on the maximal ideal even when A is the polynomial ring in two variables and hence these modules are no longer free right modules. There are thus many interesting problems that arise from passing to the higher dimensional cases.

6.5.9 Local Weyl modules have also been studied in positive characteristic in [129],[130]. As in the case of simple Lie algebras, the definition needs modification, the most crucial one being that the conditions on $L(\mathfrak{h})$ must be replaced with the conditions on the integral imaginary root vectors defined by Garland. The authors also study the question of base change, i.e., the functor of extension of scalars from the category of finite-dimensional $\mathbf{U}(L(\mathfrak{g}))_{\mathbb{K}}$-modules to that of $\mathbf{U}(L(\mathfrak{g}))_{\mathbb{F}}$-modules. For simple Lie algebras, the functor induces a bijection from the set of isomorphism classes of highest-weight modules for $\mathbf{U}(\mathfrak{g})_{\mathbb{K}}$ to that of isomorphism classes of highest-weight modules for $\mathbf{U}(\mathfrak{g})_{\mathbb{F}}$, where $\mathbb{F}$ is an algebraic extension of $\mathbb{K}$. In particular, the characters of the irreducible modules and the Weyl modules for the hyperalgebras $\mathbf{U}(\mathfrak{g})_{\mathbb{L}}$ do not depend on the choice of the field $\mathbb{L}$, but only on its characteristic. In the case of hyper loop algebras, the story is more complicated since the functor does not send irreducible modules to irreducible modules and, hence, does not preserve the length of a module. For more details on this functor in this case see [131].

6.5.10 An important family of quotients of the Weyl modules are the Kirillov–Reshetikhin (KR)–modules. In the quantum case, these modules are very important, since they admit a crystal basis. Most finite–dimensional representations do not admit crystal bases. These modules first arose from the study of solvable lattice models and a good reference is [119] and again more recent references can be found in [63]. The results of [75] show that one can put $q = 1$ and get modules for the loop algebra. The KR–modules, for our purposes, can be viewed as quotients of the local Weyl modules $W(m\omega_i, \chi_{m\omega_i,a})$, where $i \in I$, $m \in \mathbb{N}$ and $a \in \mathbb{C}^\times$. They are obtained by imposing a single additional relation:

$$x_{\alpha_i}^- \otimes (t - a)w_{m\omega_i} = 0.$$

Amongst other things, one of the interesting problems is to compute the $\mathfrak{g}$–character of this module and this is done in [69] for the classical Lie algebras and for some nodes of the exceptional Lie algebras. These modules come equipped with a $\mathbb{N}$–grading and, in fact, the graded character is computed in [69] and shown to coincide with the conjectural character formulae in [119]. The relationship between the grading and the parameter q that appears in [119], though, is quite mysterious. An obvious question from a mathematical point of view, is what happens when $m\omega_i$ is replaced with an arbitrary weight $\lambda \in P^+$. Partial answers to this question can be found in [177]. An alternative way to think of KR-modules is as projective modules in a suitable subcategory and this is being explored in ongoing work with Jacob Greenstein.

6.6 Koszul algebras, quivers, and highest weight categories

In this section we discuss an approach, developed jointly with Jacob Greenstein in [61], [62], to the graded representation theory of the algebras $\mathfrak{g}\otimes\mathbb{C}[t]$. We follow the axiomatic approach of Cline, Parshall and Scott which has been successful in both the study of $\mathcal{O}$ and representations in characteristic p. The basic idea is to look at subcategories, each of which has finitely many simple objects and enough projectives (or injectives), and to relate it to the module category of a finite–dimensional algebra. One can then also study the quiver with relations attached to such a subcategory. It is usual to assume that the subcategory is a block and also that one has a natural ordering of the simple objects in the block. This is usually given by the Bruhat order on the Weyl group, which in turn is related to extensions between Verma modules.

However, in our case, as we have seen in Section 6.4.7, the blocks of the finite–dimensional representations are very large and this remains true if we restrict our attention to graded representations of $\mathfrak{g}\otimes\mathbb{C}[t]$. Nevertheless, we are able to identify interesting subcategories, which contain finitely many irreducibles and the corresponding (generalized) Kirillov–Reshetikhin modules discussed in Section 6.5.10. We are also able to define an order on the simple modules which comes from an understanding of the extensions between simple modules. The endomorphism algebra of such a subcategory is a finite–dimensional algebra given by a quiver with relations. Under further restrictions the endomorphism algebra has a natural grading and using the results of [22] we prove that the grading is Koszul. In many cases (in fact for an infinite family) it is possible [117] to compute the relations explicitly and we give some concrete examples below. The twisted analogs of these results and in fact further generalizations may be found in [66].

6.6.1 The algebra $\mathfrak{g}[t]=\mathfrak{g}\otimes\mathbb{C}[t]$ is naturally graded by $\mathbb{N}$ with the rth–graded piece being $\mathfrak{g}\otimes t^r$ for $r\in\mathbb{N}$. The enveloping algebra $\mathbf{U}(\mathfrak{g}[t])$ acquires a well–defined natural grading by $\mathbb{N}$ as well: we say that an element of $\mathbf{U}(\mathfrak{g}[t])$ has grade s if it is a linear combination of elements of the form $(x_1\otimes t^{s_1})\cdots(x_k\otimes t^{s_k})$ where $k\in\mathbb{N}$, $s_j\in\mathbb{N}$, $x_j\in\mathfrak{g}$, $1\le j\le k$ and $s=s_1+\cdots+s_k$. Let $\mathfrak{g}[t]_+=\mathfrak{g}\otimes t\mathbb{C}[t]$, which is clearly a Lie ideal of $\mathfrak{g}[t]$. Note that $\mathbf{U}(\mathfrak{g}[t])\cong\mathbf{U}(\mathfrak{g}[t]_+)\otimes\mathbf{U}(\mathfrak{g})$ and that the graded pieces $\mathbf{U}(\mathfrak{g}[t]_+)[r]$ are finite dimensional $\mathfrak{g}$-modules. Hence for all $\lambda\in P^+$, we see that $\mathbf{U}(\mathfrak{g}[t])[s]\otimes_{\mathbf{U}(\mathfrak{g})}V(\lambda)$ is a finite–dimensional $\mathfrak{g}$–module.

Let $\mathcal{G}$ be the category whose objects are $\mathbb{N}$–graded modules V with finite dimensional graded pieces,

$$V=\bigoplus_{r\in\mathbb{N}}V[r],\quad \text{all }\dim V[r]<\infty,$$
$$(\mathfrak{g}\otimes t^s)V[r]\subseteq V[r+s],\quad r,s\in\mathbb{N},$$

and the morphisms are given by

$$\operatorname{Hom}_{\mathcal{G}}(V,W)=\{f\in\operatorname{Hom}_{\mathfrak{g}[t]}(V,W):f(V[r])\subseteq W[r],\quad r\in\mathbb{N}\}.$$

For $r\in\mathbb{N}$ let τ_r be the grading shift functor $\mathcal{G}\to\mathcal{G}$, that is for all $V\in\operatorname{Ob}\mathcal{G}$, τ_rV is isomorphic to V as a $\mathfrak{g}[t]$–module and $(\tau_rV)[s]=V[r+s]$.

6.6.2 Let $\mathrm{ev}_0 : \mathfrak{g}[t] \to \mathfrak{g}$, given by $x \otimes t^r \to \delta_{r,0}x$, be the evaluation at zero. For $\lambda \in P^+$ we can regard the $\mathfrak{g}[t]$–module $\mathrm{ev}_0^* V(\lambda)$ as an object of $\mathcal{G}$ by considering it as being concentrated in grade zero. Set $V(\lambda, r) = \tau_r \mathrm{ev}_0^* V(\lambda)$. The projective cover of $V(\lambda, r)$ in $\mathcal{G}$ is

$$P(\lambda, r) = \tau_r \mathbf{U}(\mathfrak{g}[t]) \otimes_{\mathbf{U}(\mathfrak{g})} V(\lambda), \quad P(\lambda, r)[s] = \mathbf{U}(\mathfrak{g}[t])[s-r] \otimes_{\mathbf{U}(\mathfrak{g})} V(\lambda), \qquad s \geq r.$$

The following is not hard to prove.

Proposition *Any irreducible object in $\mathcal{G}$ is isomorphic to $V(\lambda, r)$ for a unique $(\lambda, r) \in P^+ \times \mathbb{N}$. Moreover, $P(\lambda, r)$ is an indecomposable projective object of $\mathcal{G}$ and has $V(\lambda, r)$ as its unique irreducible quotient via a map $\tau_r \pi_\lambda$ which maps $1 \otimes v_\lambda \to v_\lambda$.*

Since $\mathfrak{g} = [\mathfrak{g}, \mathfrak{g}]$ one deduces that the kernel $K(\lambda, r)$ of the canonical surjection $P(\lambda, r) \twoheadrightarrow V(\lambda, r)$ is generated by $\tau_r \mathbf{U}(\mathfrak{g}[t])[1] \otimes_{\mathbf{U}(\mathfrak{g})} V(\lambda)$, and if we note that $\mathbf{U}(\mathfrak{g}[t])[1] \cong (\mathfrak{g} \otimes t)\mathbf{U}(\mathfrak{g})$, we get

$$K(\lambda, r) = \tau_r \left(\mathbf{U}(\mathfrak{g}[t])(\mathfrak{g} \otimes t)\right) \otimes V(\lambda).$$

A little more work proves that

$$\mathrm{Hom}_{\mathcal{G}}(K(\lambda, r), V(\mu, s)) \cong \begin{cases} 0, & s \neq r+1, \\ \mathrm{Hom}_{\mathfrak{g}}(\mathfrak{g} \otimes V(\lambda), V(\mu)), & s = r+1. \end{cases}$$

We note the following

Corollary *For all $(\lambda, r), (\mu, s) \in P^+ \times \mathbb{N}$ we have*

$$\mathrm{Ext}^1_{\mathcal{G}}(V(\lambda, r), V(\mu, s)) = 0, \qquad s \neq r+1,$$
$$\mathrm{Ext}^1_{\mathcal{G}}(V(\lambda, r), V(\mu, r+1)) \cong \mathrm{Hom}_{\mathfrak{g}}(\mathfrak{g} \otimes V(\lambda), V(\mu)).$$

Motivated by the previous corollary, we define a partial order $\preceq$ on the index set $P^+ \times \mathbb{N}$ of the simple modules in $\mathcal{G}$ by extending the cover relation $(\lambda, r) \prec (\mu, s)$ if and only if $s = r+1$ and $\lambda - \mu \in \Phi \sqcup \{0\}$. Since a pair (μ, s) can cover only finitely many elements it follows that there exist only finitely many elements less than a given element. A subset Γ of the poset $P^+ \times \mathbb{N}$ is called interval closed if for all $(\mu, s) \prec (\nu, k) \prec (\lambda, r)$ with $(\mu, s), (\lambda, r) \in \Gamma$ we have $(\nu, k) \in \Gamma$.

6.6.3 Even though an object V of $\mathcal{G}$ could be infinite dimensional, one can still talk about the multiplicity of $V(\lambda, r)$ in V by setting

$$[V : V(\lambda, r)] = [V_{\leq r} : V(\lambda, r)], \qquad V_{\leq r} = \frac{V}{\bigoplus_{s>r} V[s]},$$

where the right hand side of the preceding equality makes sense since $V_{\leq r}$ is a finite–dimensional object of $\mathcal{G}$. If $[V : V(\lambda, r)] > 0$, then we call $V(\lambda, r)$ a *composition factor* of V. It follows from the definition of $P(\lambda, r)$ that

$$[P(\lambda, r) : V(\mu, s)] = \dim \mathrm{Hom}_{\mathfrak{g}}(\mathbf{U}(\mathfrak{g}[t]_+)[s-r] \otimes V(\lambda), V(\mu)).$$

Moreover, one can describe the $\mathfrak{g}$-module structure of $\mathbf{U}(\mathfrak{g}[t]_+)$ in terms of symmetric powers of the adjoint representation of $\mathfrak{g}$.

Given a finite interval closed subset Γ of $P^+ \times \mathbb{N}$, let $\mathcal{G}[\Gamma]$ be the full subcategory of $\mathcal{G}$ consisting of modules V whose composition factors all lie in Γ, that is, $V \in \mathrm{Ob}\,\mathcal{G}[\Gamma]$ if and only if

$$[V : V(\mu, s)] > 0 \implies (\mu, s) \in \Gamma.$$

Clearly, $\mathcal{G}[\Gamma]$ is closed under extensions, submodules and quotients. Given $V \in \operatorname{Ob}\mathcal{G}$, let V_Γ be the maximal $\mathfrak{g}[t]$–submodule of V such that

$$[V_\Gamma : V(\mu, s)] > 0 \implies (\mu, s) \notin \Gamma,$$

and set $V^\Gamma = V/V_\Gamma$. It is clear from the definition that $V^\Gamma \in \mathcal{G}[\Gamma]$ but it is generally not true that $V_\Gamma \in \mathcal{G}[\Gamma^c]$, where Γ^c is the complement of Γ in $P^+ \times \mathbb{N}$. However, as the following proposition shows, it is true for modules $P(\mu, s)$, with $(\mu, s) \in \Gamma$, and it is this fact that makes it possible to compute the ext quiver of the endomorphism algebra of the projective generator of the category $\mathcal{G}[\Gamma]$ for finite, interval closed Γ.

Proposition *Let Γ be an interval closed subset of $\mathcal{G}$ and let $(\lambda, r), (\mu, s) \in \Gamma$. Then,*

$$[P(\lambda, r) : V(\mu, s)] = [P(\lambda, r)^\Gamma : V(\mu, s)],$$

or equivalently

$$[P(\lambda, r)_\Gamma : V(\mu, s)] = 0$$

and hence $P(\lambda, r)_\Gamma \in \mathcal{G}[\Gamma^c]$.

6.6.4 We continue to assume that Γ is finite and interval closed and choose $N \in \mathbb{N}$ so that $(\mu, s) \in \Gamma$ only if $s < N$. Set

$$P(\Gamma) = \bigoplus_{(\lambda,r)\in\Gamma} P(\lambda, r), \qquad \mathfrak{A}(\Gamma) = \operatorname{End}_{\mathcal{G}}(P(\Gamma)).$$

We have

$$\mathfrak{A}(\Gamma) \cong \operatorname{End}_{\mathcal{G}}(P(\Gamma)_{\le N}) \cong \operatorname{End}_{\mathcal{G}[\Gamma]}(P(\Gamma)^\Gamma),$$

where the first isomorphism follows easily from the definitions and the second is an application of Proposition 6.6.3. One then uses standard arguments to prove that $\mathfrak{A}(\Gamma)$ is a finite–dimensional basic algebra whose left module category is equivalent to $\mathcal{G}[\Gamma]$. Moreover, one can also prove that $\mathfrak{A}(\Gamma)$ is a directed quasi–hereditary algebra and an algorithm to compute the Ext-quiver of this algebra along with the number of relations is given in [61]. As Γ varies over interval closed sets, one gets many interesting quivers and the algebras $\mathfrak{A}[\Gamma]$ have varying representation type.

We give one specific example, motivated by the study of Kirillov-Reshetikhin modules, where the algebra is tame. Let $\mathfrak{g}$ be of type D_n with $n \geq 6$ and use the standard Bourbaki numbering of the Dynkin diagram. In particular, 4 is not a spin node. Let

$$\Gamma = \{(2\omega_4, 0), (\omega_2 + \omega_4, 1), (2\omega_2, 2), (\omega_4, 2), (\omega_1 + \omega_3, 2), (\omega_2, 3), (0, 4)\}.$$

A simple exercise shows that Γ is interval closed. With some further work, the results discussed so far can be used to prove that the category $\mathcal{G}[\Gamma]$ is equivalent to the module category of the path algebra of the following quiver

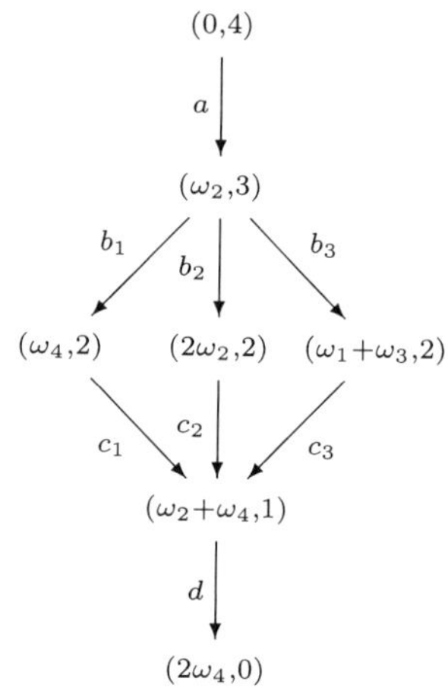

with relations

$$b_3a = 0, \qquad dc_3 = 0, \qquad c_1b_1 + c_2b_2 + c_3b_3 = 0.$$

In particular, this algebra is quadratic. It can be shown to be tame and of global dimension 2.

6.6.5 We now further focus our study on finite–dimensional associative algebras which arise from the full subcategory $\mathcal{G}_2$ which consists of objects in $\mathcal{G}$ satisfying

$$(\mathfrak{g}\otimes t^r)V = 0, \quad r \geq 2.$$

One could define $\mathcal{G}_s$ for $s \in \mathbb{N}$ in the obvious way. For $s = 1$ it is easy (and has been said elsewhere in different forms throughout these notes) to see that the category $\mathcal{G}_1$ is semi-simple. The irreducible objects are just $\mathrm{ev}_0^*\,V(\lambda)$ for $\lambda \in P^+$. Thus $\mathcal{G}_2$ is the first interesting case which is not semisimple, making it reasonable to limit our attention to this category. It is worth mentioning that restricted Kirillov-Reshetikhin modules in classical types are objects in this category. An object in $\mathcal{G}_2$ is actually a representation of the graded truncated Lie algebra $\mathfrak{g}\otimes(\mathbb{C}[t]/(t^2))$. This algebra is isomorphic to the semi–direct product $\mathfrak{g}\ltimes\mathfrak{g}_{\mathrm{ad}}$ where $\mathfrak{g}_{\mathrm{ad}}$ is the adjoint representation of $\mathfrak{g}$. Although this algebra is supported only in grades zero and one, the universal enveloping algebra is $\mathbb{N}$–graded and we have an isomorphism of associative algebras

$$\mathbf{U}(\mathfrak{g}\ltimes\mathfrak{g}_{\mathrm{ad}}) \cong S(\mathfrak{g}_{\mathrm{ad}})\otimes\mathbf{U}(\mathfrak{g}),$$

and the rth graded piece is

$$\mathbf{U}(\mathfrak{g}\ltimes\mathfrak{g}_{\mathrm{ad}})[r] \cong S^r(\mathfrak{g}_{\mathrm{ad}})\otimes\mathbf{U}(\mathfrak{g}).$$

Here $S(\mathfrak{g}_{\mathrm{ad}})$ is the symmetric algebra of $\mathfrak{g}_{\mathrm{ad}}$ and $S^r(\mathfrak{g}_{\mathrm{ad}})$ is the rth–symmetric space. One can now use the Koszul complex of $S(\mathfrak{g})$ to construct an explicit projective resolution for any simple object in $\mathcal{G}_2$ and to compute all Ext spaces between simple objects in $\mathcal{G}_2$.

Proposition *For all $(\lambda, r), (\mu, s) \in P^+\times\mathbb{N}$, we have*

$$\mathrm{Ext}^j_{\mathcal{G}_2}(V(\lambda,r),V(\mu,s)) \cong \begin{cases} 0, & j \neq s-r, \\ \mathrm{Hom}_{\mathfrak{g}}(\bigwedge^j\mathfrak{g}_{\mathrm{ad}}\otimes V(\lambda), V(\mu)), & j = s-r. \end{cases}$$

In particular, $\mathrm{Ext}^j_{\mathcal{G}_2}(M, N) = 0$ *if* M, N *are finite dimensional and* $j > \dim \mathfrak{g}$.

This proposition strongly suggests that there is an associative algebra with a Koszul grading in the background and we now describe how to find this algebra.

6.6.6 We first need to refine the partial ordering $\preceq$ on $P^+ \times \mathbb{N}$. Instead of allowing all possible roots in the cover relation, we restrict ourselves to particular subsets of Φ^+ obtained as follows. Given $\psi \in P$, let

$$\Phi^+_\psi = \{\alpha \in R^+ : \kappa(\psi, \alpha) \geq \kappa(\psi, \beta) \text{ for all } \beta \in \Phi\},$$

and define a cover relation by (μ, s) covers (λ, r) iff $s = r + 1$ and $\lambda - \mu \in \Phi^+_\psi$. Let $\leq_\psi$ be the corresponding partial order or, equivalently, $(\lambda, r) \leq_\psi (\mu, s)$ iff

$$\mu - \lambda = \sum_{\alpha \in \Phi^+_\psi} n_\alpha \alpha, \quad n_\alpha \in \mathbb{N}, \qquad \sum_{\alpha \in \Phi^+_\psi} n_\alpha = s - r.$$

The definition of Φ^+_ψ ensures that this order is well–defined. The following was proved in [62]:

Theorem *Let* $(\lambda, r) \in P^+ \times \mathbb{N}$ *and let* $\Gamma = \{(\mu, s) : (\mu, s) \leq_\psi (\lambda, r)$. *Then the endomorphism algebra* $\mathfrak{A}[\Gamma]$ *of the projective generator of* $\mathcal{G}_2[\Gamma]$ *admits a natural Koszul grading and is of global dimension at most the cardinality of* Φ^+_ψ. *Moreover the maximal dimension is attained for a suitable choice of* (λ, r).

One can define certain limits of the algebras as Γ varies and construct infinite–dimensional Koszul algebras $\mathfrak{A}_\psi$ of left global dimension $|\Phi^+_\psi|$. The Koszul dual of these algebras are also studied in [62], although one does not understand the module category of the dual in the context of representations of Lie algebras. We conclude these notes with examples of the quivers associated to the infinite–dimensional Koszul algebras arising from our study. Further details may be found in [117].

6.6.7 Suppose that $\mathfrak{g}$ is not of type A or C so that there exists a unique $i_0 \in I$ such that $\theta - \alpha_{i_0} \in \Phi^+$ and then,

$$\Phi^+_{\omega_{i_0}} = \{\theta, \theta - \alpha_{i_0}\}.$$

Then every connected subalgebra of $\mathfrak{A}_{\omega_{i_0}}$ is isomorphic to the path algebra of the translation quiver

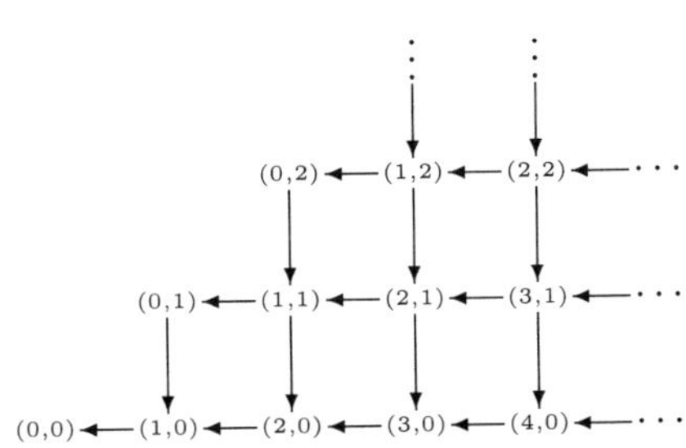

(6.16)

with the mesh relations as indicated.

6.6.8 Let $\mathfrak{g}$ be of type C. Then $\Phi^+_{\omega_2} = \{\theta, \theta - \alpha_1, \theta - 2\alpha_1\}$ and the quiver of $\mathfrak{A}_{\omega_2}$ is the (infinite, if rank of $\mathfrak{g}$ is greater than 2) union of connected components of the following two types:

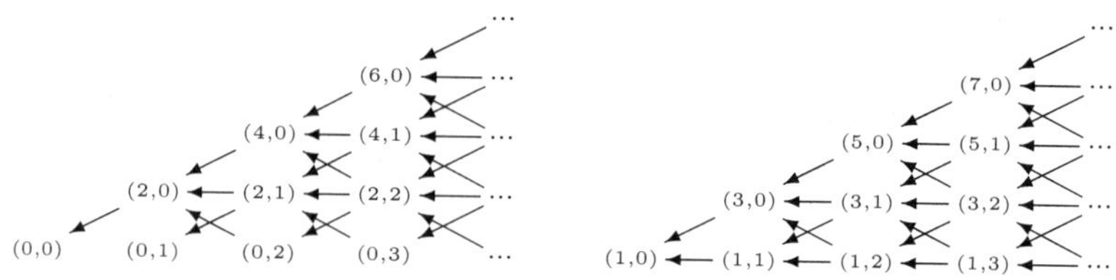

Both are translation quivers with $\tau((m,n)) = (m, n-2)$, $m > 0$, $n \geq 2$. The relations are: the commutativity relations in

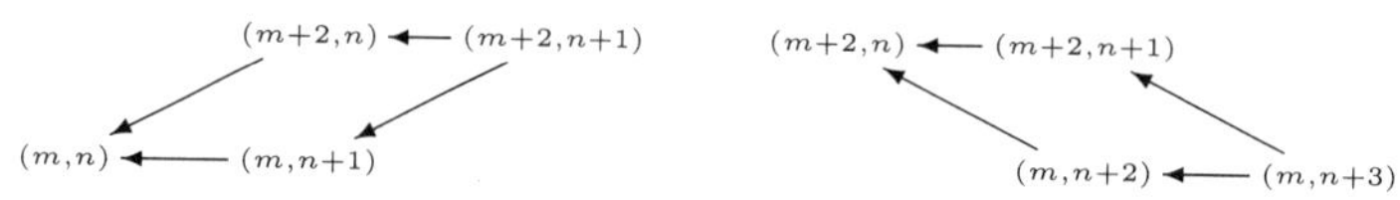

for all $m > 0$, $n \in \mathbb{Z}_+$, the zero relations $(2,n) \leftarrow (2,n+1) \leftarrow (0,n+3)$, $n \geq 0$ and

$$m^2((m,n) \leftarrow (m{+}2,n) \leftarrow (m,n{+}2)) - (m{+}2)^2((m,n) \leftarrow (m{-}2,n{+}2) \leftarrow (m,n{+}2)) \\ + (m+1)((m,n) \leftarrow (m,n+1) \leftarrow (m,n+2)), \qquad m > 1,$$

and, finally, $((1,n) \leftarrow (3,n) \leftarrow (1,n+2)) + 2((1,n) \leftarrow (1,n+1) \leftarrow (1,n+2))$. Thus, if $\ell = 2$ and $|\Phi^+_\psi| > 1$, the algebra $\mathfrak{A}_\psi$ is the direct sum of two non-isomorphic connected Koszul subalgebras of left global dimension 3.

6.6.9 Let $\mathfrak{g}$ be of type G_2 and suppose that α_1 is the long simple root. Then $\Phi^+_{\omega_1-\omega_2} = \{\alpha_1, \theta\}$ and the algebra $\mathfrak{A}_{\omega_1-\omega_2}$ is the direct sum of three isomorphic connected subalgebras with quivers

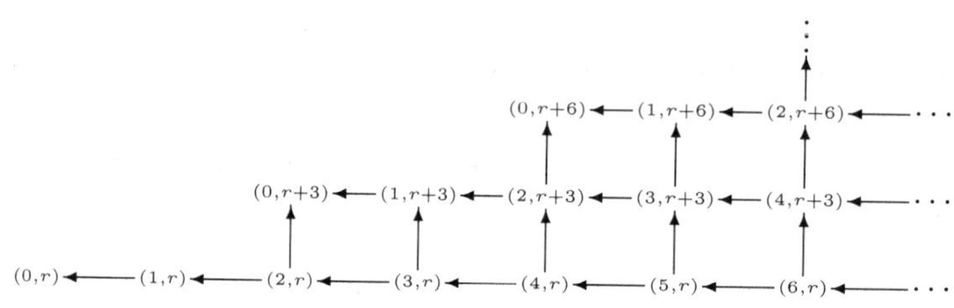

where $0 \leq r \leq 2$. The relations are again the mesh relations, the translation map being

$$\tau((m, 3k+r)) = (m+3, 3(k-1)+r), \quad m \in \mathbb{Z}_+, \quad k > 0.$$

6.6.10 In conclusion, we note that there is some work [28], [34], [67], [152], [175] on the representation theory of extended affine algebras where the full center does not act trivially. There is also the recent work [99] in which the authors give a representation theoretic way to interpret the third cohomology classes of the double loop algebras. It is perhaps clear by now, that there are many interesting avenues to pursue in the representation theory of affine, toroidal and extended affine Lie algebras and while the references are by no means anywhere near exhaustive they should provide the reader with some pointers to the literature.

Vyjayanthi Chari
Department of Mathematics
University of California
Riverside, CA 92521 USA
chari@math.ucr.edu

Bibliography

[1] Allison, B., Azam, S., Berman, S., Gao, Y., and Pianzola, A., Extended affine Lie algebras and their root systems, *Mem. Amer. Math. Soc.* **126** (1997), no. 603, x+122.

[2] Allison, B., Benkart, G., and Gao, Y., Central extensions of Lie algebras graded by finite root systems, *Math. Ann.* **316** (2000), 499–527.

[3] Allison, B., Benkart, G., and Gao, Y., Lie algebras graded by the root systems BC_r, $r \geq 2$, *Mem. Amer. Math. Soc.* **158** (2002), no. 751, x+158.

[4] Allison, B., Berman, S., Faulkner, J., and Pianzola, A., Realization of graded-simple algebras as loop algebras, *Forum Math.* **20** (2008), 395–432.

[5] Allison, B., Berman, S., Faulkner, J., and Pianzola, A. Muliloop realization of extended affine Lie algebras and Lie tori, *Trans. Amer. Math. Soc.* **361** (2009), 4807–4842.

[6] Allison, B., Berman, S., Gao, Y., and Pianzola, A., A characterization of affine Kac-Moody Lie algebras, *Comm. Math. Phys.* **185** (1997), 671–688.

[7] Allison, B., Berman, S., and Pianzola, A., Covering algebras. I. Extended affine Lie algebras, *J. Algebra* **250** (2002), no. 2, 485–516.

[8] Allison, B., Berman, S., and Pianzola, A., Covering algebras. II. Isomorphism of loop algebras, *J. Reine Angew. Math.* **571** (2004), 39–71.

[9] Allison, B., Berman, S., and Pianzola, A., Iterated loop algebras, *Pacific J. Math.* **227** (2006), 1–41.

[10] Allison, B., Berman, S., and Pianzola, A., Multiloop algebras, iterated loop algebras, and extended affine Lie algebas of nullity 2, arXiv:1002.2674v1 [math.RA].

[11] Allison, B. and Faulkner, J., Isotopy for extended affine Lie algebras and Lie tori, arXiv:0709.1181 [math.RA].

[12] Allison, B. and Gao, Y.,The root system and the core of an extended affine Lie algebra, *Selecta Math. (N.S.)* **7** (2001), 149–212.

[13] Arakawa, T., Representation theory of W-algebras, *Invent. Math.* **169** (2007), 219–320.

[14] Arakawa, T. and Suzuki, T., Duality between $\mathfrak{sl}_n(\mathbb{C})$ and the degenerate affine Hecke algebra, *J. Algebra* **209** (1998), 288–304.

[15] Assem, I., Simson, D., and Skowroński, A., Elements of the representation theory of associative algebras. In *London Mathematical Society Student Texts*, vol.1, vol.65, *Techniques of Representation Theory.* Cambridge University Press, Cambridge, 2006.

[16] Azam, S., Generalized reductive Lie algebras: connections with extended affine Lie algebras and Lie tori, *Canad. J. Math.* **58** (2006), 225–248.

[17] Azam, S., Derivations of tensor product algebras, *Comm. Algebra* **36** (2008), 905–927.

[18] Batra, P. and Pal, T., Representations of graded multi-loop Lie algebras, *Comm. Algebra* **38** (2010), 49–67.

[19] Baur, K. and Wallach, N., Nice parabolic subalgebras of reductive Lie algebras, *Represent. Theory* **9** (2005), 1–29.

[20] Beck, J. and Nakajima, H., Crystal bases and two-sided cells of quantum affine algebras, *Duke Math. J.* **123** (2004), no. 2, 335–402.

[21] Beilinson, A. and Bernstein, J., Localisation de $\mathfrak{g}$–modules, *C. R. Math. Acad. Sci. Paris* **292** (1981), 15–18.

[22] Beilinson, A., Ginzburg, V., and Soergel, W., Koszul duality patterns in representation theory, *J. Amer. Math. Soc.* **9** (1996), no. 2, 473-527.

[23] Beilinson, A., Lusztig, G., and MacPherson, R., A geometric setting for the quantum deformation of GL_n. *Duke Math. J.* **61** (1990), no. 2, 655–677.

[24] Benkart, G., Derivations and invariant forms of Lie algebras graded by finite root systems, *Canad. J. Math.* **50** (1998), 225–241.

[25] Benkart, G. and Moody, R., Derivations, central extensions, and affine Lie algebras, *Algebras Groups Geom.* **3** (1986), 456–492.

[26] Benkart, G. and Neher, E., The centroid of extended affine and root graded Lie algebras, *J. Pure Appl. Algebra* **205** (2006), 117–145.

[27] Benkart, G. and Smirnov, O., Lie algebras graded by the root system BC_1, *J. Lie Theory* **13** (2003), 91–132.

[28] Berman, S., Billig, Y., and Szmigielski, J., Vertex operator algebras and the representation theory of toroidal algebras. In *Recent Developments in Infinite-Dimensional Lie Algebras and Conformal Field Theory* (Charlottesville, VA, 2000), pp. 1-26, *Contemp. Math.* **297**, Amer. Math. Soc., Providence, RI, 2002.

[29] Berman, S., Gao, Y., and Krylyuk, Y., Quantum tori and the structure of elliptic quasi-simple Lie algebras, *J. Funct. Anal.* **135** (1996), 339–389.

[30] Berman, S., Gao, Y., Krylyuk, Y., and Neher, E., The alternative torus and the structure of elliptic quasi-simple Lie algebras of type A_2, *Trans. Amer. Math. Soc.* **347** (1995), 4315–4363.

[31] Berman, S. and Moody, R., Lie algebras graded by finite root systems and the intersection matrix algebras of Slodowy, *Invent. Math.* **108** (1992), 323–347.

[32] Bernstein, J., Gelfand, I., and Gelfand, S., A certain category of g-modules, (Russian) *Funkcional. Anal. i Prilozen.* **10** (1976), no. 2, 1-8.

[33] Bernšteĭn, I. N., Gelfand, I. M., and Ponomarev, V. A., Coxeter functors, and Gabriel's theorem, *Uspehi Mat. Nauk* **28** (1973), 19–33.

[34] Billig, Yu., *Representations of toroidal extended affine Lie algebras*, *J. Algebra* **308** (2007), 252-269.

[35] Brylinski, J. L. and Kashiwara, M., Kazhdan–Lusztig conjecture and holonomic systems, *Invent. Math.* **64** (1981), 387–410.

[36] Borcherds, R. E., Vertex algebras, Kac–Moody algebras, and the Monster, *Proc. Natl. Acad. Sci. USA* **83** (1986), 3068–3071.

[37] Bordemann, M., Nondegenerate invariant bilinear forms on nonassociative algebras, *Acta Math. Univ. Comenian. (N.S.)* **66** (1997), 151–201.

[38] Bourbaki, N., *Groupes et Algèbres de Lie*, Ch. 1, Hermann Paris 1971.

[39] Bourbaki, N., *Groupes et Algèbres de Lie*, Ch. 4–6, Masson 1981.

[40] Bourbaki, N., *Groupes et Algèbres de Lie*, Ch. 7–8, Hermann Paris 1975.

[41] Bouwknegt, and K. Schoutens, K., $\mathcal{W}$-symmetry, *Adv. Series Math. Phys.* **22**, Singapore, World Scientific, 1995.

[42] Braverman, A., Gaitsgory, D., and Vybornov, M., Relation between two geometrically defined bases in representations of GL_n. In *Quantum groups*, Volume 433 of *Contemp. Math.*, pp. 103–113. Amer. Math. Soc., Providence, RI, 2007.

[43] Braverman, A. and Gaitsgory, D., On Ginzburg's Lagrangian construction of representations of $GL(n)$, *Math. Res. Lett.* **6** (1999), no. 2, 195–201.

[44] Brown, J., Twisted Yangians and finite W-algebras, *Transform. Groups* **14** (2009), 87–114.

[45] Brown, J. and Brundan, J., Elementary invariants for centralizers of nilpotent matrices, *J. Aust. Math. Soc.* **86** (2009),1–15.

[46] Brundan, J. and Goodwin, S., Good grading polytopes, *Proc. London Math. Soc.* **94** (2007), 155–180.

[47] Brundan, J., Goodwin, S., and Kleshchev, A., Highest weight theory for finite W-algebras, *Internat. Math. Res. Notices* **15** (2008), Art. ID rnn051.

[48] Brundan, J. and Kleshchev, A., Shifted Yangians and finite W-algebras, *Adv. Math.* **200** (2006), 136–195.

[49] Brundan, J. and Kleshchev, A., Representations of shifted Yangians and finite W-algebras, *Mem. Amer. Math. Soc.* **196** (2008).

[50] Brundan, J. and Kleshchev, A., Schur-Weyl duality for higher levels, *Selecta Math.* **14** (2008), 1–57.

[51] Brundan, J. and Stroppel, C., Highest weight categories arising from Khovanov's diagram algebra III: category $\mathcal{O}$, arXiv:0812.1090.

[52] Bump, D., Lie groups. In *Graduate Texts in Mathematics*, Volume 225. Springer-Verlag, New York, 2004.

[53] Carter, R., *Finite Groups of Lie Type: Conjugacy Classes and Complex Characters*, Pure and Applied Math., John Wiley & Sons, Inc., New York, 1985.

[54] Carter, R. W., *Lie Algebras of Finite and Affine Type*, Cambridge Studies in Advanced Mathematics, 2005.

[55] Capparelli, S., Lepowsky, J., and Milas, A., The Rogers-Ramanujan recursion and intertwining operators, *Commun. Contemp. Math* **5** (2003), 947–966.

[56] Chari, V., Integrable representations of affine Lie-algebras, *Invent. Math.* **85** (1986), 317-335.

[57] Chari, V., Fourier, G., and Senesi, P., Weyl modules for the twisted loop algebras, *J. Algebra* **319** (2008), no. 12, 5016-5038.

[58] Chari, V., Fourier, G., and Khandai, T., A categorical approach to Weyl modules, arXiv:0906.2014.

[59] Chari V. and Greenstein, J., An application of free Lie algebras to current algebras and their representation theory, *Contemporary Mathematics*, **392** (2005) 15–31.

[60] Chari V. and Greenstein, J., Graded level zero integrable representations of affine Lie algebras, *Trans. of the AMS* **360** (2008), no. 6, 2923–2940.

[61] Chari V. and Greenstein, J., Current algebras, highest weight categories and quivers, *Adv. in Math.* **216** (2007), no. 2, 811–840.

[62] Chari V. and Greenstein, J., A family of Koszul algebras arising from finite-dimensional representations of simple Lie algebras, *Adv. in Math.* **220** (2009), no. 4, 1193–1221.

[63] Chari, V. and Hernandez, D., Beyond Kirillov Reshetikhin modules, *Contemp. Math.*, **506** (2010), 49–81.

[64] Chari, V. and Ilangovan, S., On the Harish–Chandra Homomorphism for Infinite-Dimensional Lie Algebras, *J. Algebra* (1984), 476–490.

[65] Chari, V. and Jing, N., Realization of level one representations of $U_q(\hat{g})$ at a root of unity, *Duke Math. J.* **108** (2001), no. 1, 183–197.

[66] Chari, V., Khare, A., and Ridenour, T. B. in preparation.

[67] Chari, V. and Le, T., Representations of double affine Lie algebras, A tribute to C. S. Seshadri (Chennai, 2002), *Trends Math.*, pp. 199-219. Birkhäuser, Basel, 2003.

[68] Chari, V. and Loktev, S., Weyl, Demazure and fusion modules for the current algebra of sl_{r+1}, *Adv. in Math.* **207** (2006), no. 2, 928–960.

[69] Chari, V. and Moura, A., Spectral characters of finite-dimensional representations of affine algebras, *J. Algebra* **279** (2004), no. 2, 820-839.

[70] Chari, V. and Moura, A., The restricted Kirillov-Reshetikhin modules for the current and twisted current algebras, *Commun. Math. Phys.* **266** (2006) 431–454.

[71] Chari, V. and Pressley, A., New unitary representations of loop groups, *Math. Ann.* **275** (1986), 87-104.

[72] Chari, V. and Pressley, A., Integrable representations of twisted affine Lie algebras, *J. Alg.* **113** (1988), no. 2, 438–464.

[73] Chari, V. and Pressley, A., Quantum affine algebras, *Comm. Math. Phys.* **142** (1991), no. 2, 261–283.

[74] Chari, V. and Pressley, A., Quantum affine algebras and their representations. In Representations of Groups (Banff, AB, 1994), *CMS Conf. Proc.*, pp. 59–78, vol. 16, Amer. Math. Soc., Providence, RI, 1995.

[75] Chari, V. and Pressley, A., Weyl modules for classical and quantum affine algebras, *Represent. Theory* **5** (2001), 191–223 (electronic).

[76] Cherednik, I., A new interpretation of Gelfand-Tzetlin bases, *Duke Math. J.* **54** (1987), 563–577.

[77] Chriss, N. and Ginzburg, V., *Representation Theory and Complex Geometry*. Birkhäuser, Boston, MA, 1997.

[78] Chuang, J. and Rouquier, R., Derived equivalences for symmetric groups and $\mathfrak{sl}_2$-categorification, *Annals Math.* **167** (2008), 245–298.

[79] Cline, E. T., Parshall, B. J., and Scott, L. L., Finite–dimensional algebras and highest weight categories, *J. Reine. Angew. Math.* **391** (1988), 85–99.

[80] Collingwood, D. and McGovern, W., *Nilpotent Orbits in Semisimple Lie Algebras*. Van Nostrand Reinhold Mathematics Series, Van Nostrand Reinhold Co., NY, 1993.

[81] D'Andrea, A., De Concini, D., De Sole, A., Heluani, R., and Kac, V., Three equivalent definitions of finite W-algebras, appendix to [84].

[82] de Boer, J. and Tjin, T., Quantization and representation theory of finite W-algebras, *Comm. Math. Phys.* **158** (1993), 485–516.

[83] Deodhar, V. V., Gaber, O., and Kac, V., Structure of some categories of representations of infinite–dimensional Lie algebras, *Adv. in Math.* **45** (1982), 92–116.

[84] De Sole, A. and Kac, V., Finite vs affine W-algebras, *Japan J. Math.* **1** (2006), 137–261.

[85] de Vos, K. and van Driel, P., The Kazhdan–Lusztig conjecture for finite W-algebras, *Lett. Math. Phys.* **35** (1995), 333–344.

[86] Dimitrov, I., Futorny, V., and Penkov, I., A reduction theorem for highest weight modules over toroidal Lie algebras, *Comm. Math. Phys.* **250** (2004), 47–63.

[87] Dixmier, J., Enveloping algebras. In *Graduate Studies in Mathematics*, volume 11, revised reprint of the 1977 translation, American Mathematical Society, Providence, RI, 1996.

[88] Drinfel′d, V. G., Hopf algebras and the quantum Yang-Baxter equation, *Soviet Math. Dokl.* **32** (1985), no. 1, 254–258.

[89] Drinfeld, V. G., Degenerate affine Hecke algebras and Yangians, *Func. Anal. Appl.* **20** (1986), 56–58.

[90] Drinfeld, V. G., A new realization of Yangians and of quantum affine algebras, *Dokl. Akad. Nauk SSSR* **296** (1987), no. 1, 13–17.

[91] Eswara Rao, S. and Moody, R., Vertex representations for n-toroidal Lie algebras and a generalization of the Virasoro algebra, *Comm. Math. Phys.* **159** (1994), 239–264.

[92] Elashvili, P. and Kac, V., Classification of good gradings of simple Lie algebras. In *Lie Groups and Invariant Theory*, (E. B. Vinberg ed.), pp. 85–104, *Amer. Math. Soc. Transl.* **213**, AMS, 2005.

[93] Etingof, P. and Moura, A., Elliptic central characters and blocks of finite dimensional representations of quantum affine algebras, *Represent. Theory* **7**, (2003), 346-373.

[94] Fehér, L., O'Raifeartaigh, L., Ruelle, P., Tsutsui, I., and Wipf, A., Generalized Toda theories and W-algebras associated with integral gradings, *Ann. Phys.* **213** (1992), 1–20.

[95] Fehér, L., O'Raifeartaigh, L., Ruelle, P., Tsutsui, I., and Wipf, A., On Hamiltonian reductions of the Wess-Zumino-Novikov-Witten theories, *Phys. Rep.* **222** (1992), 1–64.

[96] Fourier, G. and Littelmann, P., Weyl modules, Demazure modules, KR-modules, crystals, fusion products and limit constructions, *Advances in Mathematics* **211** (2007) 566-593.

[97] Frenkel, E., Kac, V., Radul, A., and Wang, W., $\mathcal{W}_{1+\infty}$ and $\mathcal{W}(gl_N)$ with central charge N, *Comm. Math. Phys.* **170** (1995), 337–357.

[98] Frenkel, E., Kac, V., and Wakimoto, M., Characters and fusion rules for W-algebras via quantized Drinfeld-Sokolov reduction, *Commun. Math. Phys.* **147** (1992), 295–328.

[99] Frenkel, E. and Zhu, X., *Gerbal Representations of Double Loop Groups* arXiv:0810.1487 (2008)

[100] Frenkel, I. B. and Jing, N., Vertex representations of quantum affine algebras, *Proc. Nat. Acad. Sci. U.S.A.* **85** (1988), no. 24, 9373–9377.

[101] Frenkel, I. B. and Kac, V. G., Basic representations of affine Lie algebras and dual resonance models, *Invent. Math.* **62** (1980), 23–66.

[102] Frenkel, I., Lepowsky, J., and Meurman, A., Vertex operator algebras and the monster, Academic Press, New York, 1988.

[103] Friedlander, E. and Parshall, B., Support varieties for restricted Lie algebras, *Invent. Math.* **86** (1986), 553–562.

[104] Fulton, W. and Harris, J., Representation theory In *Graduate Texts in Mathematics*, vol. 129. A first course, Readings in Mathematics. Springer-Verlag, New York, 1991.

[105] Gabriel, P., Unzerlegbare Darstellungen. I, *Manuscripta Math.* **6** (1972) 71–103. Correction, ibid. 6 (1972), 309.

[106] Gabriel, P., Indecomposable representations. II. In *Symposia Mathematica, Vol. XI (Convegno di Algebra Commutativa, INDAM, Rome, 1971)*, pp. 81–104. Academic Press, London, 1973.

[107] Gan, W. L. and Ginzburg, V., Quantization of Slodowy slices, *Internat. Math. Res. Notices* **5** (2002), 243–255.

[108] García, E. and Neher, E., Tits-Kantor-Koecher superalgebras of Jordan superpairs covered by grids, *Comm. Algebra* **31** (2003), 3335–3375.

[109] Garland, H., The arithmetic theory of loop algebras, *J. Algebra* **53**, (1978), 480-551.

[110] Garland, H., The arithmetic theory of loop groups, *Inst. Hautes Études Sci. Publ. Math.* **52** (1980), 5–136.

[111] Ginzburg, V., Harish–Chandra bimodules for quantized Slodowy slices, *Representation Theory* **13** (2009), 236–271.

[112] Ginzburg, V., Perverse sheaves on a loop group and Langlands duality, `math.AG/9511007` (1995).

[113] Ginzburg, V., Geometric methods in the representation theory of Hecke algebras and quantum groups. In *Representation Theories and Algebraic Geometry (Montreal, PQ, 1997), NATO Adv. Sci. Inst. Ser. C Math. Phys. Sci.*, vol. 514, pp. 127–183. Kluwer Acad. Publ., Dordrecht, 1998. Notes by Vladimir Baranovsky [V. Yu. Baranovskiĭ].

[114] Goodwin, S. Translation for finite W-algebras, arXiv:0908.2739.

[115] Goodwin, S., Röhrle, G., and Ubly, G., On 1-dimensional representations of finite W-algebras associated to simple Lie algebras of exceptional type, arXiv:0905.3714.

[116] Gordon, I. and Rumynin, D., Subregular representations of $\mathfrak{sl}_n$ and simple singularities of type A_{n-1}, *Compositio Math.* **138** (2003), 337–360.

[117] Greenstein, J., Quivers with relations arising from Koszul algebras of g-invariants, *J. of Algebra* **322** (2009), no. 12, 4430–4478.

[118] Guay, N., Affine Yangians and deformed double current algebras in type A, *Adv. Math.* **211** (2007), 436–484.

[119] Hatayama, G., Kuniba, A., Okado, M., Takagi, T., and Yamada, Y., Remarks on fermionic formula. In *Recent Developments in Quantum Affine Algebras and Related Topics* (Raleigh, NC, 1998), pp. 243–291, *Contemp. Math.* **248** (1999). Amer. Math. Soc., Providence, RI.

[120] Helgason, S., *Differential Geometry, Lie Groups, and Symmetric Spaces*, Pure and Applied Mathematics **80**, Academic Press, New York 1978.

[121] Høegh-Krohn, R. and Torrésani, B., Classification and construction of quasisimple Lie algebras, *J. Funct. Anal.* **89** (1990), 106–136.

[122] Hong, J. and Kang, S.-J., *Introduction to Quantum Groups and Crystal Bases*, Graduate Studies in Mathematics, vol. 42. American Mathematical Society, Providence, RI, 2002.

[123] Hong, J., Kang, S.-J., and Lee, H., Young wall realization of crystal graphs for $U_q(C_n^{(1)})$, *Comm. Math. Phys.* **244** (2004), no. 1, 111–131.

[124] Humphreys, J. E., *Introduction to Lie Algebras and Representation Theory*, Springer-Verlag, New York, 1968.

[125] Humphreys, J. E., *Linear Algebraic Groups*, Graduate Texts in Mathematics, vol. 21. Springer-Verlag, New York, 1975.

[126] Humphreys, J. E., *Introduction to Lie Algebras and Representation Theory*, 2nd ed. Springer-Verlag, New York, 1978.

[127] Humphreys, J. E., *Representations of Semisimple Lie Algebras in the BGG Category $\mathcal{O}$*, Graduate Studies in Mathematics, vol. 94, American Mathematical Society, 2008.

[128] Jacobson, N., *Lie algebras, Interscience Tracts in Pure and Applied Mathematics* **10**, Interscience Publishers (a division of John Wiley & Sons), New York-London 1962.

[129] Jakelic, D. and Moura, A. A., Finite-dimensional representations of hyper loop algebras, *Pacific J. Math.* **233** (2007), 371–402.

[130] Jakelic, D. and Moura, A. A., Finite-dimensional representations of hyper loop algebras over non algebraicaly closed fields, *Algebras and Representation Theory* published online first DOI 10.1007/s10468-008-9122-5.

[131] Jakelic, D. and Moura, A. A., On multiplicity problems for finite-dimensional representations of hyper loop algebras, *Contemp. Math.* **483** (2009), 147–159.

[132] Jantzen, J. C., Representations of algebraic groups, In *Pure and Applied Mathematics*. Academic Press Inc., Boston, MA, 1987.

[133] Jantzen, J. C., Representations of Lie algebras in prime characteristic, *NATO Adv. Sci. Inst. Ser. C Math. Phys. Sci.* **514**, Representation theories and algebraic geometry (Montreal, PQ, 1997), pp. 185–235, Kluwer Acad. Publ., Dordrecht, 1998.

[134] Jantzen, J. C., Nilpotent orbits in representation theory, pp. 1–211, *Progr. Math.* **228**, Birkhäuser, Boston, MA, 2004.

[135] Jimbo, M., A q-difference analogue of $U(\mathfrak{g})$ and the Yang-Baxter equation, *Lett. Math. Phys.* **10** (1985), no. 1, 63–69.

[136] Kac, V. G., Vertex algebras for beginners. In *University Lecture Series*, vol. 10, Amer. Math. Soc., Providence, RI, 1996; 2nd ed., 1998.

[137] Kac, V. G., *Infinite Dimensional Lie Algebras*, 3rd ed., Cambridge University Press, Cambridge, 1990.

[138] Kac, V., Roan, S., and Wakimoto, M., Quantum reduction for affine superalgebras, *Comm. Math. Phys.* **241** (2003), 307–342.

[139] Kang, S.-J., Crystal bases for quantum affine algebras and combinatorics of Young walls, *Proc. London Math. Soc.* (3) **86** (2003), no. 1, 29–69.

[140] Kang, S.-J., Kashiwara, M., Misra, K. C., Miwa, T., Nakashima, T., and Nakayashiki, A., Vertex models and crystals, *C. R. Acad. Sci. Paris Sér. I Math.* **315** (1992), no. 4, 375–380.

[141] Kashiwara, M., On crystal bases of the q-analogue of universal enveloping algebras, *Duke Math. J.* **63** (1991), 465–516.

[142] Kashiwara, M., The crystal base and Littelmann's refined demazure character formula, *Duke Math. J.* **71** (1993), no. 3, 839–858.

[143] Kashiwara, M. and Nakashima, T., Crystal graphs for representations of the q-analogue of classical Lie algebras, *J. Algebra* **165** (1994), no. 2, 295–345.

[144] Kashiwara, M. and Saito, Y., Geometric construction of crystal bases, *Duke Math. J.* **89** (1997), no. 1, 9–36.

[145] Kassel, C., Kähler differentials and coverings of complex simple Lie algebras extended over a commutative algebra, In *Proceedings of the Luminy conference on algebraic K-theory* (Luminy, 1983), *J. Pure Appl. Algebra* **34** (1984), 265–275.

[146] Kazhdan, D. and Lusztig, G., Representations of Coxeter groups and Hecke algebras, *Invent. Math.* **53** (1979), 165–184.

[147] Kleshchev, A., *Linear and Projective Representations of Symmetric Groups*, Cambridge Tracts in Mathematics, p. 163. Cambridge University Press, Cambridge, 2005.

[148] Kodera, R., Extensions between finite-dimensional simple modules over a generalized current Lie algebra, arXiv:0908.3738.

[149] Kostant, B., Groups over Z, algebraic groups and discontinuous subgroups, In *Proc. Symp. Pure Math. IX*, Providence, AMS, 1966.

[150] Kostant, B., On Whittaker modules and representation theory, *Invent. Math.* **48** (1978), 101–184.

[151] Lauda, A. and Vazirani, M., Crystals from categorified quantum groups, arXiv:0909.181 (2009).

[152] Larsson, T., Extended diffeomorphism algebras and trajectories in jet space, *Commun. Math. Phys.* **214** (2000), 469-491.

[153] Lau, M., Representations of multi–loop algebras, math. RT/ arXiv:0811.2011v2.

[154] Lepowsky, J. and Wilson, R. L., Construction of the affine Lie algebra $A_1^{(1)}$, *Commun. Math. Phys.* **62** (1978), 43–53.

[155] Loos, O., Spiegelungsräume und homogene symmetrische Räume, *Math. Z.* **99** (1967), 141–170.

[156] Loos, O. and Neher, E., Reflection systems and partial root systems, Jordan Theory Preprint Archives, paper #182, to appear in *Forum Math.*

[157] Losev, I., Quantized symplectic actions and W-algebras, *J. Amer. Math. Soc.* **23** (2010), 35–59.

[158] Losev, I., Finite dimensional representations of W-algebras, arXiv:0807.1023.

[159] Losev, I., On the structure of the category $\mathcal{O}$ for W-algebras, arXiv:0812.1584.

[160] Losev, I., 1-dimensional representations and parabolic induction for W-algebras, preprint, arXiv:0906.0157.

[161] Losev, I., Finite W-algebras, ICM talk 2010, arXiv:1003.5811.

[162] Lurie, J., A proof of the Borel-Weil-Bott theorem, *http://www.math.harvard.edu/~lurie/papers/bwb.pdf.*

[163] Lusztig, G., Singularities, character formulas, and a q-analog of weight multiplicities. In *Analysis and Topology on Singular Spaces, II, III (Luminy, 1981), Astérisque*, vol. 101, pp. 208–229. Soc. Math. France, Paris, 1983.

[164] Lusztig, G., Quantum deformation of certain simple modules over enveloping algebras, *Adv. Math.* **70** (1988), 237–249.

[165] Lusztig, G., Canonical bases arising from quantized enveloping algebras, *J. Amer. Math. Soc.* **3** (1990), no. 2, 447–498.

[166] Lynch, T., Generalized Whittaker vectors and representation theory, *Ph.D. dissertation*, M.I.T., 1979.

[167] Malkin, A., Tensor product varieties and crystals: GL case, *Trans. Amer. Math. Soc.* **354**(2) (2002), 675–704 (electronic).

[168] Malkin, A., Tensor product varieties and crystals: the ADE case, *Duke Math. J.* **116**(3) (2003), 477–524.

[169] Mirković, I. and Vilonen, K., Geometric Langlands duality and representations of algebraic groups over commutative rings, *Ann. of Math. (2)* **166**(1) (2007), 95–143.

[170] Mirković, I. and Vybornov, M., On quiver varieties and affine Grassmannians of type A, *C. R. Math. Acad. Sci. Paris* **336**(3) (2003), 207–212.

[171] Misra, K. and Miwa, T., Crystal base for the basic representation of $U_q(\hat{\mathfrak{sl}}(n))$., *Comm. Math. Phys.* **134** (1990), no. 1, 79–88.

[172] Mitzman, D., Integral bases for affine Lie algebras and their universal enveloping algebras, *Contemp. Math* **40** (1985).

[173] Moeglin, C., Modèles de Whittaker et idéaux primitifs complètement premier dans les algèbres enveloppantes I, *C. R. Acad. Sci. Paris* **303** (1986), 845–848; II, *Math. Scand.* **63** (1988), 5–35.

[174] Moody, R. and Pianzola, A., Lie algebras with triangular decompositions, *Can. Math. Soc.* series of monographs and advanced texts, John Wiley, 1995.

[175] Moody, R. V., Rao, S. E., and Yokonuma, T., Toroidal Lie algebras and vertex representations, *Geom. Dedicata* **35** (1990), 283–307.

[176] Morita, J. and Yoshii, Y., Locally extended affine Lie algebras, *J. Algebra* **301** (2006), 59–81.

[177] Moura, A., Restricted limits of minimal affinizations, *Pacific J. Math.* **244** (2010), no. 2, 359–397.

[178] Nakajima, H., Instantons on ALE spaces, quiver varieties, and Kac-Moody algebras. *Duke Math. J.* **76**(2) (1994), 365–416.

[179] Nakajima, H., Quiver varieties and finite-dimensional representations of quantum affine algebras, *J. Amer. Math. Soc.* **14** (2001), no. 1, 145–238 (electronic).

[180] Nakajima, H., Quiver varieties and tensor products, *Invent. Math.* **146**(2) (2001), 399–449.

[181] Nakajima, H., Quiver varieties and Kac-Moody algebras, *Duke Math. J.* **91**(3) (1998), 515–560.

[182] Naoi, K., Multiloop Lie algebras and the construction of extended affine Lie algebras, arXiv:0807.2019.

[183] Neher, E., Lie algebras graded by 3-graded root systems and Jordan pairs covered by a grid, *Amer. J. Math* **118** (1996), 439–491.

[184] Neher, E., An introduction to universal central extensions of Lie superalgebras, In *Groups, Rings, Lie and Hopf Algebras* (St. John's, NF, 2001), *Math. Appl.* vol. 555, pp. 141–166. Kluwer Acad. Publ., Dordrecht, 2003.

[185] Neher, E., Lie tori, *C. R. Math. Acad. Sci. Soc. R. Can.* **26** (2004), 84–89.

[186] Neher, E., Extended affine Lie algebras, *C. R. Math. Acad. Sci. Soc. R. Can.* **26** (2004), 90–96.

[187] Neher, E., Extended affine Lie algebras and other generalizations – a survey, In *Trends and Developments in Infinite Dimensional Lie Theory, Progress in Mathematics,* Birkhäuser, to appear in 2010.

[188] Neher, E., Universal central extensions of graded Lie algebras, preprint 2010.

[189] Neher, E., Savage, A., and Senesi, P., Irreducible finite-dimensional representations of equivariant map algebras, arXiv:0906.5189 (to appear in *Trans. Amer. Math. Soc.*).

[190] Neher, E. and Tocón, M., Lie tori of type B_2 and graded-simple Jordan structures covered by a triangle, preprint 2009.

[191] Neher, E. and Yoshii, Y., Derivations and invariant forms of Jordan and alternative tori, *Trans. Amer. Math. Soc.* **355** (2003), 1079–1108.

[192] Panyushev, D., Premet, A., and Yakimova, O., On symmetric invariants of centralisers in reductive Lie algebras, *J. Algebra* **313** (2007), 343–391.

[193] Premet, A., Irreducible representations of Lie algebras of reductive groups and the Kac-Weisfeiler conjecture, *Invent. Math.* **121** (1995), 79–117.

[194] Premet, A., Special transverse slices and their enveloping algebras, *Adv. in Math.* **170** (2002), 1–55.

[195] Premet, A., Enveloping algebras of Slodowy slices and the Joseph ideal, *J. Eur. Math. Soc.* **9** (2007), 487–543.

[196] Premet, A., Primitive ideals, non-restricted representations and finite W-algebras, *Mosc. Math. J.* **7** (2007), 743–762.

[197] Premet, A., Commutative quotients of finite W-algebras, preprint, arXiv:0809.0663.

[198] Ragoucy, E. and Sorba, P., Yangian realisations from finite W-algebras, *Comm. Math. Phys* **203** (1999), 551–572.

[199] Ramos, E., Sah, C.-H., and Shrock, R. E., Algebras of diffeomorphisms of the N-torus, *J. Math. Phys.* **31** (1990), 1805–1816.

[200] Rao, S. E., On representations of loop algebras, *Comm. Algebra* **21** (1993), 2131-2153.

[201] Rao, S. E., Classification of irreducible integrable modules for multiloop algebras with finite-dimensional weight spaces, *Jour. Alg.* **246**, (2001), 215-225.

[202] Saito, Y., Crystal bases and quiver varieties, *Math. Ann.* **324** (2002), no. 4, 675–688.

[203] Savage, A., A geometric construction of crystal graphs using quiver varieties: extension to the non-simply laced case, In *Infinite-Dimensional Aspects of Representation Theory and Applications (Charlottesville, VA, 2004), Contemp. Math.*, vol. 392, pp. 133–154. Amer. Math. Soc., Providence, RI, 2005.

[204] Savage, A., Geometric and combinatorial realizations of crystal graphs, *Algebr. Represent. Theory* **9**(2) (2006), 161–199.

[205] Savage, A., On two geometric constructions of $U(\mathfrak{sl}_n)$ and its representations, *J. Algebra* **305**(2) (2006), 664–686.

[206] Senesi, P., The block decomposition of finite-dimensional representations of twisted loop algebras, *Pacific J. Math.* **244** (2010), no. 2, 335–357.

[207] Sevostyanov, A., Regular nilpotent elements and quantum groups, *Comm. Math. Phys.* **204** (1999), 1–16.

[208] Skryabin, S., A category equivalence, Appendix to [194].

[209] Skryabin, S., Representations of the Poisson algebra in prime characteristic, *Math. Z.* **243** (2003), 563–597.

[210] Slodowy, P., *Simple singularities and simple algebraic groups*, *Lect. Notes. in Math.* **815**. Springer, Berlin, 1980.

[211] Soergel, W., Kategorie O, perverse Garben und Moduln über den Koinvarianten zur Weylgruppe, *J. Amer. Math. Soc.* **3** (1990), no. 2, 421-445.

[212] Stroppel, C., Category O: Quivers and endomorphism rings of projectives, *Represent. Theory* **7** (2003), 322-345 (electronic).

[213] van der Kallen, W., *Infinitesimally Central Extensions of Chevalley Groups*, In *Lecture Notes in Mathematics*, vol. 356. Springer-Verlag, Berlin, 1973.

[214] Wang, W., Dimension of a minimal nilpotent orbit, *Proc. Amer. Math. Soc.* **127** (1999), 935–936.

[215] Wang, W., Lagrangian construction of the $(\mathrm{gl}_n, \mathrm{gl}_m)$-duality, *Commun. Contemp. Math.* **3**(2) (2001), 201–214.

[216] Wang W. and Zhao, L., Representations of Lie superalgebras in prime characteristic I, *Proc. London Math. Soc.* **99** (2009), 145–167.

[217] Wang W. and Zhao, L., Representations of Lie superalgebras in prime characteristic II: The queer series, arXiv:0902.2758.

[218] Webster, B., Singular blocks of parabolic category $\mathcal{O}$ and finite W-algebras, Preprint, arXiv:0909.1860.

[219] Weibel, C., An introduction to homological algebra, In *Cambridge Studies in Advanced Mathematics*, vol. 38. Cambridge University Press, 1994.

[220] Weisfeiler, B. and Kac, V., On irreducible representations of Lie p-algebras, *Func. Anal. Appl.* **5** (1971), 111–117.

[221] Wilson, R., Euclidean Lie algebras are universal central extensions, In *Lie Algebras and Related Topics* (New Brunswick, NJ, 1981), Lecture Notes in Math., vol. 933, pp. 210–213. Springer, Berlin, 1982.

[222] Yoshii, Y., Coordinate algebras of extended affine Lie algebras of type A_1, *J. Algebra* **234** (2000), 128–168.

[223] Yoshii, Y., Root-graded Lie algebras with compatible grading, *Comm. Algebra* **29** (2001), 3365–3391.

[224] Yoshii, Y., Root systems extended by an abelian group and their Lie algebras, *J. Lie Theory* **14** (2004), 371–394.

[225] Yoshii, Y., Lie tori—a simple characterization of extended affine Lie algebras, *Publ. Res. Inst. Math. Sci.* **42** (2006), 739–762.

[226] Zhao, L., Representations of Lie superalgebras in prime characteristic III, arXiv:0910.2077, *Pacific J. Math* (to appear).

[227] Zhu, Y., Modular invariance of characters of vertex operator algebras, *J. Amer. Math. Soc.* **9** (1996), 237–302.

Index

Titles in This Series